ELECTRICAL ENGINEERING

ELECTRICAL ENGINEERING

C L Wadhwa

Formerly-Professor
Department of Electrical Engineering
Delhi College of Engineering, Delhi
Formerly-Dean, Faculty of Technology
University of Delhi, Delhi

NEW AGE

NEW AGE INTERNATIONAL (P) LIMITED, PUBLISHERS
New Delhi • Bangalore • Chennai • Cochin • Guwahati
Hyderabad • Jalandhar • Kolkata • Lucknow • Mumbai • Ranchi

Published by New Age International (P) Ltd., Publishers
First Edition: 2004
Reprint: 2013

BRANCHES

- **Bangalore** 37/10, 8th Cross (Near Hanuman Temple), Azad Nagar, Chamarajpet, Bangalore-560 018
  Tel.: (080) 26756823, Telefax: 26756820, E-mail: bangalore@newagepublishers.com
- **Chennai** 26, Damodaran Street, T. Nagar, Chennai-600 017
  Tel.: (044) 24353401, Telefax: 24351463, E-mail: chennai@newagepublishers.com
- **Cochin** CC-39/1016, Carrier Station Road, Ernakulam South, Cochin-682 016
  Tel.: (0484) 2377004, Telefax: 4051303, E-mail: cochin@newagepublishers.com
- **Guwahati** Hemsen Complex, Mohd. Shah Road, Paltan Bazar, Near Starline Hotel, Guwahati-781 008
  Tel.: (0361) 2513881, Telefax: 2543669, E-mail: guwahati@newagepublishers.com
- **Hyderabad** 105, 1st Floor, Madhiray Kaveri Tower, 3-2-19, Azam Jahi Road, Nimboliadda, Hyderabad-500 027
  Tel.: (040) 24652456, Telefax: 24652457, E-mail: hyderabad@newagepublishers.com
- **Kolkata** RDB Chambers (Formerly Lotus Cinema) 106A, 1st Floor, S.N. Banerjee Road, Kolkata-700 014
  Tel.: (033) 22273773, Telefax: 22275247, E-mail: kolkata@newagepublishers.com
- **Lucknow** 16-A, Jopling Road, Lucknow-226 001
  Tel.: (0522) 2209578, 4045297, Telefax: 2204098, E-mail: lucknow@newagepublishers.com
- **Mumbai** 142C, Victor House, Ground Floor, N.M. Joshi Marg, Lower Parel, Mumbai-400 013
  Tel.: (022) 24927869, Telefax: 24915415, E-mail: mumbai@newagepublishers.com
- **New Delhi** 22, Golden House, Daryaganj, New Delhi-110 002
  Tel.: (011) 23262368, 23262370, Telefax: 43551305, E-mail: sales@newagepublishers.com

ISBN : 978-81-224-1535-3

₹ 175.00

C-12-08-6498

Printed in India at Print-O-Pack, Delhi.
Typeset at Goswami Associates, Delhi.

PUBLISHING FOR ONE WORLD
NEW AGE INTERNATIONAL (P) LIMITED, PUBLISHERS
7/30 A, Darya Ganj, New Delhi-110002
Visit us at **www.newagepublishers.com**

To

My mother, who taught me how to hold pen in my little fingers.

My father, who taught me modesty and tolerance.

My wife, a symbol of mutual trust and mutual respect.

My daughter and son, who exhibited a high degree of patience.

My students, who made me learn the subject.

The Almighty, who has created such a beautiful world.

PREFACE

Electrical Engineering has been written as a core course for all engineering students viz. Electronics and Communication, Computer Engineering, Civil, Mechanical Engineering etc. With advancement in technologies in almost all spheres of engineering, it is becoming difficult to provide more than one slot for interdisciplinary courses. However, the author feels that no engineering can work without electric energy. The basic input to all engineering is the electric energy. A basic course on Electrical Engineering is almost essential for all engineering students. Keeping this compulsion in mind and also that this course will normally be offered at the first year level of engineering, the author has made modest effort to give in a concise form various features of Electrical Engineering starting from simple ac circuits to Network theorems, measuring instruments, transformers, various dc and ac machines including stepper motor. Various physical phenomenon have been explained using simple language avoiding the rigorus of mathematics.

Chapter I deals with the steady state analysis of a.c. series and parallel circuits and series and parallel resonance.

Chapter II describes various network theorems, star-delta transformation and methods of mesh and nodal analysis for dc networks.

Balanced and unbalanced, three phase circuits have been analysed and methods of 3-phase power measurements have been discussed in Chapter III.

Measuring instruments like, ammeter, voltmeter, wattmeter and energy meter have been described in chapter IV.

Transformer is a very important electrical equipment. The construction, principle of operation, parameter calculations, efficiency and regulation of transformer have been discussed in chapter V.

Chapter VI deals in types of dc machines, their operation and characteristics, efficiency and speed control and application of these machines.

Chapter VII deals into the construction, principle of operation and applications of three phase synchronous motors and stepper motor.

Three phase induction motors form atmost 60 to 70% of the total load on power system. These motors have been discussed in chapter VIII including types, construction, principle of operation, methods of starting, torque-slip characteristic, application and single phasing operation. Single phase induction motors find wide application in Office and domestic appliances. Construction, principle of operation and application of these motors have been described in chapter IX.

Power systems is the most capital intensive and the most complex system ever developed by man. Chapter X gives in brief, various supply and distribution systems and suggests methods for improvement of power factor.

A large number of problems have been solved to help understand the theory. At the end of each chapter unsolved problems with their answers have been suggested for further practice. At the end, a large number of objective type questions have been added to help the reader to test himself.

Any constructive suggestion for the improvement of the book will be gratefully acknowledged.

Last but not the least, I with to express my gratitude to my wife Usha, daughter Meenu and son Sandeep for their patience and encouragement during the preparation of the book.

—C.L. Wadhwa

CONTENTS

CHAPTER

1 A-C Circuits

1.1 INTRODUCTION

Network theory, in general, deals into three components, the network, usually consisting of a suitable combination of R, L and C, the source and the response. Mostly, it is desirable that whatever is the waveform of the source, same waveform should appear in the response (Exceptions to be ignored). We have various waveforms viz. square, triangular, trapezoidal, sinusoidal etc. Out of all these, if the circuit has resistance only, one can use any of the waveforms mentioned above and the response (v-i relation) will have the same waveform as the excitation (source). However, with L and C as the element, it is only sinusoidal waveform which satisfies the requirement of identical waveforms for source and response. If the source is voltage source the current through the inductor is proportional to the integral of the voltage wave and for the capacitor, it is the derivative of the voltage wave. It is a very useful and important property of sine waves that both it's derivative and integrals are also sinusoidal. The basis for standardization on the sine wave thus comes from both physics and calculus. Also, it is very easy to generate the sine waves using rotating machines and transmit and distribute electric energy using transformers. We next illustrate that a uniform circular motion or simple harmonic motion is a sinusoidal variation.

Sine wave is the basic waveform of a.c. circuit theory and it is necessary to learn how to handle such waves.

Let us relate a uniform circular motion or rotation to a sinusoidal wave form. We consider a rotating voltage/current phasor that has a length V_m/I_m and is coincident along x-axis Fig. 1.1. We start counting time and rotate the phasor anti-clockwise and at various instant of time we take the projection of the rotating phasor along y-axis.

At $t = 0$, the projection is zero and hence we plot a point on the right hand side along $t(\theta)$ axis. After certain time say t the phasor rotates through an angle θ and there is some finite projection along y-axis, mark this on the right hand side along $t(\theta)$ axis as shown in Fig. 1.1. As we rotate the phasor further when $\theta = 90°$ the value of the projection is V_m/I_m. Mark it on the right hand side along $t(\theta)$ axis as shown. Rotate phasor further and obtain the projection along y-axis, the projections decreases beyond 90° and continues to be along + y axis till θ = 180° when projection is zero.

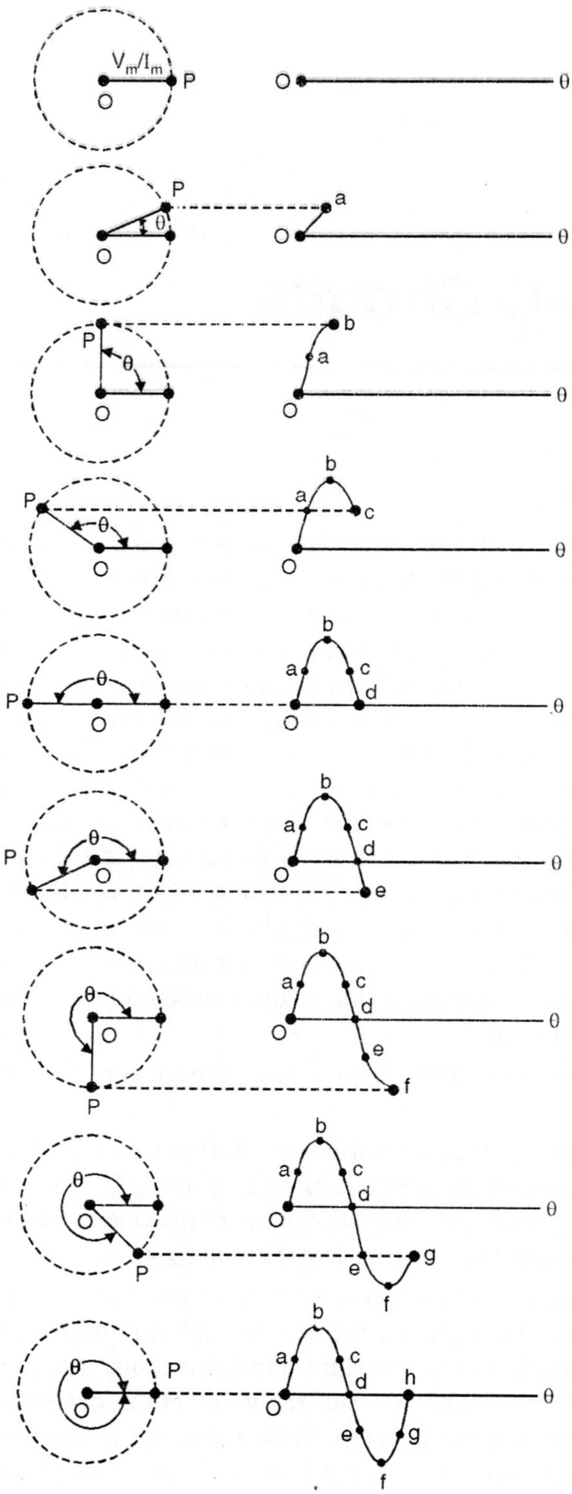

Fig. 1.1

Similarly if we rotate the phasor further through 2π radian or 360° or one cycle, we obtain a complete sine wave as shown in Fig. 1.1. Thus, if the speed of rotation is f revolutions/sec the phasor has a frequency of f hertz (Hz). Hence the angular velocity is $\omega = 2\pi f$ rad/sec.

If the instantaneous value of time at any instant of rotation is t seconds the total angle in radians from time zero upto that point is

$$2\pi ft = \omega t$$

The concept of rotating phasor is a function of time, we may substitute $2\pi ft$ or ωt for θ in the equation.

$$V = V_m \sin \theta$$

and now therefore, the instantaneous voltage in terms of time is

$$V = V_m \sin 2\pi ft = V_m \sin \omega t \qquad ...(1.1)$$

when equation 1.1 is used, the axis of the waveform is not θ but time t.

1.2 PHASE

A sinusoidal waveform having a displacement of θ to the left is shown in Fig. 1.2(a) which means that at $t = 0$ the phasor has an initial displacement of θ in counterclockwise direction. Therefore, the initial instantaneous value of voltage/current at time $t = 0$ is

$V = V_m \sin \theta$ putting $\omega t = 0$. This shift of the waveform is called a phase displacement and the angle θ is called phase angle. Fig. 1.2 (b-d) show different phase displacements of the same sine wave. The sine wave at 'a' is said to be leading the wave at 'b' as

$$V_a = V_m \sin (\omega t + \theta) \text{ and at } b$$

$$V_b = V_m \sin (\omega t - \theta)$$

Example 1.1: The equation for a voltage wave is $v = 0.02 \sin (4000t + 30°)$. Find the frequency, the instantaneous voltage when $t = 320\ \mu$ sec. What is the time represented by 30° phase difference?

Solution: The general expression for voltage is given as

$$V = V_m \sin (2\pi ft + \theta)$$

here $$2\pi f = 4000 \quad \text{or} \quad f = \frac{4000}{2\pi} = 637 \text{ Hz.}$$

When $$t = 320\mu \text{ sec.}$$

$$v = 0.02 \sin \left(4000 \times 320 \times 10^{-6} \times \frac{180}{\pi} + 30°\right)$$

$$v = 0.02 \sin (73.3 + 30)$$

$$= 0.02 \sin (103.3)$$

$$= 0.0195 \text{ volt}$$

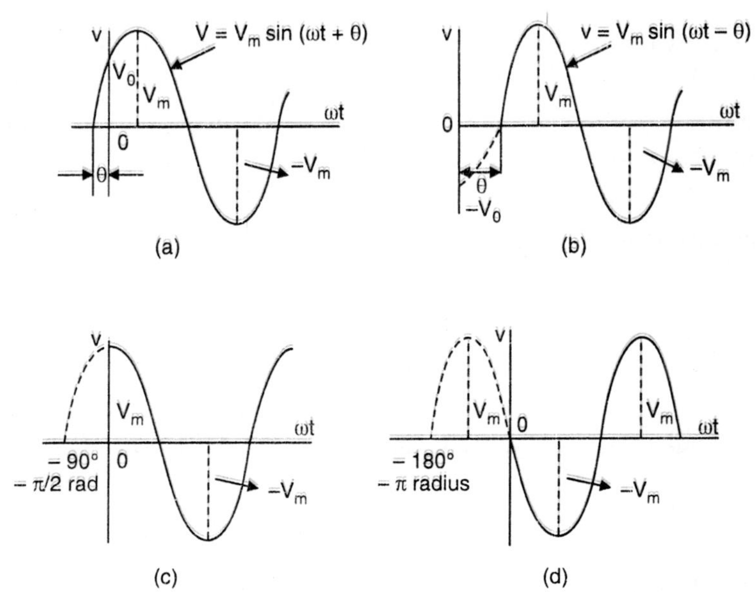

Fig. 1.2 Displaced sinusoidal voltage waveforms:
(a) displacement of θ to the left, (b) Displacement of θ to the right,
(c) Displacement to the left of 90° (π/2 radians), (d) Displacement of 180° (π radians).

The time of 1 cycle is

$$\frac{1}{f} = \frac{1}{637}$$

which is equal to 360° electrical degrees. Hence for 30° the time taken is

$$\frac{30}{360} \times \frac{1}{637} \text{ sec} = 0.131 \text{ m sec} \quad \textbf{Ans.}$$

1.3 THE AVERAGE VALUE OF A WAVEFORM

In a.c. circuit applications we are interested in finding out the average value of a waveform, that wave could be sinusoidal, triangular trapezoidal or any other shape.

The average value of a cycle of a waveform is the area under the waveform divided by the length of one cycle.

Mathematically,

$$V_{av} = \frac{1}{T}\int_0^T v\,dt = \frac{1}{2\pi}\int_0^{2\pi} v\,dt \qquad ...(1.2)$$

where T is the length of one cycle.

We first consider average value of sinusoidal waves for which we consider the following different shapes.

These are all sine waves In (a) since it is a full sine wave, the total area under the curve is zero as half of it is above ωt axis and equal half is below ωt-axis and hence the average area of the curve is zero and hence over the complete sine wave the average value is zero.

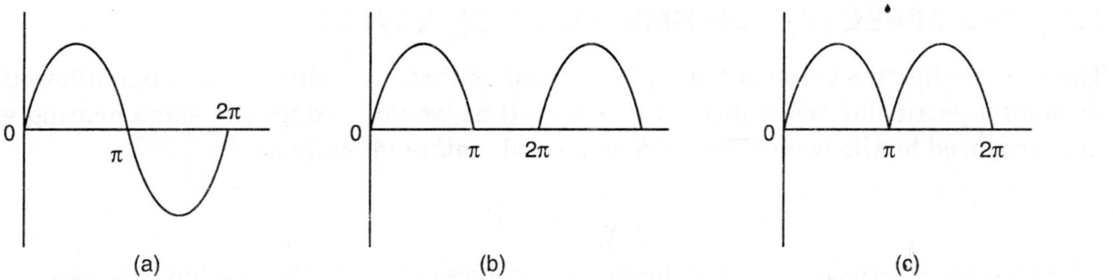

Fig. 1.3 (a) Full sine wave (b) Half-rectified wave (c) Full rectified wave

However, in Fig. 1.3(*b*) the average value is

$$V_{av} = \frac{1}{2\pi}\int_0^{\pi} V_m \sin \omega t \, d\omega t$$

$$= \frac{1}{2\pi}\Big[-V_m \cos \omega t\Big]_0^{\pi}$$

$$= \frac{-V_m}{2\pi}[\cos \pi - \cos 0]$$

$$= \frac{V_m}{\pi}$$

For Fig. 1.3(*c*) the average value is just double of what we have for a half wave rectified sine wave as over the same period 2π, the area under the curve is double of what we have for the half wave.

i.e., For full rectified wave, average value is $\frac{2V_m}{\pi}$

Consider a non-sinusoidal wave shown in Fig. 1.4 and find the average value of the wave for one half cycle.

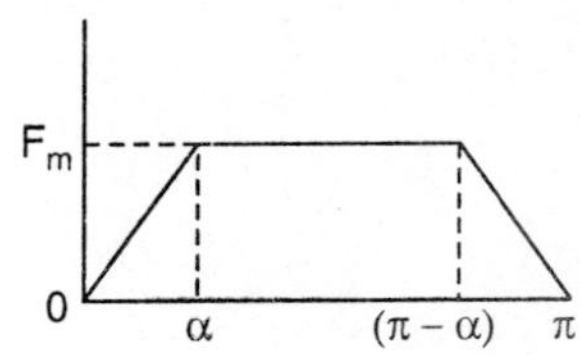

Fig. 1.4 Trapezoidal wave

Since the wave has a trapezoidal shape the area under the curve is

$$\frac{1}{2}(\pi + \pi - 2\alpha) F_m$$

Hence Average Value = $\frac{2(\pi - \alpha)}{2\pi} F_m$

If $\alpha = 0$

Average Value is F_m

If $\alpha = \frac{\pi}{6}$

Average Value is $\frac{(\pi - \pi/6)}{\pi} F_m = \frac{5}{6} F_m$

Similarly if $\alpha = \frac{\pi}{2}$, the average value is

$$\frac{\pi - \pi/2}{\pi} F_m = \frac{F_m}{2}$$

1.4 THE EFFECTIVE OR RMS VALUE OF A WAVE

The *rms* or effective value of a wave is defined as that d.c. value which when allowed to flow through a particular resistance for a certain time would produce the same heating effect as that produced by the wave. This is represented mathematically as

$$I_{rms} = I_{eff} = \sqrt{\frac{1}{T}\int_0^T i^2\, dt}$$

As the name *rms* (root of mean of the squared values) suggests that we have to take the square of the various ordinates and find the area of the squared wave over the period T and find the mean value of this quantity and then obtain its square root. This is the effective value.

We again take first, various sinusoidal waves as shown in Fig. 1.3(*a*) to (*c*) and find out their *rms* values Fig. 1.3(*a*).

As per the definition we have first to express the wave as per its instantaneous value as every instantaneous value is to be squared and then area under this curve is to be obtained. Now, for this sine wave

$$i = I_m \sin \omega t$$

Since the wave is to be squared, the negative part of the wave during π to 2π would also become positive and hence *rms* value over the complete cycle will not be zero.

Hence

$$I_{rms} = \sqrt{\frac{1}{2\pi}\int_0^{2\pi} i^2\, d\omega t}$$

$$= \sqrt{\frac{1}{2\pi}\int_0^{2\pi} I_m^2 \sin^2 \omega t\, d\omega t}$$

$$= \sqrt{\frac{I_m^2}{2\pi}\int_0^{2\pi} \frac{1-\cos 2\omega t}{2}\, d\omega t}$$

$$= I_m \sqrt{\frac{1}{4\pi}\left(\omega t - \frac{\sin 2\omega t}{2}\right)_0^{2\pi}}$$

$$= I_m \sqrt{\frac{2\pi}{4\pi}} = \frac{I_m}{\sqrt{2}}$$

For Fig. 1.3(*b*) where it is a half rectified sine wave, the *rms* value over the complete cycle will be

$$\sqrt{\frac{I_m^2}{2\pi}\int_0^{\pi}\left(\frac{1-\cos 2\omega t}{2}\right) d\omega t}$$

$$= \frac{I_m}{2}$$

i.e. *rms* value of half rectified sine wave is $\frac{1}{2}$ of its peak value.

For Fig. 1.3(*c*) since it is a full rectified wave and the square of each ordinate will be similar to the full sine wave, the *rms* value will be same i.e. $\frac{I_m}{\sqrt{2}}$.

Let us find out *rms* value of the wave shown in Fig. 1.4. The wave can be expressed mathematically as

$$f(\theta) = \frac{F_m}{\alpha}\theta \qquad 0 < \theta \le \alpha$$
$$= F_m \qquad \alpha \le \theta \le \pi - \alpha$$
$$= -\frac{F_m}{\alpha}\theta \quad (\pi - \alpha) \le \theta \le \pi$$

Since the area under the squared ordinate in the interval 0 to α and $(\pi - \alpha)$ to α is same we have the *rms* value as

$$i_{rms} = \sqrt{\frac{1}{\pi}\left[2\int_0^{\alpha} \frac{F_m^2}{\alpha^2}\theta^2\, d\theta + \int_{\alpha}^{\pi-\alpha} F_m^2\, d\theta\right]}$$
$$= \sqrt{\frac{F_m^2}{\pi}\left[2\left(\frac{\theta^3}{3\alpha^2}\right)_0^{\alpha} + (\theta)_{\alpha}^{\pi-\alpha}\right]}$$
$$= \sqrt{\frac{F_m^2}{\pi}\left[\frac{2\alpha}{3} + \pi - \alpha - \alpha\right]}$$
$$= \sqrt{\frac{F_m^2}{\pi}\left(\pi - \frac{4}{3}\alpha\right)}$$
$$= F_m\sqrt{\left(\pi - \frac{4}{3}\alpha\right)\Big/\pi}$$

now when $\alpha = 0$ i.e. it is *a* d.c. wave

$i_{rms} = F_m$ which is an expected value as the *rms* value of *a* d.c. wave is the d.c. value itself. Similarly the average value of *a* d.c. wave is the value of the d.c. wave.

If $\qquad \alpha = \frac{\pi}{6}$

$$i_{rms} = F_m\sqrt{\left(\pi - \frac{4}{3}\frac{\pi}{6}\right)\Big/\pi}$$
$$= \sqrt{\frac{7}{9}}\,F_m$$

when $\qquad \alpha = \frac{\pi}{2}$

$$i_{rms} = F_m\sqrt{\left(\pi - \frac{4}{3}\cdot\frac{\pi}{2}\right)\Big/\pi} = \frac{F_m}{\sqrt{3}}$$

1.4.1 Form Factor

It is defined as the ratio of *rms* value of a wave and the average value of the wave.

i.e. $$\text{Form factor} = \frac{rms \text{ value of the wave}}{\text{Average value of the wave}}$$

For a sinusoidal wave

$$\text{Form factor} = \frac{V_m}{\sqrt{2}} \cdot \frac{\pi}{2V_m}$$
$$= 1.11$$

1.4.2 Peak Factor or Crest Factor

It is the ratio of the peak value of the wave to its *rms* value.

For a sinusoidal wave

$$\text{Peak factor} = \frac{V_m}{V_m/\sqrt{2}} = \sqrt{2} = 1.414$$

It is sometimes misunderstood that the operating voltages e.g. 110 V, 220 V, 132 KV are used because of form factor which is 1.11 i.e. multiple of 11. In fact these voltages have been standardized as standardization has its own economic and operating advantages. In fact there are many other operating voltages e.g. 400 V, 400 KV, 765 KV which are not multiple of 11. So it is the standardization which has selected certain voltages as operating voltages.

Form factor does not have much of significance except when two wave forms one with higher form factor when applied to magnetic materials, produces higher hysteresis loss per loop as compared to one with lower form factor.

Crest factor has an indication of electric stresses developed in insulating material as the stresses are a function of the peak value of the wave.

1.5 PHASOR DIAGRAM

Consider Fig. 1.5 which shows time waves of a sinusoidal current waveform lagging a sinusoidal voltage waveform by θ degrees. The same aspect can be represented by two phasors rotating counter clockwise at the angular velocity ω.

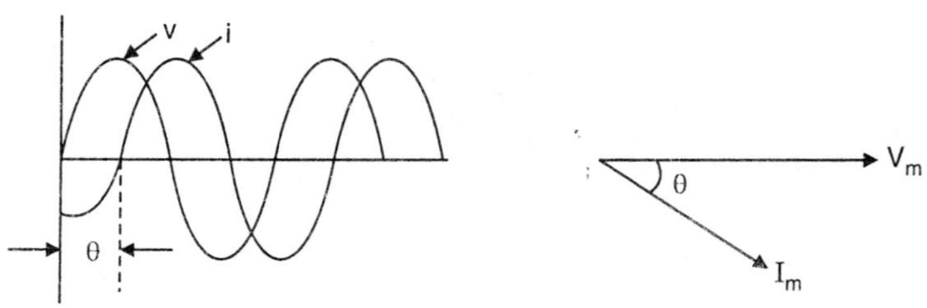

(a) (b)

Fig. 1.5 (a) *V-i* sinusoidal wave (b) Phasor representation of (a)

At each point on the time waveform, the angle of current lag is θ because the angle between the phasors V_m and I_m is at all time θ. Therefore, either the time waveform of the rotating phasors or the phasor diagram, can be used to describe the system. Since both the diagrams, the time diagram and the phasor diagram convey the same information, the phasor diagram being much more simpler, it is used for an explanation in circuit theory analysis. Since electrical data is given in terms of *rms* value, we draw phasor diagram with phasor values as *rms* rather than peak value used so far.

1.5.1 Phasor Algebra

Consider a phasor V in Fig. 1.6. This can be represented mathematically in the following ways :

$$V = V_x + jV_y \text{ (Cartesian Co-ordinates)}$$

$$V = |V| \angle\theta \text{ in polar coordiantes.}$$

Here V_x is the component of V along x-axis and V_y is the component of V along y-axis and j is an operator which when multiplied to a phasor rotates the phasor 90° anticlockwise and

$$j = \sqrt{-1} \quad \text{or} \quad j^2 = -1$$

Addition and subtraction of phasors is very simple in cartesian coordinates as each phasor is to be decomposed into two components, one along x-axis and the other along y-axis and algebraic addition or subtraction can be carried out with ease.

Multiplication and division is very simple in polar coordinates. As the multiplication of various phasor equals the multiplication of magnitude of individual phasor and the angle is the algebraic sum of the phase angle of individual phasors i.e.

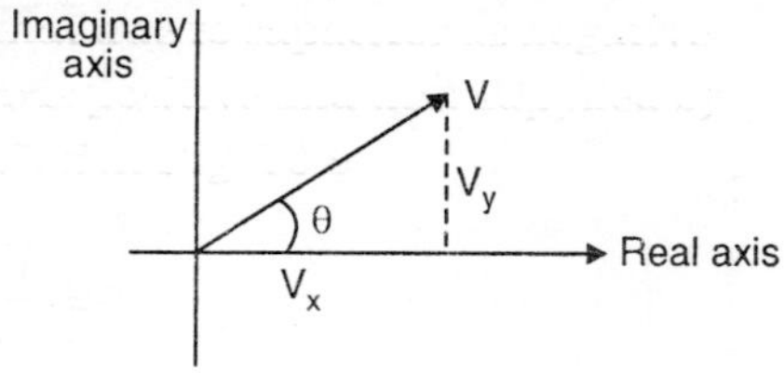

Fig. 1.6 Representation of phasor

$$V = |V_1| \angle\theta_1 \; |V_2| \angle\theta_2 \; |V_3| \angle -\theta_3$$

$$= |V_1| \; |V_2| \; |V_3| \angle\theta_1 + \theta_2 - \theta_3$$

Similarly, division of two phasors is the division of the magnitude of the two phasors and the angle is the difference of the angles of the two phasors e.g.

$$V \angle\theta = \frac{V_1 \angle\theta_1}{V_2 \angle -\theta_2} = \left|\frac{V_1}{V_2}\right| \angle\theta_1 + \theta_2$$

Let us now consider phase relations between voltages and currents when sinusoidal voltages are applied to R, L and C elements.

1.6 RESISTANCE CONNECTED ACROSS A SINUSOIDAL VOLTAGE

Fig. 1.7 shows resistance connected to a sinusoidal source of V_m as peak voltage

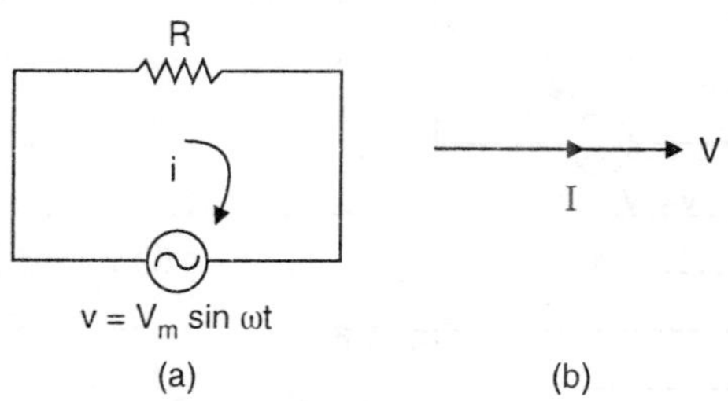

Fig. 1.7 Resistance across a c supply.

The instantaneous value of current in the circuit is

$$i = \frac{v}{R} = \frac{V_m}{R} \sin \omega t$$

$$= I_m \sin \omega t$$

Since the two phasors voltage and current have exactly similar representation except for the magnitude, the two phasors are in phase i.e. they are represented as in Fig. 1.7(*b*) where *V* and *I* are the *rms* values of voltage and current.

Now power

$$p = \frac{1}{2\pi}\int_0^{2\pi} vi\, d\omega t$$

$$= \frac{1}{2\pi}\int_0^{2\pi} I_m V_m \sin \omega t \sin \omega t\, d\omega t$$

$$= \frac{V_m I_m}{2\pi}\int_0^{2\pi} \sin^2 \omega t\, d\omega t$$

$$= \frac{V_m I_m}{2\pi}\int_0^{2\pi} \frac{1 - \cos 2\omega t}{2}\, d\omega t$$

$$= \frac{V_m I_m}{4\pi}\left[\omega t - \frac{\sin 2\omega t}{2}\right]_0^{2\pi}$$

$$= \frac{V_m I_m}{4\pi} . 2\pi$$

$$= \frac{V_m I_m}{4\pi}$$

$$= VI$$

Hence average power in a resistive element equals the product of the *rms* voltage and current across and through the resistance respectively.

1.6.1 Inductor Across Sinusoidal Source

Whenever an *ac* source is connected across an inductor a voltage is induced in the inductor and the instantaneous value of this voltage is given by

$$v_L = -L\frac{di}{dt}$$

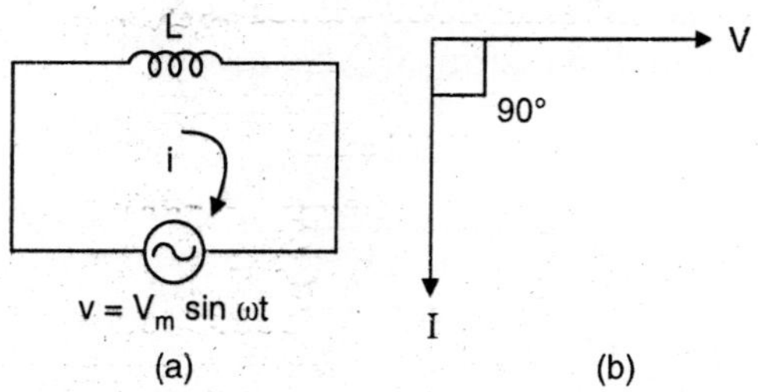

Fig. 1.8

and this voltage opposes the increase in current in the inductor and hence

$$v_L = -v$$

Hence $$v = L\frac{di}{dt}$$

or $$di = \frac{v}{L}dt$$

$$i = \frac{1}{L}\int V_m \sin \omega t \, dt$$

$$= -\frac{V_m}{L}\frac{\cos \omega t}{\omega}$$

$$= -\frac{V_m}{\omega L}\cos \omega t$$

$$= \frac{V_m}{\omega L}\sin(\omega t - 90°)$$

Since $\frac{V_m}{\omega L}$ dimensionally should be current and it is the ratio of voltage to ωL, therefore, the denominator should be something like resistance in d.c. circuit. The term ωL is known as inductive reactance as it is due to an inductance and we see that the two phasors v and i for an inductor are not identical. If we compare expression for v and i we find that current i lags the voltage exactly by 90° as shown in Fig. 1.8(b).

Now $$I = \frac{I_m}{\sqrt{2}} \text{ and } V = \frac{V_m}{\sqrt{2}}$$

Hence the ratio of the two phasors

$$\frac{V\angle 0}{I\angle -90°} = \frac{V_m/\sqrt{2}}{\frac{V_m}{\sqrt{2}\omega L}\angle -90°} = \omega L \angle 90°$$

$$= j\omega L$$

$$= jX_L$$

The average power in the circuit.

$$p = \frac{1}{2\pi}\int_0^{2\pi} vi \, d\omega t$$

$$= \frac{1}{2\pi}\int_0^{2\pi} V_m \sin \omega t \, I_m \sin(\omega t - 90°)\, d\omega t$$

$$= \frac{1}{2\pi}\int_0^{2\pi} V_m I_m \sin \omega t \cos \omega t \, d\omega t$$

$$= \frac{1}{2\pi}\cdot V_m I_m \int_0^{2\pi} \frac{\sin 2\omega t}{2} d\omega t = 0$$

This shows that the power consumed in a purely inductive circuit is zero.

1.6.2 Capacitor Connected Across Sinusoidal Source

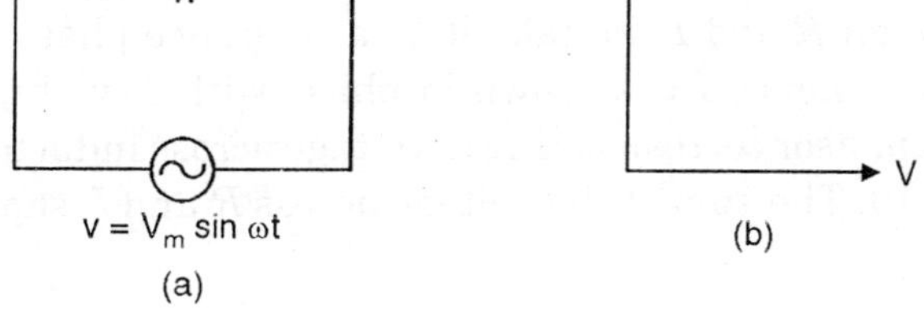

Fig. 1.9

The voltage and current in a capacitor are related as

$$i = C\frac{dv}{dt}$$
$$= CV_m\,\omega\cos\omega t$$
$$= V_m\,\omega C\sin(\omega t + 90)$$
$$= \frac{V_m}{1/\omega C}\sin(\omega t + 90)$$

This shows that in a purely capacitive circuit the current through the capcitor leads the voltage across it as shown in Fig. 1.9(*b*). Here $\frac{1}{\omega C}$ represents reactance of the capacitor. Taking the ratio of the voltage phasor and current phasor we have

$$\frac{V\angle 0}{I\angle 90} = \frac{(V_m/\sqrt{2})\angle 0}{\frac{V_m}{\sqrt{2}\,.\,1/\omega C}\angle 90°} = \frac{1}{\omega C}\angle -90°$$
$$= \frac{-j}{\omega C} = -jX_C$$

The average power in a capacitor

$$p = \frac{1}{2\pi}\int_0^{2\pi} V_m I_m \sin\omega t\sin(\omega t + 90°)\,d\omega t$$
$$= \frac{1}{2\pi}\int_0^{2\pi} V_m I_m \sin\omega t\cos\omega t\,d\omega t$$
$$= \frac{1}{2\pi}\int_0^{2\pi} V_m I_m \frac{\sin 2\omega t}{2}\,d\omega t$$
$$= 0$$

This again shows that the power consumed in a purely capacitive circuit is zero.

1.6.3 Series R-L Circuit Connected to Sinusoidal Source

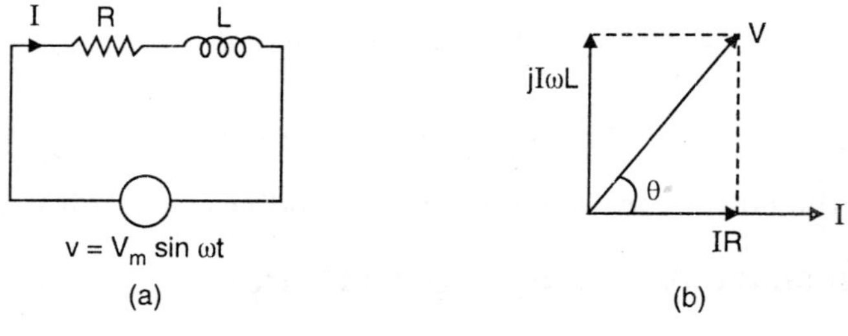

Fig. 1.10

Since current is same through R and L we take it as a reference phasor. The voltage across resistance is in phase with I, hence IR is drown in phase with I, voltage across inductor is leading the current hence a phasor corresponding to voltage across inductor should lead phasor I by 90° as shown in Fig. 1.10. The sum of the voltage across R and L should equal the supply voltage V.

From the phasor diagram

$$V = IR + jI\omega L$$

or $$\frac{V}{I} = R + j\omega L$$

From the phasor diagram it is seen that the phase angle between V and I is neither 0° to represent the ratio of $\frac{V}{I}$ as resistance nor it is 90° to represent the ratio $\frac{V}{I}$ as reactance Hence we call this ratio something different from resistance or reactance, we call it impedance. It is to be noted that if we divide the phasor diagram by I we will have another diagram as shown in Fig. 1.10.1.

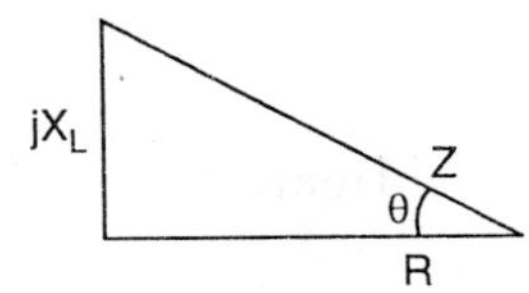

Fig. 1.10.1

Impedance even though is a ratio of two phasors it is not a phasor by itself as R and X_L are not function of time and hence while drawing impedance triangle, arrows should not be attached with R and X_L. Angle θ between R and Z is same as the angle between I and V and hence is the phase angle of the impedance and impedance is to be represented by $Z = |Z| \angle\theta$ if it is an inductive circuit and $Z = |Z| \angle -\theta$ if it is a capacitive circuit.

From Fig. 1.10.1 it is seen that

$$R = Z\cos\theta, \quad X_L = Z\sin\theta$$

and $$Z = \sqrt{R^2 + X_L^2}$$

and $$\theta = \tan^{-1}\frac{X_L}{R}$$

If V and I correspond to the same circuit and θ is the angle between the two phasors then power in the circuit is given by $P = VI\cos\theta$ where θ is known as power factor angle and cos θ is known as power factor of the circuit. The product of V and I is known as apparent power. Hence power factor can be defined as

$$\cos\theta = \frac{P}{VI} = \frac{\text{Real Power}}{\text{Apparent Power}}$$

The component of current along the voltage phasor ($I\cos\theta$) gives real power or actual power whereas the component of current at right angle to the voltage phasor gives what is known as reactive power i.e. the power which does not do any useful work but is present in the system. In fact reactive power is associated with the energy storing elements L and C. The analogy for this can be drawn between force applied to an object. The component of the force along the displacement produces useful work whereas the force applied at right angle to the displacement produces no work.

1.6.4 RC Series Circuit Connected to Sinusoidal Source

Here again since current is same through both the elements, it is taken as the reference phasor. The voltage across R is in phase with I and voltage across C lags the current by 90° as shown in Fig. 1.11(b). The resultant phasor sum of IR and $-jIX_C$ is the supply voltage V.

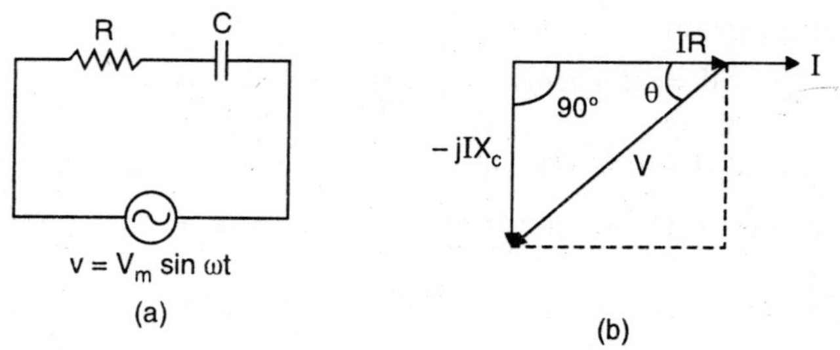

Fig. 1.11

$$V = IR - jIX_C$$

$$\frac{V}{I} = (R - jX_C) = Z$$

Hence $\quad Z = R - jX_C = \sqrt{R^2 + X_C^2}$

$$R = Z\cos\theta, \quad X_C = Z\sin\theta \quad \text{and} \quad \tan\theta = \frac{-X_C}{R}$$

Example 1.2: The impedance of a circuit placed across a 120*V*, 50 Hz source is (10 + *j*20). Find current and the power.

Solution: Taking voltage as the reference we have $V = (120 + j0)$

Hence current $\quad I = \frac{V}{Z} = \frac{120 + j0}{(10 + j20)}$

or $\quad I = \frac{(120 + j0)(10 - j20)}{(10 + j20)(10 - j20)} = \frac{12 - j24}{5}$

$$= 2.4 - j4.8 \text{ Amp.}$$

This shows that the current lags the voltage by an angle

$$\theta = \tan^{-1}\frac{-4.8}{2.4} = -63.4°$$

$$I = \sqrt{2.4^2 + 4.8^2}\ \angle -63.4$$

$$= 5.36\ \angle -63.4$$

Power $\quad P = VI\cos\theta = 120 \times 5.36\cos 63.4$

$$= 288 \text{ watt}$$

The same problem can be solved using polar coordinates as follows :

Taking voltage as reference

$$V = 120\ \angle 0$$

Impedance $\quad Z = \sqrt{10^2 + 20^2}\ \tan^{-1}\frac{20}{10}$

$$= 22.36\ \angle 63.4$$

$$I = \frac{V}{Z} = \frac{120\ \angle 0}{22.36\ \angle 63.4}$$

$$= 5.37\ \angle -63.4$$

Power $P = 120 \times 5.37 \cos 63.4 = 288$ watts

Power can also be calculated using the relation

$$P = \text{Real } [VI^*]$$

In Cartesian co-ordinates

$$V = 120 + j0$$

$$I = 2.4 - j4.8$$

Hence $I^* = 2.4 + j4.8$

and

$$P = Re\ [(120 + j0)(2.4 + j4.8)]$$

$$= 120 \times 2.4$$

$$= 288 \text{ watts}$$

Example 1.3: A 15 *V* source is applied to a capacitive circuit that has an impedance of $(10 - j25)\ \Omega$. Determine the current and the power in the circuit.

Solution: Taking voltage as reference and using *j*-notation method we have

$$I = \frac{(15 + j0)(10 + j25)}{(10 - j25)(10 + j25)}$$

$$= \frac{150 + j375}{10^2 + 25^2} = \frac{6 + j15}{29}$$

$$= 0.2 + j0.517 \text{ A}$$

This means the current leads the voltage

$$I = 0.55 \tan^{-1} \frac{0.517}{0.2} = 0.55 \angle 68.85$$

which means the current leads the supply voltage by 68.85° which, in this case is also the angle of the impedance $(10 - j25)$.

$$\tan \theta = \frac{-25}{10}, \qquad \theta = -68.2°$$

This being capacitive circuit the impedance triangle is as shown here in Fig. E1.3.

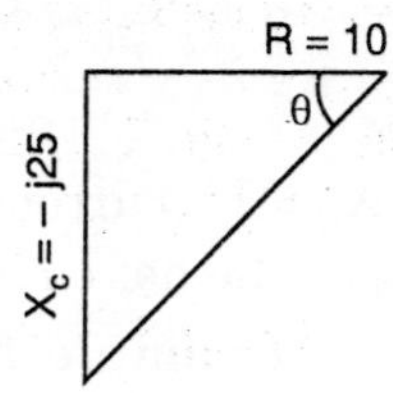

Fig. E1.3

The power loss $Re\ [VI^*]$

$$= \{15[0.2 - j0.517]\}$$

$$= 3 \text{ watts}$$

This can also be obtained using the formula

$$P = V I \cos \theta$$

$$= 15 \times 0.55 \cos 68.8$$

$$= 3 \text{ watts}$$

Since it is a series circuit and the current through the circuit is 0.55 A and resistance is 10 Ω, the power consumed by the circuit is I^2R

$$0.55^2 \times 10 = 3 \text{ watts}$$

The reactive power $Q = VI \sin \theta = 15 \times 0.55 \sin (-68.8)$

$$= -7.7 \text{ Var}$$

Here reactive power is negative as it is a capacitive circuit. However, for inductive circuits the reactive power is positive.

This is because an inductor absorbs reactive power whereas a capacitor supplies reactive power. This is why, capacitor is also known as reactive power generator.

The current can also be obtained using polar co-ordinates.

$$\text{Impedance } z = \sqrt{10^2 + 25^2} \tan^{-1} \frac{-25}{10}$$

$$= 26.95 \angle -68.2°$$

$$\text{Hence current } I = \frac{15}{26.95 \angle -68.2}$$

$$= 0.55 \angle 68.2$$

1.6.5 R-L-C Series Circuit Connected to Sinusoidal Source

Refer to Fig. 1.12

Taking current as reference, the phasor diagram for the circuit is shown in Fig. *b* below. Depending upon which reactance is larger, the current will lead or lag the supply voltage

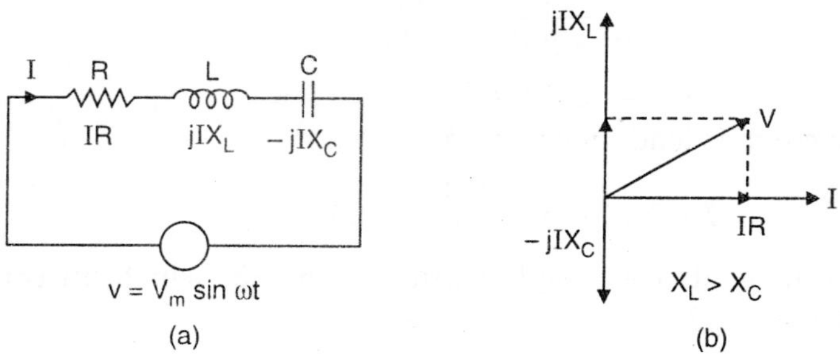

Fig. 1.12

If $X_C > X_L$, current will lead the supply voltage as the net circuit is capacitive otherwise it will lag as the net circuit becomes inductive.

Example 1.4: A 100 Ω resistor, 20 μF capacitor and a 2H inductor are connected in series. At what frequency is the phase angle 45°.

Solution: There are two possibilities for angle to be 45°, one when the total circuit is inductive i.e. when $X_L > X_C$ and the other when $X_C > X_L$.

For θ = 45°, tan θ = 1 for inductance load and when

$$\theta = -45°, \tan\theta = -1$$

Now $$\tan\theta = \frac{X_L}{R} = 1$$

Therefore $$X_L = R$$

where X_L is the net inductive reactance

$$X_L = X_{LT} - X_C$$

$$= \omega L - \frac{1}{\omega C}$$

$$100 = 2\pi f_1(2) - \frac{10^6}{2\pi f_1 \times 20}$$

$$= 12.56 f_1 - \frac{7.96 \times 10^3}{f_1}$$

or $$f_1^2 - 7.96\, f_1 - 633 = 0$$

or $$f_1 = 29.5 \text{ Hz}$$

For the second case

$$X_C - X_{LT} = 100$$

$$\frac{10^6}{2\pi f_2 \times 20} - 2\pi f_2(2) = 100$$

or $$f_2^2 + 7.96\, f_2 - 633 = 0$$

or $$f_2 = 21.5 \text{ Hz}$$

One can check that by using these frequencies the difference of the two reactances in both the cases comes out to be 100 Ω.

Example 1.5: A 120 V, 500 W lamp is used with a series choke on a 230 V, 60 Hz line, what is the inductance of the required choke if the choke has a Q of 2?

Solution: Here we have a 500 W lamp rated at 120 volt whereas the supply voltage is 230 V. Naturally we have to connect some circuit element R, L or C so that the bulb gets 120 V. The resistance of the lamp is given by

$$\frac{V^2}{R} = 500 = \frac{120^2}{R}$$

or $$R = \frac{120^2}{500} = 28.8\ \Omega$$

Here it is suggested to connect a choke which has both resistance as well inductance. It is also given that Q of the coil is 2

i.e. $\frac{\omega L}{R} = 2$ where R is the resistance of the choke itself.

Now $\omega = 2\pi \times 60 = 376.8$ rad/sec.

$$376.8L = 2R$$

or $$R = 188.4\, L$$

Now current in the circuit is

$$120 \times I = 500$$

or $$I = \frac{500}{120} = 4.167A$$

Hence impedance of the circuit is

$$\frac{230}{4.167} = 55.2\ \Omega$$

Also that impedance is

$(28.8 + 188.4L)^2 + (367.8L)^2 = 55.2^2$

The solution gives $L = 87$ mH.

1.7 PARALLEL R-L CIRCUIT ACROSS SINUSOIDAL SUPPLY

Refer to Fig. 1.13

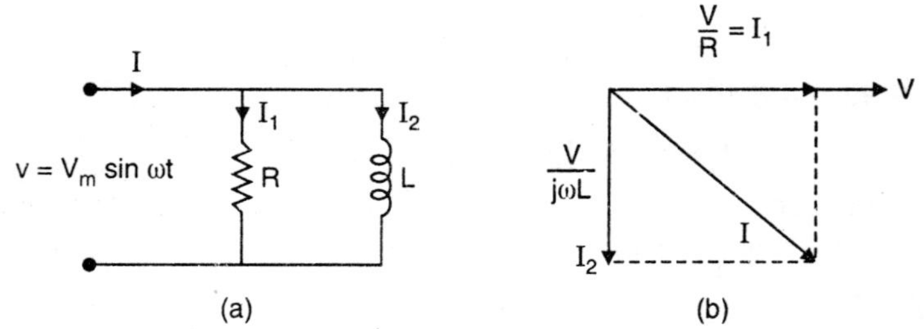

Fig. 1.13

We have taken here V as the reference quantity as it is the voltage which is common to both the elements R and L and not the current. Now current through R will be in phase with the voltage across it whereas current through inductor will lag by 90° the voltage across it. Hence, the phasor diagram in Fig. (*b*) follows and current I supplied by the source equals the phasor sum of the current I_1 and I_2.

Now $$I = I_1 + I_2 = \frac{V}{R} + \frac{V}{j\omega L}$$

Here $\frac{1}{R}$ is the reciprocal of resistance and is termed as conductance whereas $\frac{1}{j\omega L}$ is the reciprocal of the inductive reactance and is known as inductive susceptance.

From the equation above we have

$$\frac{I}{V} = \frac{1}{R} + \frac{1}{j\omega L} \quad \text{...(1.3)}$$

Now dimensionally $\frac{I}{V}$ is reciprocal of impedance and is known as admittance. The above equation can be rewritten as

$Y = G - jB$ where Y is admittance G the conductance and B the susceptance.

$$Y = \frac{1}{Z}, G = \frac{1}{R} \quad \text{and} \quad B = \frac{1}{\omega L}$$

It is to be noted that whereas for an inductive circuit $z = R + j\omega L$, the admittance is

$$Y = G - jB$$

whereas for a capacitive circuit.

$Z = R - jX_C$ the admittance is

$$Y = G + j\omega C = G + jB$$

and hence ωC is the capacitive susceptance. The same observation can also be made as follows : By definition

$$Y = \frac{1}{Z} = \frac{1}{R + jX} \text{ or inductive circuit}$$

$$= \frac{R - jX}{R^2 + X^2} = \frac{R}{R^2 + X^2} - j\frac{X}{R^2 + X^2}$$

where R and X are connected in series.

The units of admittance is *mho* or siemens and denoted by ℧ or S respectively.

The power consumed by the circuit is again $P = VI \cos \theta$ where I is the total current and θ is the angle between V and I.

From equation 1.3 we have

$$\frac{1}{Z} = \frac{1}{R} + \frac{1}{j\omega L}$$

if $Z_1 = R$ and $Z_2 = j\omega L$ we have

$$\frac{1}{Z} = \frac{1}{Z_1} + \frac{1}{Z_2} \quad ...(1.4)$$

or

$$Z = \frac{Z_1 Z_2}{Z_1 + Z_2}$$

i.e. if there are two impedances Z_1, Z_2 connected in parallel, their equivalent impedance equals the ratio of the product of the two impedance to the sum of the two impedances. Also from equation (1.4)

$$Y = Y_1 + Y_2$$

i.e. if there are two admittances Y_1 and Y_2 connected in parallel, the equivalent admittance is the sum of the two admittances.

1.7.1 Parallel R-C Circuit Across Sinusoidal Source

Refer to Fig. 1.14

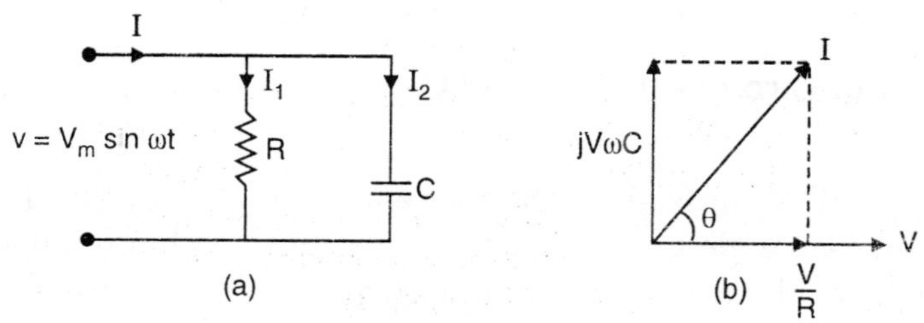

Fig. 1.14

From the phasor diagram

$$I = \frac{V}{R} + jV\omega C$$

or

$$\frac{I}{V} = \frac{1}{R} + j\omega C$$

$$Y = G + jB$$

Example. 1.6: The impedance of a parallel *RC* circuit is (10 – *j*30) Ω at 1 MHz. Determine the values of the components.

Solution: The admittance of the circuit is

$$Y = \frac{1}{Z} = \frac{1(10+j30)}{(10-j30)(10+j30)} = \frac{10+j30}{1000} = \frac{1}{100} + j\frac{3}{100}$$

i.e. $$G = \frac{1}{R} = \frac{1}{100} \quad \text{and} \quad B = \omega C$$

i.e. $$R = 100\ \Omega \quad \text{and} \quad B = \frac{3}{100} = 10^6\ C\ .\ 2\pi$$

or $$C = \frac{3}{100} \times \frac{1}{2\pi \times 10^6} = \frac{3}{200\pi}\ \mu F = 4.7\ nF.$$

Example 1.7: A 4700 Ω resistor and a 2 μ*F* capacitor are connected in parallel across a 240 *V* 60 Hz source. Determine the circuit impedance and the line current.

Solution:

Current $$I_R = \frac{240}{4700} = 0.051\ A$$

$$I_C = j\omega CV$$

$$= j2\pi \times 60 \times 2 \times 10^{-6} \times 240$$

$$= j0.1808$$

Total current $$= I_R + I_C = 0.051 + j0.1808$$

$$= \frac{V}{R} + jV/1/\omega C$$

Therefore, impedance of the circuit is $\frac{V}{I}$

$$= \frac{240}{0.188\ \angle 74.3}$$

$$= 1276\ \angle -74.3\ \Omega$$

1.7.2 Series Parallel Circuits of R, L and C

Refer to Fig. 1.15

The current I_C will of course-lead the reference voltage *V* by 90°. Regarding *RL* series circuit, let the impedance angle be θ. Since it is an inductive circuit, current I_L will lag the voltage across it (supply voltage here) by angle θ. Hence total line current *I* is phasor sum of I_L and I_C as shown in Fig. 1.15(*b*).

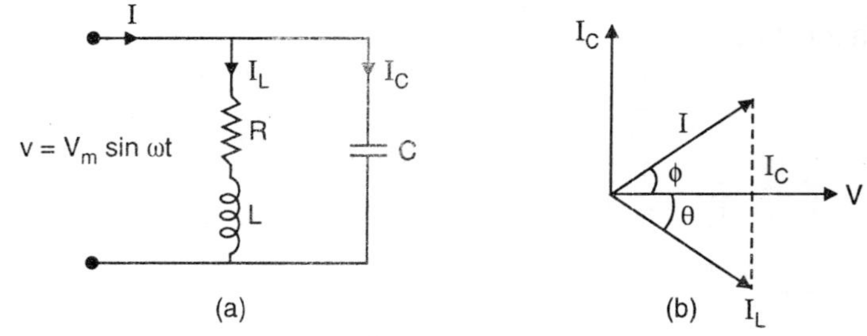

Fig. 1.15

Example 1.8: Determine the line current and the total impedance and admittance of the circuit shown in Fig. E 1.8.

Solution:

$$Z_1 = 25 + j50 = 55.9 \angle 63.4$$

$$Z_2 = 40 \angle -90$$

Hence $$Z_{eq} = \frac{Z_1 Z_2}{Z_1 + Z_2} = \frac{55.9 \angle 63.4 \; 40 \angle -90}{25 + j50 - j40}$$

$$= \frac{2236 \angle -26.6}{26.9 \angle 21.8}$$

$$= 83.12 \angle -48.4°$$

Admittance $$Y = \frac{1}{Z} = \frac{1}{83.12 \angle -48.4}$$

$$= 1.2 \times 10^{-2} \angle 48.4°$$

Total current $$I = VY = 100 \times 1.2 \times 10^{-2} \angle 48.4°$$

$$= 1.2 \angle 48.4° \text{A}$$ **Ans.**

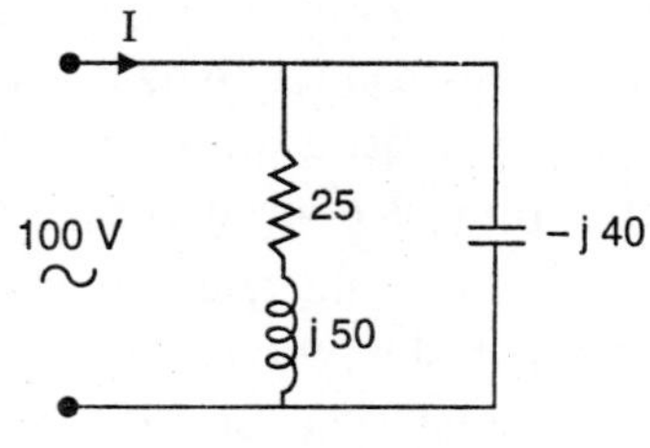

Fig. E 1.8

1.8 RESONANCE

Whenever the natural frequency of oscillation of a system (could be electrical, mechanical or a civil structure or a hydraulic) coincides with the frequency of the driving force (a voltage source in an electric circuit or a wind force in a civil structure etc.), the two system resonate with respect to each other and the system has maximum response to a fixed magnitude of driving force. This phenomenon is known as resonance. This phenomenon may be useful under certain conditions and sometimes it may prove to be disastrous for the system.

There are many engineering applications of resonance. In a refrigerator, the reciprocating compressor is mounted on a support designed to minimize the vibrations transmitted to the cabinet. A suspension bridge in Washington showed tendencies to oscillate up and down during construction and only a few months after construction, it began to build up oscillations under a moderate wind and within an hour the multi-billion dollar bridge was reduced to pieces. This is a typical example of designing a bridge ignoring the possibility of phenomenon of resonance on the bridges.

In case of a series compensated power system if the difference between the normal power frequency and the natural frequencies of oscillation of the system after compensation coincides with one of the natural torsional frequencies of the machine's shaft system (Turbine, or combination of turbine and generator), torsional oscillation may be excited which may build up sufficiently to break the shaft of these machines. This is known as subsynchronous resonance as the resonance occurs at a frequency less than the normal supply frequency.

A power transformer is normally operated at the knee-point of its B-H curve due to economic reasons. If some how the system voltage increases due to Ferranti effect or due to rejection of load, the transformer operates along the saturation region of the B-H curve where the inductance offered by the winding decreases and keeps on changing depending upon the dynamic behaviour of increase of system voltage. Saturation of transformers subjected to over voltages can produce high currents rich in harmonics and in the presence of sufficient capacitance there is risk of phenomenon of resonance taking place which may result in further increase in system voltages which may prove disastrous for the system insulation. This is known as Ferro-resonance as it is due to the non-linear behaviour of the iron core of the transformer. In such a situation it may be necessary to disconnect shunt capacitors very quickly to reduce the chances of occurrences of Ferro-resonance.

In an electric circuit with inductance and capacitor in series, there is always a frequency at which the two reactances just cancel resulting in the minimum impedance (resistive circuit) characteristic of series resonance. In electronic circuits many a times resonance condition is desired for maximum response for a given magnitude of excitation.

In general there are two types of resonance in the electric circuits :

(*a*) Series resonance,

(*b*) Parallel resonance.

We now study basic features and characteristic of these phenomena.

1.9 SERIES RESONANCE

Consider a series *RLC* circuit connected to a variable frequency voltage source as shown in Fig. 1.16.

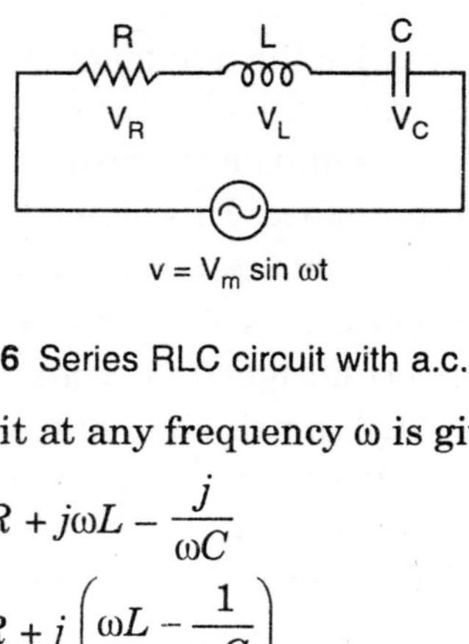

Fig. 1.16 Series RLC circuit with a.c. source.

The impedance of the circuit at any frequency ω is given as

$$Z = R + j\omega L - \frac{j}{\omega C}$$

$$= R + j\left(\omega L - \frac{1}{\omega C}\right) \quad \text{...(1.5)}$$

Since resistance is independent of frequency the circuit will have minimum impedance at some frequency.

When $$Z = R \quad \text{...(1.6)}$$

or when $$\omega_0 L - \frac{1}{\omega_0 C} = 0$$

or $$\omega_0 = \frac{1}{\sqrt{LC}}$$

$$f_0 = \frac{1}{2\pi\sqrt{LC}} \quad \text{...(1.7)}$$

Here f_0 is the frequency of resonance i.e. if a circuit has fixed values of *R*, *L* and *C*, resonance will take place if the supply frequency

$f_0 = \dfrac{1}{2\pi\sqrt{LC}}$ when the impedance of the circuit is purely resistive i.e. the power factor of the circuit is unity, the supply voltage and current are in phase. However, it is to be noted that the phase relations between the voltage and current in the individual elements R, L and C are not same. The current in the inductor lags its voltage by 90° and in the capacitor it leads its voltage by 90°.

The variation of impedance of the three elements as a function of frequency is shown in Fig. 1.17.

Since R is independent of frequency it is shown by a horizontal line $Z = R$.

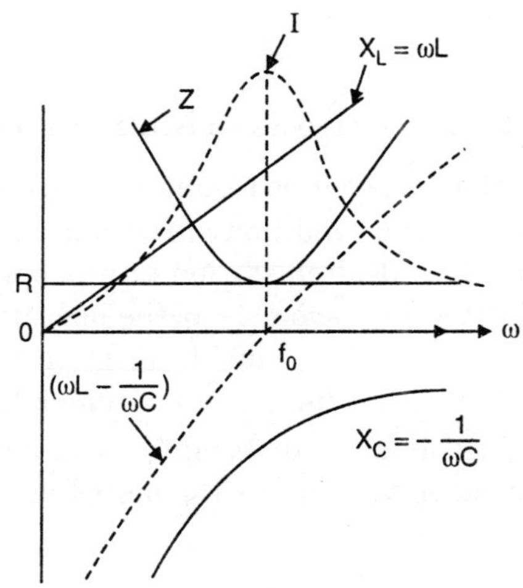

Fig. 1.17 $ZV_s\,\omega$ for RLC series circuit.

Also $X_L = \omega L$ the inductive reactance is a straight line passing through the origin and inductive reactance is taken as +ve, where as $X_C = \dfrac{1}{\omega C}$ the capacitive reactance as a function of ω is a rectangular hyperbola and the reactance is taken as –ve. The net impedance is shown a positive quantity. The resonance frequency is f_0 where $\left(\omega L - \dfrac{1}{\omega C}\right)$ is zero and at this frequency the impedance curve has minimum value equal to R. The variation of current is also shown in Fig. 1.17 as a function of frequency and is maximum at f_0 whereas on either side the current decreases. It is to be noted that at $\omega = 0$, the current in the RLC series circuit is zero as the capacitor reactance is infinite and, therefore, the graph starts from origin whereas it is again zero at $\omega = \infty$ and hence the graph should not be passing through zero rather it should have some finite value as indicated in the diagram.

Again, it can be seen that the series circuit is capacitive for all frequency $\omega < \omega_0$ and at $\omega = \omega_0$ the circuit is resistive. Fig. 1.18 shows the current response of the RLC series circuit for certain voltage V when one of the three parameters is changed at a time.

It is seen that the current response curve must always start from origin and the frequency of resonance can be varied by varying either L, or C or both. The steepness of the response curve can be varied by varying the resistance of the circuit.

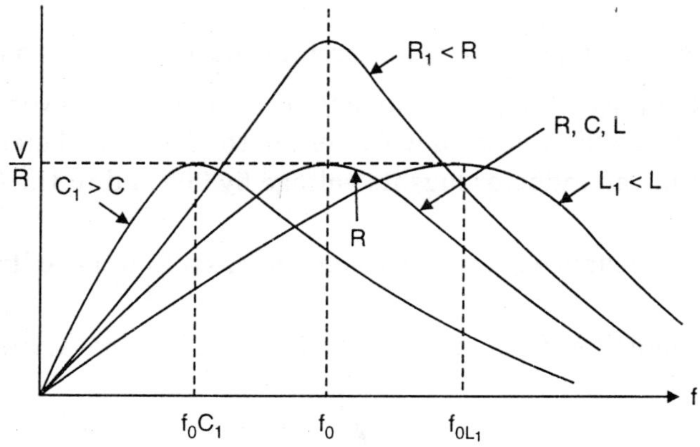

Fig. 1.18 Current response for RLC series circuit.

There are various applications of a series resonant circuit where the frequency is fixed and either L or C is varied to obtain the condition of resonance. A typical example is that of tuning a radio receiver to a particular desired station that is operating at a fixed frequency. Here a circuit or L and C is adjusted to resonance at the operating frequency of the desired station. The capacitor C (parallel plate) is variable in most portable radio receivers and the inductance of the coil is usually varied in tuning of an automobile radio receiver.

Example 1.9: For the Fig. E 1.9 shown determine the maximum current, the frequency at which it occurs and the resulting voltage across the inductance and capacitance.

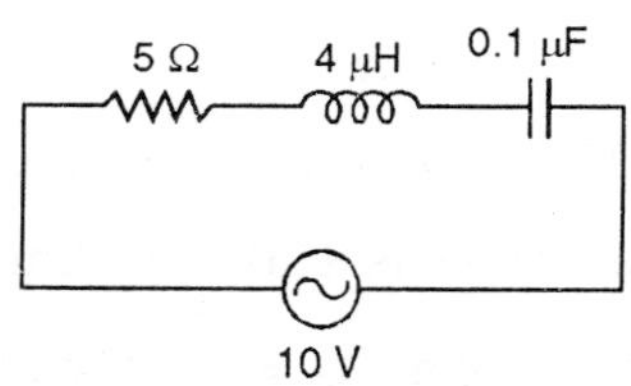

Fig. E1.9

Solution: The frequency at which the current is maximum is

$$\omega_0 = \frac{1}{\sqrt{4 \times 10^{-3} \times 0.1 \times 10^{-6}}}$$

$$= 5 \times 10^4 \text{ rad/sec.}$$

The current at resonance is $\frac{V}{R} = \frac{10}{5} = 2\text{A}$

The voltage across the inductance

$$= \omega_0 LI = 5 \times 10^4 \times 4 \times 10^{-3} \times 2$$

$$= 400 \ V$$

and voltage magnitude across the capacitor is

$$= \frac{1}{\omega_0 C} = \frac{2}{5 \times 10^4 \times 0.1 \times 10^{-6}}$$

$$= 400 \ V$$

It is seen that the voltage across L or C under resonance is greater than the voltage across all the three elements i.e. the supply voltage.

Fig. E 1.9 shows phasor diagram of the RLC series circuit corresponding to $\omega < \omega_0$, $\omega = \omega_0$ and $\omega > \omega_0$. We have taken current as the reference phasor as it is a series circuit and current is same through all the elements.

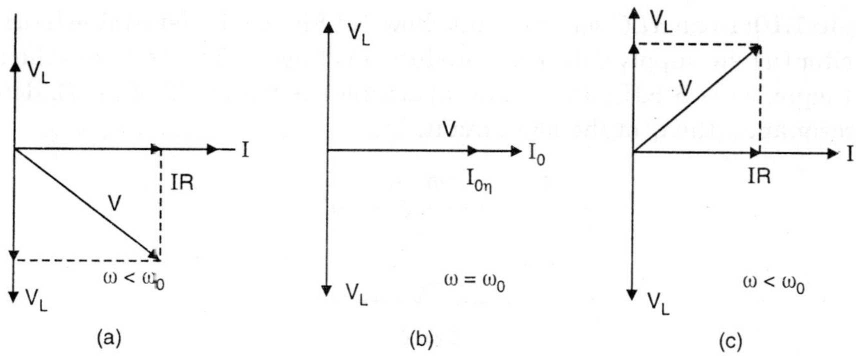

Fig. E1.9 Phasor diagram of RLC series circuit (a) $\omega < \omega_0$ (b) $\omega = \omega_0$ (c) $\omega > \omega_0$.

In Fig. 1.9(a) since $\omega < \omega_0$, the circuit is predominantly capacitive hence the net supply voltage lags behind the circuit current. However the voltage across the inductor leads the current by 90°, the voltage across the resistor is in phase with the circuit current whereas voltage across the capacitor lags behind the current by 90°. The voltage across inductor and capacitor are 180° out of phase with reference to each other. Since the circuit at this frequency is capacitive the net voltage lags behind the circuit current by some angle.

In Fig. E 1.9(b) since $\omega = \omega_0$ the circuit is resistive as the voltage drop across inductance exactly cancels the voltage across the capacitance as these are equal in magnitude but opposite in phase relation. It is to be seen that under this condition voltage across L or C individually may far exceed the supply voltage.

Next we find expression for voltage drops across individual elements under resonance condition.

Under resonance

$$I_0 = \frac{V}{R} \qquad \text{...(1.8)}$$

Therefore,

$$V_R = I_0 R = \frac{V}{R} \cdot R = V \text{ the supply voltage} \qquad \text{...(1.9)}$$

$$V_L = I_0 \omega L = \frac{V}{R}\, \omega_0 L = V \cdot \frac{\omega_0 L}{R}$$

$$= VQ \qquad \text{...(1.10)}$$

where Q is known as the quality factor of the series network. Usually $Q >> 1$, hence it is also known as voltage gain, as the voltage across the inductor is much greater than the supply voltage.

Also $$V_C = \frac{I_0}{\omega_0 C} = V \cdot \frac{1}{\omega_0 CR}$$
$$= VQ \qquad \text{...(1.11)}$$

Hence $$V_L = V_C >> V$$

Therefore, extreme care must be taken when working on series circuits that may become resonant when connected to power line sources.

Example 1.10: In an *RLC* series circuit shown in Fig. E1.10 determine (*i*) the necessary value of capacitor (*ii*) the supply voltage to produce a voltage of 5 volts across the capacitance if resonance frequency is 5 KHz (*iii*) If the capacitance is made 1/2 of at (*i*), determine the frequency of resonance, the *Q* of the new circuit.

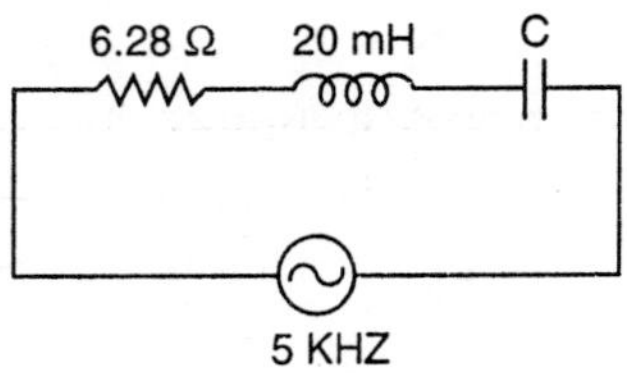

Fig. E1.10

Solution:

$$5000 = \frac{1}{2\pi\sqrt{20 \times 10^{-3} \times C \times 10^{-6}}}$$
$$C = 0.0507\ \mu F$$
$$V_C = 5 = V \cdot \frac{1}{2\pi \times 5 \times 10^3 \times 0.0507 \times 10^{-6} \times 6.28}$$

or $$V = 0.05 \text{ volt}$$

(*iii*) If the capacitance is halved, the frequency of resonance would be $\sqrt{2}\, f_0$
$$= \sqrt{2} \times 5000 = 7071 \text{ Hz}$$

and *Q* of the coil $$= \frac{2\pi \times 7071 \times 20 \times 10^{-3}}{6.28}$$
$$= 141.4 \quad \textbf{Ans.}$$

1.9.1 Quality Factor Q

In a practical circuit *R* is essentially resistance of the coil since practical capacitors have very low loss in comparison to practical inductor. Hence *Q* is a measure of the energy storage property (LI^2) in relation to the energy dissipation property (I^2R) of a coil or a circuit. The *Q* is, therefore, defined as

$$Q = 2\pi \cdot \frac{\text{maximum energy stored}}{\text{energy dissipated per cycle}} \qquad \text{...(1.12)}$$

In electric circuit energy is stored in the form of electromagnetic field in the inductance where as in electrostatic form of energy across a capacitance. It can be proved that at any instant at a certain frequency the sum of energy stored by the inductor and the capacitor is constant. At the extreme situation when the current through the inductor is maximum, the voltage across the capacitor is zero hence the total energy is

$$\frac{1}{2}L(\sqrt{2}I)^2 = LI^2 \qquad ...(1.13)$$

where $\sqrt{2}\,I$ is the instantaneous maximum value of the current. At this since V_C is zero, therefore maximum energy stored is LI^2. The power consumed per cycle is the energy per sec divided by f_0 under resonance condition. Therefore $P_R = \dfrac{I^2 R}{f_0}$

Hence
$$Q = \frac{2\pi LI^2 f_0}{I^2 R} = \frac{\omega_0 L}{R} \qquad (1.14)$$

Q can also be looked as the ratio of

$$\frac{\text{Time rate of change of energy stored}}{\text{Time rate of change of energy dissipated}}$$

$$= \frac{\text{Reactive power absorbed by the inductor}}{\text{Active power consumed by the resistor}} \qquad ...(1.15)$$

The great advantage of this definition of Q is that it is also applicable to more complicated lumped circuits, to distributed circuits such as transmission lines and to non-electrical circuits.

Q is also a measure of the frequency selectivity of the circuit. A circuit with high Q will have a very sharp current response curve as compared to one which has a low value of Q. To understand this let us consider Fig. 1.19. Here we find that the current response is maximum at f_0 and on either side of f_0, the current decreases sharply.

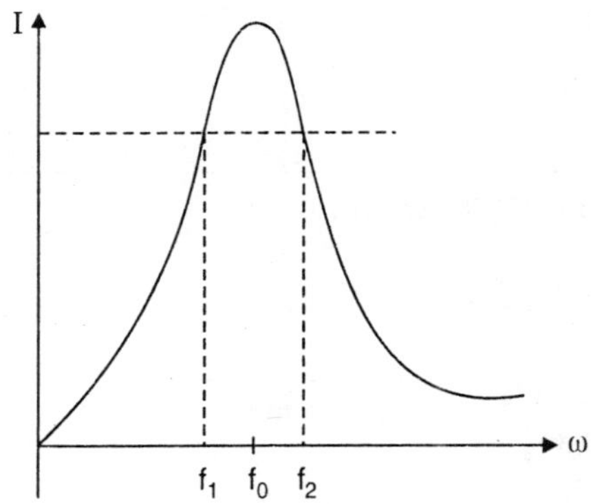

Fig. 1.19 Frequency selectivity.

In order to obtain quantitative analysis of this reduction in current, we specify two frequencies f_1 and f_2 at which the magnitude $(X_L - X_C)$ is equal to R.

Since at f_1 the circuit is capacitive X_C is greater than X_L, therefore, at f_1,

$$X_C - X_L = R \qquad ...(1.16)$$

and at f_2

$$X_L - X_C = R \qquad ...(1.17)$$

The corresponding impedances are :

$$Z_1 = \sqrt{R^2 + R^2} \angle \tan^{-1} \frac{-R}{R} = \sqrt{2} R \angle -45° \quad ...(1.18)$$

and

$$Z_2 = \sqrt{R^2 + R^2} \angle \tan^{-1} \frac{-R}{R} = \sqrt{2} R \angle 45° \quad ...(1.19)$$

The current at these two frequencies are :

$$I_1 = \frac{V \angle 45°}{\sqrt{2}R} \quad ...(1.20)$$

and

$$I_2 = \frac{V \angle -45°}{\sqrt{2}R} \quad ...(1.21)$$

Since $\frac{V}{R} = I_0$ the current at resonance at f_0.

$$I_1 = 0.707\, I_0 \angle 45° \quad ...(1.22)$$

and

$$I_2 = 0.707\, I_0 \angle -45° \quad ...(1.23)$$

Frequencies f_1 and f_2 are known as half-power frequencies as at these frequencies the power dissipated by the circuit is half of that dissipated at f_0.

The band width of a resonant circuit is defined as the frequency range between the 70.7% current points.

i.e.

$$BW = f_2 - f_1 \quad ...(1.24)$$

At f_1

$$X_C - X_L = R \quad ...(1.25)$$

$$\frac{1}{2\pi f_1 C} - 2\pi f_1 L = R \quad ...(1.26)$$

and at f_2

$$X_L - X_C = R \quad ...(1.27)$$

$$2\pi f_2 L - \frac{1}{2\pi f_2 C} = R \quad ...(1.28)$$

On subtracting (1.28) from (1.26) we have

$$\left(\frac{1}{\omega_1} + \frac{1}{\omega_2}\right)\frac{1}{C} - (\omega_1 + \omega_2)L = 0$$

or

$$\frac{\omega_2 + \omega_1}{\omega_1 \omega_2} = LC\,(\omega_1 + \omega_2)$$

or

$$\frac{1}{\omega_1 \omega_2} = LC = \frac{1}{\omega_0^2}$$

or

$$\omega_0 = \sqrt{\omega_1 \omega_2} \quad ...(1.29)$$

This means the resonance frequency is the geometric mean of the half-power frequencies.

Now on addition of (1.26) and (1.28) we have

$$\frac{1}{\omega_1 C} - \frac{1}{\omega_2 C} + \omega_2 L - \omega_1 L = 2R$$

or $$\frac{\omega_2 - \omega_1}{\omega_1\omega_2 C} + (\omega_2 - \omega_1)L = 2R$$

or $$(\omega_2 - \omega_1)\left(\frac{1}{\omega_0^2 C} + L\right) = 2R$$

or $$(\omega_2 - \omega_1)\left(\frac{2}{\omega_0^2 C}\right) = 2R$$

or $$(\omega_2 - \omega_1) = \omega_0^2 CR$$

$$\frac{\omega_2 - \omega_1}{\omega_0} = \omega_0 CR = \frac{1}{Q}$$

$$f_2 - f_1 = \frac{f_0}{Q} \qquad ...(1.30)$$

Bandwidth is thus given by the ratio of the frequency of resonance to the quality factor and selectivity is defined as the ratio of resonant frequency to the bandwidth $f_0/(f_2 - f_1)$. This, therefore, shows that the larger the value of Q the smaller is $(f_2 - f_1)$ and hence sharper is the current response.

Example 1.11: For the circuit shown determine the value of inductance for resonance if $Q = 50$ and $f_0 = 175$ KHZ. Also find the circuit current the voltage across the capacitor and the bandwidth of the circuit.

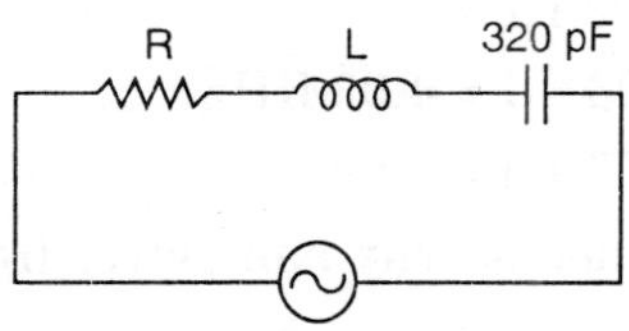

Fig. E1.11

Solution: $$f_0 = \frac{1}{2\pi\sqrt{LC}}$$

$$175 \times 10^3 = \frac{1}{2\pi\sqrt{320 \times 10^{-12}\, L}}$$

or $$L = 2.58 \text{ mH}$$

The reactance of the coil at resonance is

$2\pi \times 175 \times 10^3 \times 2.58 \times 10^{-3} = 2840\ \Omega$

Since $$Q = \frac{\omega_0 L}{R}$$

or $$R = \frac{\omega_0 L}{Q} = \frac{2840}{50} = 56.8\ \Omega$$

The impedance of the circuit at resonance is

$$Z = R = 56.8\ \Omega$$

Therefore, current $I_0 = \dfrac{0.85}{56.8} = 14.96$ mA

The voltage across the capacitor = QV

$$= 50 \times 0.85 = 42.5 \text{ volt}$$

The bandwidth of the circuit is $\frac{f_0}{Q} = \frac{175 \times 10^3}{50}$

$$\frac{f_0}{Q} = 3.5 \times 10^3 \text{ Hz}$$

$$= 3.5 \text{ KHz.}$$

Example 1.12: For the circuit shown, determine the impedance magnitude at resonance, at 1 KHz below resonance and, at 1 KHz above resonance.

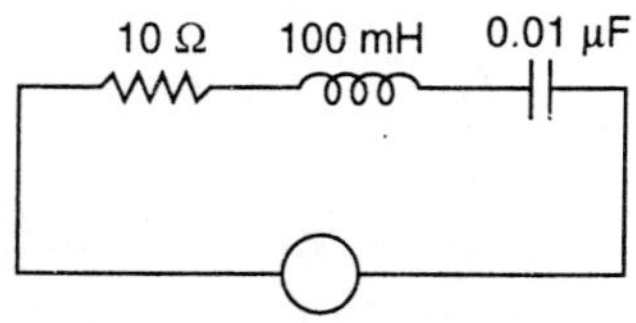

Fig. E1.12

Solution:

$$f_0 = \frac{1}{2\pi\sqrt{100 \times 10^{-3} \times 0.01 \times 10^{-6}}}$$

$$= 5.03 \text{ KHz}$$

Frequency $f_1 = 5.03 - 1 = 4.03$ KHz

and $f_2 = 5.03 + 1 = 6.03$ KHz

Note these f_1 and f_2 are not necessarily half power frequencies and hence Z_1 need not necessarily equal Z_2.

$$Z_1 = \sqrt{10^2 + (2.53 - 3.95) \times 10^6}$$

Since 10 is negligibly small as compared to X_L and X_C which are in $K\,\Omega$, therefore,

$$Z_1 = -j1.42 \text{ K}\Omega$$

$$Z_2 = j3.79 - j2.64 = j1.15 \text{ K}\Omega \quad \textbf{Ans.}$$

Before we move to parallel resonance, let us summarize the characteristics of series resonance circuit.

1. At resonance power factor of the circuit is unity.
2. Therefore, supply voltage and current are in phase.
3. The reactive component of the input impedance is zero and hence the circuit is resistive and hence current drawn by the circuit is maximum.
4. The frequency of resonance is given by

$$f_0 = \frac{1}{2\pi\sqrt{LC}}.$$

5. Even there is voltage across the individual reactive element but across both the elements the net voltage is zero as the two voltages are equal in magnitude but 180° out of phase. The voltage across each reactive element is QV where Q is the quality factor of the element and V is the supply voltage.

6. The quality factor

$$Q = \frac{\omega_0 L}{R} = \frac{1}{\omega_0 CR} \quad \text{...(1.31)}$$

7. For $\omega < \omega_0$ the circuit is capacitive
   $\omega = \omega_0$ the circuit is resistive
   $\omega > \omega_0$ circuit is inductive.

8. The band width $f_2 - f_1 = \dfrac{f_0}{Q}$

and $f_0 = \sqrt{f_1 f_2}$

Example 1.13: For the circuit shown in Fig. E1.13 $R_1 = 0.5\ \Omega$, $R_2 = 1.5\ \Omega$, $R_3 = 0.5\ \Omega$ $C_1 = 6\ \mu F$ and $C_2 = 12\ \mu F$ $L_1 = 25$ mH and $L_2 = 15$ mH. Determine (*i*) the frequency of resonance (*ii*) Q of the circuit (*iii*) Q of coil 1 and coil 2 individually

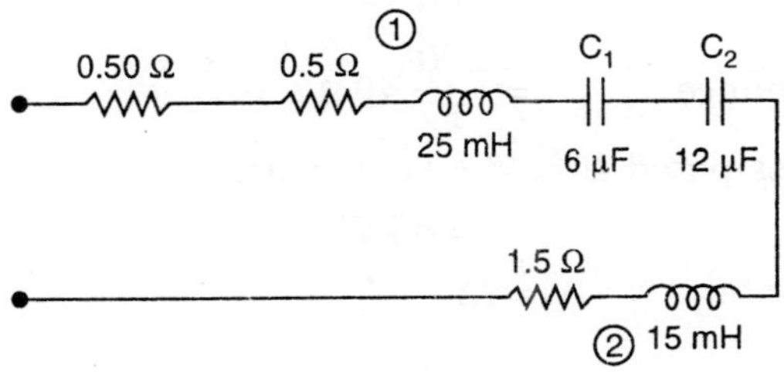

Fig. E1.13

Solution:

Total inductance of the circuit 25 + 15 = 40 mH

Total capacitance of the circuit $\dfrac{6 \times 12}{18} = 4\mu F$

$$\text{The frequency of resonance} = \frac{1}{2\pi\sqrt{40 \times 10^{-3} \times 4 \times 10^{-6}}}$$

$$= \frac{10^4}{8\pi} \text{ Hz} \quad \text{or} \quad \omega_0 = 2.5 \times 10^3 \text{ rad/sec}$$

$$Q \text{ of the circuit } = \frac{\omega_0 L_{eq}}{R_{eq}} = \frac{2.5 \times 10^3 \times 40 \times 10^{-3}}{2.5}$$

$= 40$ **Ans.**

$$Q \text{ of coil 1 } \frac{\omega_0 L_1}{R_1} = \frac{2.5 \times 10^3 \times 25 \times 10^{-3}}{0.5} = 125$$

$$Q \text{ of coil 2 } \frac{\omega_0 L_2}{R_2} = \frac{2.5 \times 10^3 \times 15 \times 10^{-3}}{1.5} = \frac{2.5 \times 15}{1.5}$$

$= 25$ **Ans.**

$$\text{Bandwidth of the circuit} = \frac{f_0}{Q}$$

$$= \frac{10^4}{8\pi \times 40} = 9.95 \text{ Hz.}$$

Example 1.14: A coil having a 5 ohm resistor is connected in series with a 50 μF capacitor. The circuit resonates at 100 Hz (*a*) Determine the inductance of the coil. (*b*) If the circuit is connected across a 200 V 100 Hz a.c. source, determine the power delivered to the coil (*c*) the voltage across the capacitor and the coil (*d*) the bandwidth of the circuit.

Solution: At resonance $\omega_0 = \dfrac{1}{\sqrt{LC}}$

$$2\pi \times 100 = \frac{1}{\sqrt{50 \times 10^{-6} L}}$$

or $L = \dfrac{25}{500} = 50$ mH

The current at resonance $= \dfrac{200}{5}$ $40\,A$

Therefore, power dissipated $40^2 \times 5 = 8000$ watts.

$= 8$ kW.

Voltage across capacitor $= IX_c$

$$= \frac{40}{50 \times 10^{-6} \times 100 \times 6.28}$$

$$= \frac{40 \times 10^6}{5 \times 10^3 \times 6.28} = \frac{8000}{6.28}$$

The impedance of the coil $R + j\,\omega_0 L$

$= 5 + j628 \times 50 \times 10^{-3}$

$= 5 + j31.40$

The voltage $V_L = 40(5 + j31.4)$

$\simeq 1256$ volts **Ans.**

The Q_0 of the coil $= \dfrac{31.4}{5} \simeq 6.3$

The bandwidth $= \dfrac{f_0}{Q_0} = \dfrac{100}{6.3} \simeq 16$ Hz

1.10 PARALLEL RESONANCE

Consider Fig. 1.20 where *L* and *C* are connected in parallel. In series with these elements, resistances shown are their own resistances. Coil will have its own resistance and the capacitor may have some loss component. In general there could be external resistance in series with these elements.

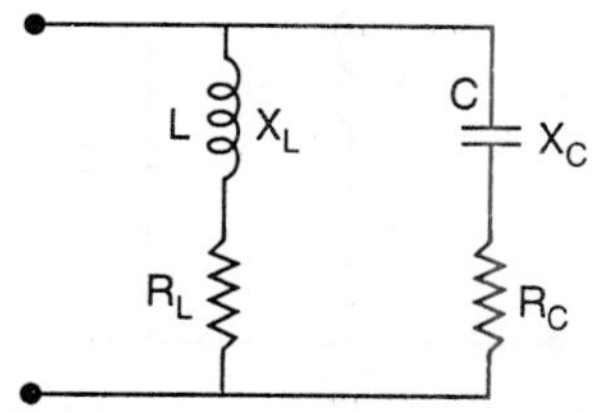

Fig. 1.20 Parallel resonance.

The input admittance is the sum of the admittances of the two branches.

$$Y_{eq} = \left(\frac{R_c}{R_c^2 + X_c^2} + j\frac{X_c}{R_c^2 + X_c^2}\right) + \left(\frac{R_L}{R_L^2 + X_L^2} - j\frac{X_L}{R_L^2 + X_L^2}\right)$$

$$= \left(\frac{R_c}{R_c^2 + X_c^2} + \frac{R_L}{R_L^2 + X_L^2}\right) + j\left(\frac{X_c}{R_c^2 + X_c^2} - \frac{X_L}{R_L^2 + X_L^2}\right)$$

Now at resonance the imaginary part (susceptance) of admittance is zero.

Therefore,

$$\frac{X_{c_0}}{R_c^2 + X_{c_0}^2} = \frac{X_{L_0}}{R_L^2 + X_{L_0}^2} \qquad ...(1.32)$$

or

$$X_{c_0}R_L^2 + X_{c_0}X_{L_0}^2 = X_{L_0}R_c^2 + X_{L_0}X_{c_0}^2$$

$$\frac{1}{2\pi f_0 C}R_L^2 + \frac{1}{2\pi f_0 C}\cdot(2\pi f_0 L)^2 = 2\pi f_0 L\ R_c^2 + \frac{2\pi f_0 L}{(2\pi f_0 C)^2}$$

Multiplying by $2\pi f_0 C$ we have

$$R_L^2 + 4\pi^2 f_0^2 L^2 = 4\pi^2 f_0^2 R_c^2 LC + \frac{L}{C}$$

or

$$4\pi^2 f_0^2 LC\left(R_c^2 - \frac{L}{C}\right) = R_L^2 - \frac{L}{C}$$

or

$$4\pi^2 f_0^2 = \frac{1}{LC}\cdot\frac{R_L^2 - L/C}{R_c^2 - L/C}$$

or

$$f_0 = \frac{1}{2\pi\sqrt{LC}}\cdot\sqrt{\frac{R_L^2 - L/C}{R_c^2 - L/C}} \qquad ...(1.33)$$

This is the frequency of resonance of the general parallel *RLC* circuit.

The equivalent input admittance is real and is given as

$$Y_{eq_0} = G_0 = \frac{R_c}{R_c^2 + X_c^2} + \frac{R_L}{R_L^2 + X_L^2} \qquad ...(1.34)$$

Here if R_c is the equivalent series resistance of the capacitance, it can be ignored as it usually is small in magnitude, the equivalent circuit then is given in Fig. 1.21.

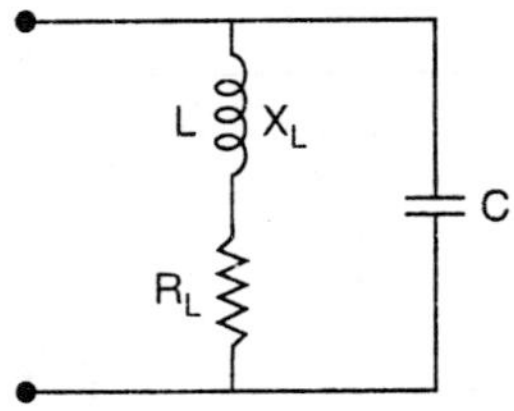

Fig. 1.21 Parallel LC circuit with $R_C = 0$.

The resonance frequency for the circuit is given by the following expression when $R_c = 0$ in the general circuit expression.

$$f_0 = \frac{1}{2\pi\sqrt{LC}} \cdot \sqrt{\frac{R_L^2 - L/C}{-L/C}}$$

$$= \frac{1}{2\pi\sqrt{LC}} \cdot \sqrt{1 - R_L^2 \frac{C}{L}} \qquad ...(1.35)$$

The general expression for Y_{in} in this case is obtained by putting $R_c = 0$ in the expression for Y_{eq} and variation of Z_{in} and I_{input} vs ω are drawn in Fig. 1.22 (*a*) and (*b*).

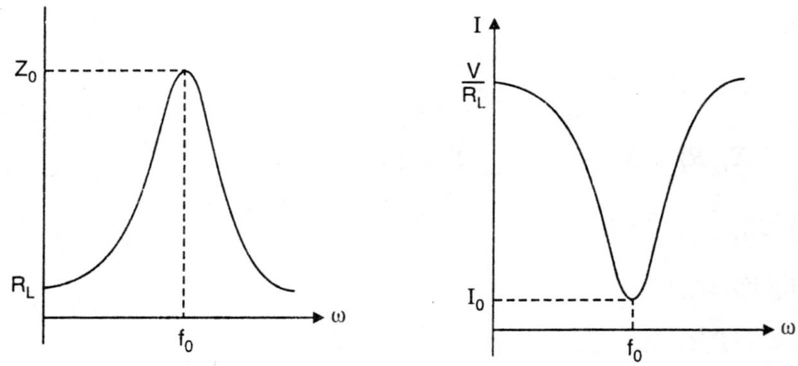

Fig. 1.22 (*a*) *Z Vs f* (*b*) *I Vs f* for parallel resonance.

The shape of curve for current and Y_{in} will be identical as $I = Y_{in} V$.

It is seen that at resonance the circuit has maximum impedance and hence the current is minimum. At this frequency since the susceptance of the circuit is zero, the power factor of the circuit is unity.

At frequency $f < f_0$ since the capacitance acts as an open circuit the net circuit is inductive and at higher frequencies $f > f_0$ the inductor acts as an open circuit the net circuit is capacitive.

If we now consider that in the general circuit if R_L is also zero besides R_c being zero, we have the equivalent circuit (Fig. 1.23).

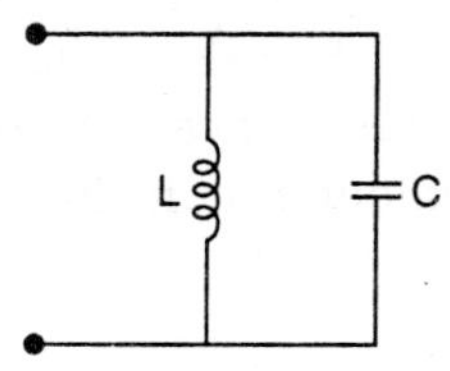

Fig. 1.23 Parallel LC circuit.

For this the frequency of resonance is

$$f_0 = \frac{1}{2\pi\sqrt{LC}} \quad ...(1.36)$$

and it can be seen that at the frequency the input impedance is

$$\frac{jX_L(-jX_c)}{jX_L - jX_c} = \frac{jX_L(-jX_c)}{0}$$

$$= \infty$$

and hence theoretically the current drawn from the source is zero. This means there will be current in each parallel path but the net current from the supply is zero i.e. there will be exchange of energy between the inductance and the capacitance of the circuit. Actually because of the small winding resistance of the coil some power is lost so the reactive currents are not exactly identical and there is a small current from the source to take into account this loss.

If we assume that

$$R_L = R_c = \sqrt{\frac{L}{C}} = \sqrt{\frac{\omega L}{\omega C}} = \sqrt{X_L X_c}$$

Therefore, $$Y_{in} = \frac{R_c}{X_L X_c + X_c^2} + \frac{R_L}{X_L X_c + X_L^2} + j\left(\frac{X_c}{X_L X_c + X_c^2} - \frac{X_L}{X_L X_c + X_L^2}\right)$$

The imaginary part

$$= \frac{X_L X_c^2 + X_c X_L^2 - X_L^2 X_c - X_L X_c^2}{(X_L X_c + X_c^2)(X_c X_L + X_L^2)}$$

$$= 0$$

Therefore, Y_{in} is resistive at all frequencies that means resonance will occur at all frequencies.

Consider Fig. 1.21. The impedance of the parallel circuit is

$$Z_T = \frac{(R + jX_L).(-jX_c)}{R + jX_L - jX_c} \quad ...(1.37)$$

Let $$X_L = X_C$$

Therefore, $$Z_T = \frac{(X_L - jR)X_c}{R} \quad ...(1.38)$$

Let $$\frac{X_L}{R} = Q \text{ the quality factor of the coil,}$$

$$Z_T = (Q - j1)X_c$$

Usually $Q >> 1$, therefore, ...(1.39)

At resonance

$$Z_T = Q\,X_{C0} \quad ...(1.40)$$

Since this impedance is effectively a pure resistance, for the high Q-circuit, resonance occurs when $X_{L0} = X_{C0}$.

and $$f_0 = \frac{1}{2\pi\sqrt{LC}}$$

These relations hold good only when $Q >> 1$ (should be greater than 5). Similarly bandwidth of the circuit is

$$BW = \frac{f_0}{Q}$$

Example 1.15: For the circuit shown in Fig. E 1.15 determine the current through the parallel branches and, the supply current.

(*b*) If the frequency is made $f_0/2$ and $2f_0$ determine the currents from the source under these conditions.

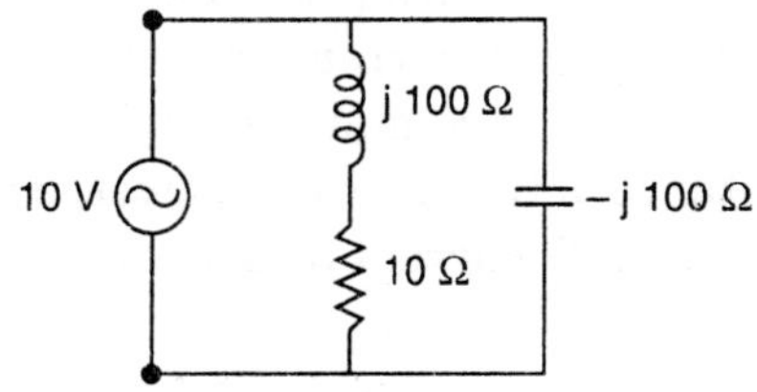

Fig. E1.15 Value at resonance.

Solution: (*a*) $$Z_0 = QX_{C0} = \frac{100}{10} \times 100 = 1000\ \Omega$$

The current $$I_0 = \frac{V}{Z_0} = \frac{10}{1000} = 10 \text{ mA}$$

The current through the capacitor branch

$$I_C = \frac{10}{-j100} = j100 \text{ mA}$$

$$I_L = \frac{10}{j100} = -j\ 100 \text{ mA}$$

The current through capacitor and inductor are QI_0. These large currents in L and C are called oscillating currents because they are confined within the L and C of the parallel resonant circuit. The supply current is 10 mA which supplies the losses of the parallel resonance circuit. The input power from the source is 10 × 10 = 100 mW.

The I^2R loss in the 10 Ω resistor is

$$(0.01)^2 \times 10 = 100 \text{ mW}$$

If however, the frequency is reduced to $f_0/2$ the inductive reactance become 50 Ω and capacitor 200 Ω.

The net impedance $$= \frac{(10 + j50)(-j200)}{10 + j50 - j200}$$

$$= 68.8 \angle 75°$$

The current $$I = \frac{10 \angle -75°}{68.8}$$

$$= 0.145 \angle -75°$$

If the frequency is doubled

$$X_L = j200 \quad \text{and} \quad X_C = -j50$$

$$Z_T = \frac{(10 + j200)(-j50)}{10 + j200 - j50}$$

$$= 66.4 \angle -89°$$

The current $$I = \frac{10\ \angle 89^\circ}{66.4} = 0.151\ \angle 89^\circ$$

This shows that impedance at resonance is maximum and the current is minimum. The currents through the parallel branches gets magnified at resonance and is QI_0. The circuit is inductive for $f < f_0$ and is capacitive for $f > f_0$.

If we load this parallel resonant circuit with some resistance R the effective resistance will be less than QX_{C0}. Let the resistance be $R' = Q' X_{C0}$. Therefore, Q' is smaller than Q and hence the bandwidth of the circuit is increased when the resonant circuit is shunted with the resistance R.

1.10.1 Q of Parallel RLC Circuit

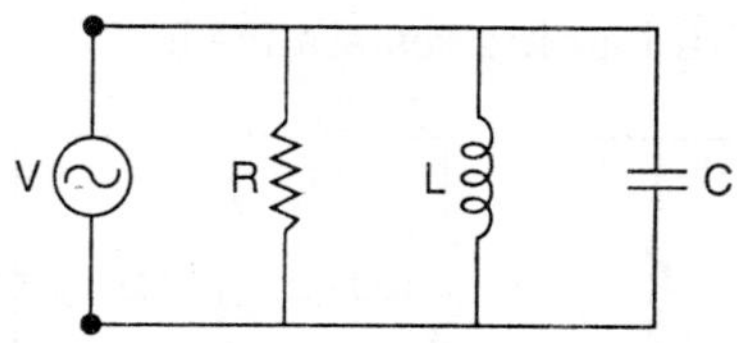

Fig. 1.24 Parallel RLC circuit.

At resonance there will be exchange of energy between the inductor and the capacitor. When the inductor is carrying maximum current, the voltage across capacitor is zero and hence the electrostatic energy is zero.

The time rate of change of energy through inductor is $\frac{V^2}{\omega_0 L}$ and the time rate of change of power is $\frac{V^2}{R}$.

Therefore, $$Q_0 = \frac{V^2}{\omega_0 L} \cdot \frac{R}{V^2} = \frac{R}{\omega_0 L}$$

$$= \frac{V^2 \omega_0 cR}{V^2} = \omega_0 CR \qquad \text{...(1.41)}$$

and bandwidth $$= \frac{f_0}{Q}$$

Example 1.16: Consider Fig. E1.16 determine the original bandwidth of the circuit and the loaded circuit bandwidth.

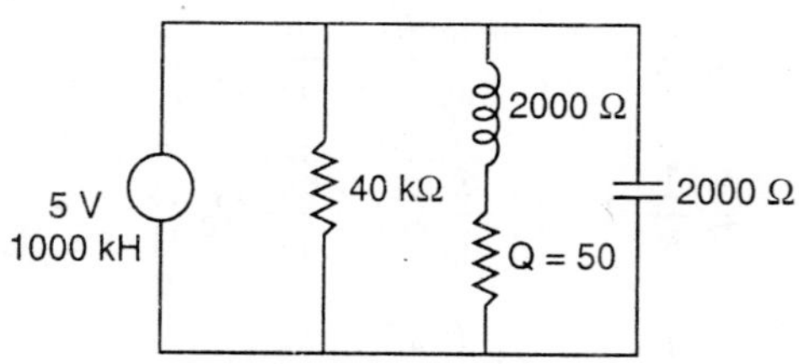

Fig. E1.16

The original equivalent circuit is

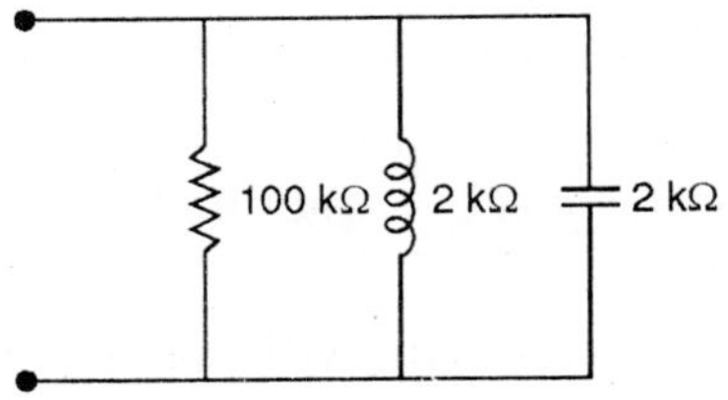

Fig. E1.16 (*a*)

$$Z_0 = Q\,X_{C0} = 50 \times 2 = 100 \text{ k}\Omega$$

The bandwidth $= \dfrac{f_0}{Q} = \dfrac{1000}{50} = 20 \text{ kHz}$

The equivalent circuit with loading resistance is

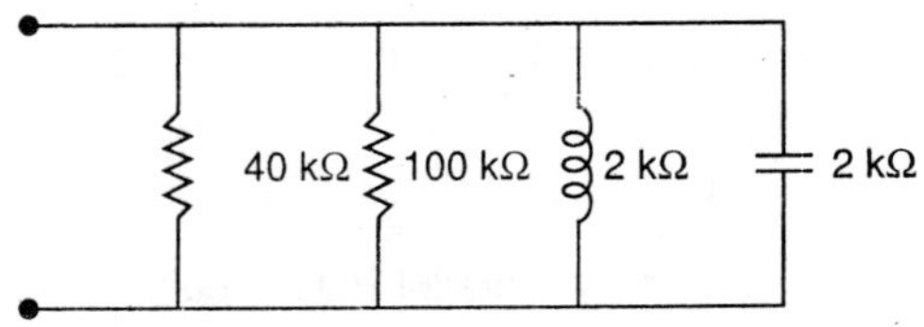

Fig. E1.16 (*b*)

$$Q'\,X_{C0} = \frac{40 \times 100}{140} = \frac{200}{7}$$

Therefore, $Q' = \dfrac{200}{7} \times \dfrac{1}{2} = \dfrac{100}{7}$

Therefore, bandwidth $= \dfrac{1000}{100} \times 7 = 70 \text{ kHz}$

It is to be noted that bandwidth in case of parallel resonance is obtained when net impedance of the circuit is 70.7% of the maximum value.

For a parallel *RLC* circuit the admittance is given as

$$Y_{in} = \frac{1}{R} + j\left(\omega C - \frac{1}{\omega L}\right) \qquad ...(1.42)$$

Since $Q_0 = \omega_0 CR$ we have

$$Y_{in} = \frac{1}{R} + j\left(\frac{\omega\omega_0 CR}{\omega_0 R} - \frac{\omega_0}{\omega\omega_0 L}\right)$$

$$= \frac{1}{R} + \frac{j}{R}\left(\frac{\omega\omega_0 CR}{\omega_0} - \frac{\omega_0 R}{\omega\omega_0 L}\right)$$

$$= \frac{1}{R}\left[1 + j\left(\frac{\omega}{\omega_0}Q_0 - \frac{\omega_0}{\omega}Q_0\right)\right]$$

$$= \frac{1}{R}\left[1 + jQ_0\left(\frac{\omega}{\omega_0} - \frac{\omega_0}{\omega}\right)\right] \qquad ...(1.43)$$

Since at resonance the equivalent admittance of the circuit is $1/R$ at 1/2 power frequencies it would be $\frac{\sqrt{2}}{R}$. Therefore, at $\frac{1}{2}$ power frequencies the imaginary quantity should be equal to unity so that

$$Y_{in} = \left|\frac{\sqrt{2}}{R}\right|$$

Therefore,

$$Q_0 \cdot \frac{\omega_1}{\omega_0} - \frac{\omega_0}{\omega_1} Q_0 = -1 \qquad \text{...(1.44)}$$

or $$\omega_1^2 - \omega_0^2 = -\frac{\omega_1\omega_0}{Q_0}$$

or $$\omega_1^2 + \frac{\omega_1\omega_0}{Q_0} - \omega_0^2 = 0$$

or $$\omega_1 = \frac{-\frac{\omega_0}{Q_0} + \sqrt{\frac{\omega_0^2}{Q_0^2} + 4\omega_0^2}}{2}$$

$$= \omega_0\left[\sqrt{\left(\frac{1}{2Q_0}\right)^2 + 1} - \frac{1}{2Q_0}\right] \qquad \text{...(1.45)}$$

Similarly at ω_2

$$Q_0 \cdot \left[\frac{\omega_2}{\omega_0} - \frac{\omega_0}{\omega_2}\right] = 1$$

This gives on solution

$$\omega_2 = \omega_0\left[\sqrt{\left(\frac{1}{2Q_0}\right)^2 + 1} + \frac{1}{2Q_0}\right] \qquad \text{...(1.46)}$$

Hence bandwidth $\Delta\omega = \omega_2 - \omega_1 = \frac{\omega_0}{Q_0}$

or $$f_2 - f_1 = \frac{f_0}{Q_0}$$

Similarly using the equations (1.45) and (1.46) it can be proved that

$$\omega_1\omega_2 = \omega_0^2 \quad \text{or} \quad \omega_0 = \sqrt{\omega_1\omega_2} \qquad \text{...(1.47)}$$

Summarising the characteristics of a parallel resonance we have

1. At resonance the input impedance is maximum or the input admittance is minimum and the circuit is purely resistive and hence power factor is unity.
2. The circuit is capacitive for frequencies $f > f_0$. It is inductive for $f < f_0$.
3. The current through the inductor and capacitor are equal in magnitude but opposite in phase and the current is amplified by the factor Q, hence the current in each branch is QI_0 where I_0 is current from the source.

4. Quality factor is given as $Q_0 = \frac{R}{\omega_0 L} = \omega_0 CR.$

5. The resonance frequency when $R_L \to 0$ and $R_C \to 0$ is given by

$$f_0 = \frac{1}{2\pi\sqrt{LC}}.$$

6. Bandwidth is given as $f_2 - f_1 = \frac{f_0}{Q_0}.$

7. $f_0 = \sqrt{f_1 f_2}$ where f_1 and f_2 are the lower and upper half power frequencies respectively.

Example 1.17: Show that the sum of energy stored by the inductor and the capacitor connected in series at resonance at any instant is constant and is given by LI^2.

Solution: Let i and v be the instantaneous current through inductor and the voltage across the capacitor at any time t.

Let $\quad i = I_m \cos \omega_0 t$

The energy stored is $\frac{1}{2} Li^2 = \frac{1}{2} LI_m^2 \cos^2 \omega_0 t$

The energy stored in the capacitor $\frac{q^2}{2C} = \frac{1}{2C}\left[\int_0^t i\,dt\right]^2$

$$= \frac{1}{2C} \cdot I_m^2 \left[\int_0^t \cos \omega_0 t\,dt\right]^2$$

$$= \frac{1}{2C} I_m^2 \left[\left(\frac{\sin \omega_0 t}{\omega_0}\right)_0^t\right]^2$$

$$= \frac{I_m^2}{2} L \sin^2 \omega_0 t$$

Therefore total energy

$$= \frac{1}{2} LI_m^2 (\cos^2 \omega_0 t + \sin^2 \omega_0 t)$$

$$= \frac{1}{2} I_m^2 L$$

$$= LI^2$$

Example 1.18: Show that the sum of energy stored by the inductor and the capacitor in a parallel RLC circuit at any instant is constant at resonant frequency and is equal to CV^2.

Solution: Let $\quad v = V_m \cos \omega_0 t$

The energy stored by the capacitor is

$$= \frac{1}{2} CV_m^2 \cos^2 \omega_0 t$$

The energy stored by the inductor is $\frac{1}{2} Li^2$

Now
$$v = L\frac{di}{dt}$$
$$di = \frac{V}{L}dt$$
$$i^2 = \left(\int_0^t \frac{v}{L}dt\right)^2$$
$$= \frac{V_m^2}{L^2}\cdot\frac{\sin^2\omega_0 t}{\omega_0^2} = \frac{V_m^2}{L^2}\ .\ \sin^2\omega_0 t\ LC$$
$$i^2 = \frac{V_m^2 C}{L^2}\sin^2\omega_0 t$$
Energy
$$= \frac{1}{2}LC^2 = \frac{1}{2}L\ .\ \frac{V_m^2 C}{L}\sin^2\omega_0 t$$
$$= \frac{C}{2}V_m^2\sin^2\omega_0 t$$
Therefore total energy $= \frac{1}{2}CV_m^2 = CV^2$. Hence proved.

Example 1.19: A coil having a resistance of 50 Ω and inductances 10 mH is connected in series with a capacitor and is supplied at constant voltage and variable frequency source. The maximum current is 1 A at 750 Hz. Determine the bandwidth and half power frequencies.

Solution: Q of the coil $= \frac{\omega_0 L}{R} = \frac{2\pi \times 750 \times 10 \times 10^{-3}}{50}$
$$= 0.9425$$
Bandwidth $= \frac{f_0}{Q} = \frac{750}{0.9425} = 795.8$ Hz

i.e. $\quad f_2 - f_1 = 795.8$ Hz

and $\quad f_0 = \sqrt{f_1 f_2}$
$$750^2 = f_1 f_2$$
$$(f_2 + f_1)^2 = 795.8^2 + 4 \times 750^2$$
$$= 1697.8^2$$
or $\quad f_2 + f_1 = 1697.8$
$$f_2 - f_1 = 795.8$$
$$2f_2 = 2493.6$$
$$f_2 = 1246.8 \text{ Hz}$$
and $\quad f_1 = 451$ Hz.

Example 1.20: Two branches of a parallel circuit have element $R_L = 6\ \Omega\ L = 1$ mH and $R_C = 4\ \Omega$ and $C = 20\ \mu F$. Determine the frequency of resonance. Also determine the maximum value of R_C for which resonance can occur.

Solution: We know that for such a circuit
$$\omega_0 = \frac{1}{\sqrt{LC}}\cdot\sqrt{\frac{R_L^2 - L/C}{R_c^2 - L/C}}$$

$$= \sqrt{\frac{1}{1 \times 10^{-3} \times 20 \times 10^{-6}}} \cdot \sqrt{\frac{6^2 - \dfrac{10^{-3}}{20 \times 10^{-6}}}{4^2 - \dfrac{10^{-3}}{20 \times 10^{-6}}}}$$

$$= \frac{10^4}{\sqrt{200}} \sqrt{\frac{14}{34}}$$

$$= 4537 \text{ rad/sec.}$$

In order for resonance to take place the quantity within the square root sign should be positive. Since $R_L^2 < \frac{L}{C}$, R_c^2 should also be less than $\frac{L}{C}$ or $R_c < \sqrt{\frac{L}{C}}$ or $R_c < 7.07\ \Omega$ **Ans.**

Example 1.21: In the problem of example 1.20 if $R_c = 5\ \Omega$ and R_L is variable, the circuit is fed from a sinusoidal source at a frequency f such that $X_c = 6\ \Omega$ and $X_L = 15\ \Omega$. Determine the value of R_L for which resonance can take place.

Solution: At resonance

$$\frac{X_c}{R_c^2 + X_c^2} - \frac{X_L}{R_L^2 + X_L^2} = \frac{5}{25 + 36} - \frac{15}{R_L^2 + 225} = 0$$

or $$\frac{5}{61} - \frac{15}{R_L^2 + 225} = 0$$

or $$R_L^2 + 225 = 183$$

or $$R_L^2 = -42$$

This means R_L should be negative which is not possible. Hence no value of R_L will be able to bring resonance in the circuit at the given condition.

Example 1.22: A coil has an inductance of 1.3 mH and resonates at 600 KHz and its $Q = 30$. If the bandwidth required is 50 KHz what resistor should be connected across the coil?

Solution: The reactance of the coil $= 2\pi (600 \times 10^3) \times 1.3 \times 10^{-3}$

$$= 4900\ \Omega$$

Since at resonance $X_{L0} = X_{C0}$

$$X_{c0} = 4900$$

The input resistance $= Z_{T0} = QX_{c0}$

$$= 30 \times 4900$$

$$= 147000\ \Omega$$

The required bandwidth is 50 kHz

Therefore, required Q of the circuit is $\frac{f_0}{50}$

$$= \frac{600}{50} = 12$$

The equivalent input resistance required is

$$Z'_{TO} = Q' X_{c0} = 12 \times 4900$$

$$= 58800$$

The resistance required for shunting is say R'

$$58800 = \frac{147000R'}{147000 + R'}$$

or $$147000 \times 58800 = 147000R' - 58800R'$$

or $$R' = \frac{147000 \times 58800}{88200}$$

$$= 98\ k\Omega. \quad \textbf{Ans.}$$

PROBLEMS

1.1. Determine the average value of the waveform shown

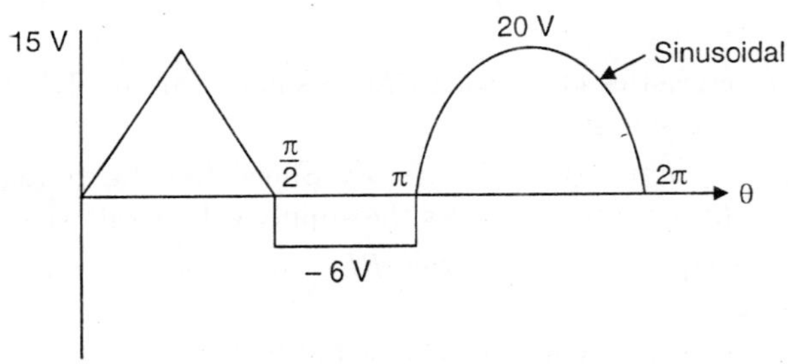

Fig. P 1.1

[**Ans.** 6.74]

1.2. Determine the effective value for the waveform shown

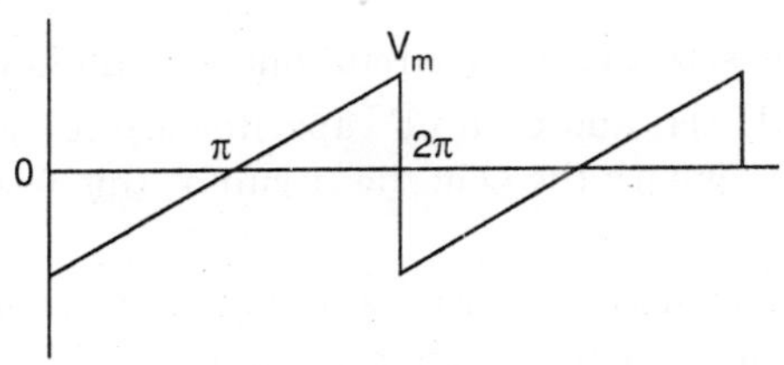

Fig. P 1.2

[**Ans.** $V_m/\sqrt{3}$]

1.3. The power in a 20 ohm load is 10 W. What is the peak to peak sinusoidal voltage across the load. [**Ans.** 40 V]

1.4. A resistor and a capacitor in series are connected to a 120 V, 60 Hz supply. The impedance of the circuit is 86 ohms and the power consumed is also 86 W. Determine the value of R and C. [**Ans.** 44.17 Ω, 35.9 μF]

1.5. A 200 ohm resistor is in series with a capacitor across 240 V, 60 Hz. Supply. The power is 100 W. Determine C and I. [**Ans.** 9.67 μF, 0.707 A]

1.6. The resistance of a coil is 140 Ω and its inductance 0.85 H. Determine the current, the p.f. and the circuit impedance when the coil is connected to 120 V, 60 Hz supply. [**Ans.** 0.344 $\angle$–66.4, 0.4, 349$\angle$66.4]

1.7. A 240 V, 50 KVA generator operates at a leading p.f. of 0.8. Determine the equivalent series impedance of the load. [**Ans.** $0.922 - j0.691$]

1.8. A 0.1 H inductor is in series with a 5 Ω resistor. The series combination is placed across a 120 V, 60 Hz supply. Determine the current in rectangular and polar coordinates.

[**Ans.** $0.415 - j3.128$ A]

1.9. Two appliances are operated in parallel from a 120 V supply. One draws 12 A at a lagging p.f. of 0.8 and the other draws 8A at 0.75 p.f. lead. Determine the total current and the power.

[**Ans.** $15.6 - j1.91$ A, 1872 W]

1.10. Two branches connected in parallel have impedances of $120 + j180$ Ω and $150 - j130$ Ω respectively. Determine the branch currents, the line current and the power delivered to the circuit if it is connected across 115 V supply. [**Ans.** $0.53 \angle{-56.3}$, $0.58 \angle 40.9$, 0.732 A, 84.2 W]

1.11. Two circuits, the impedances of which are given by $Z_1 = 10 + j15$ Ω and $Z_2 = 6 - j8$ Ω are connected in parallel. If the total current supplied is 15 A, what is the power taken by each branch ?

[**Ans.** 737 W, 1430 W]

1.12. Discuss various characteristics of a series *RLC* resonant circuit. Derive mathematical expressions in support of your discussion.

1.13. Define *Q* of a coil and *Q* of a series *RLC* circuit. Show that the voltage across the inductor or capacitor in a series *RLC* circuit is *Q* times the supply voltage at resonance.

1.14. What are half-power frequencies in a series *RLC* resonance circuit? Derive an expression for bandwidth of the circuit.

1.15. Discuss various characteristics of a general parallel *RLC* circuit at resonance. Derive resonance frequency when (*a*) coil is loss free (*b*) capacitor is loss free (*c*) both are loss free. Draw the equivalent circuit when *Q* of the circuit is high.

1.16. Derive an expression for *Q* of a parallel resonance circuit from first principles. A given parallel circuit has certain bandwidth. If it is required to obtain a lower bandwidth what measure should be taken to achieve this?

1.17. Explain what you mean by selectivity of a circuit and explain how it can be improved.

1.18. A 70 ohm resistor, a 5 mH coil and a 15 pF capacitor are in series across a 110 volt supply. Determine the resonant frequency the *Q* of the circuit at this frequency, the voltage across the capacitor at resonance. [**Ans.** 581 kHz, 261, 28.71 kV]

1.19. A variable inductor is connected in series with a resistor and a capacitor. The circuit is connected to a 200 volt 50 Hz supply and the maximum current obtainable by varying inductance is 0.314 A. The voltage across the capacitor then is 300 V. Determine the circuit constants.

[**Ans.** 3.3 μF, 637 Ω, 3.04 H]

1.20. A coil of 20 Ω resistance and 0.2 H inductance is connected in parallel with a capacitor of 100 μF capacitance. Find the frequency of resonance and the effective impedance at resonance.

[**Ans.** 31.8 Hz, 100 Ω]

1.21. A series circuit consists of a 115 Ω resistor, a 0.024 μF capacitor and coil of inductance *L*. If the resonant frequency of the circuit is 1000 Hz determine *L* and the bandwidth.

[**Ans.** 1.055 H, 17.3 Hz]

CHAPTER

2 Network Theory

2.1 INTRODUCTION

Kirchhoff's postulated two basic laws way back in 1845 which are used for writing network equations. These laws concern the algebraic sum of voltages around a loop and currents entering or leaving a node. The word algebraic is used to indicate that summation is carried out taking into account the polarities of voltages and direction of currents. While traversing a loop we will take voltage drops as positive and voltage rises as negative. Also while considering currents at a node, the currents entering the node will be taken as positive and those leaving would be taken as negative.

2.2 KIRCHHOFF'S LAWS

Kirchhoff's voltage law usually abbreviated as KVL is stated as follows :

The algebraic sum of all branch voltage around any closed loop of a network is zero at all instants of time. Alternatively, Kirchhoff's voltage law can be stated in terms of voltage drops and rises as follows: The sum of voltage rises and drops in a closed loop at any instant of time are equal. KVL is a consequence of law of conservation of energy as voltage is energy or work per unit charge. If we start from one node in a loop and move along the closed loop and come-back to the same node, obviously the total potential difference or sum of potential rises must equal the total sum of potential falls. Just as, if we start from one point on the surface of the earth and after travelling through valleys and hills come back to the same point, the total displacement is zero. We talk elevation's and depression on the earth with respect to the sea level. Similarly, in case of voltages we take ground as the reference which is shown in Fig. 2.1(a). Here potential of node A is above the ground and that of B is below the ground potential.

We know that electronic current flows form negative potential to positive potential, the conventional current flows from positive potential to negative potential. Therefore, when current i flows in the circuit of Fig. 2.1(b) it produces voltages polarities in various elements as shown in the Fig. Applying Kirchhoff's voltage law

$$v_1 + v_2 + v_3 - v = 0$$

Or in terms of voltage drops and voltage rises

$$v_1 + v_2 + v_3 = v \qquad ...(2.1)$$

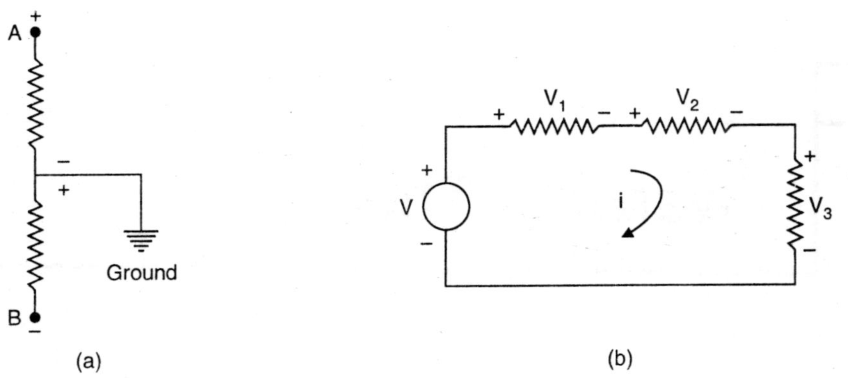

Fig. 2.1 (a) Potential reference, (b) KVL application.

Kirchhoff's current law states that the algebraic sum of all currents terminating at a node equals zero at any instant of time. Alternatively, this states that sum of all currents entering a node equals the sum of currents leaving the same node at any instant of time.

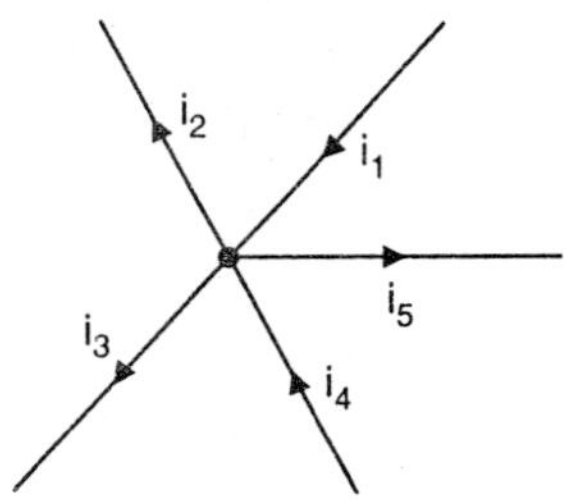

Fig. 2.2 One node with terminating branches.

Fig. 2.2 shows one node along with various branches terminating in it, in a network and the currents with directions in the various branches are shown by the arrows. Applying Kirchhoff's current law abbreviated by KCL we get

$$i_1 + i_4 - i_2 - i_3 - i_5 = 0$$

or

$$i_1 + i_4 = i_2 + i_3 + i_5 \qquad ...(2.2)$$

KCL is a consequence of law of conservation of charge. The charge that enters a node, must leave that node as it can't be stored there. Since the algebraic sum of charges at a node must be zero, it's time derivative must also be zero at any instant of time.

The two basic laws by Kirchhoff's can be applied to solve any network irrespective of its complexity. One typical application could be to find out equivalence between two networks or given a network in some configurations, how to find its equivalent so that it could be used more conveniently.

By definition, two networks are said to be equivalent at a pair of terminals if the voltage current relationships for the two networks are identical at these terminals. Consider a network having two resistances in series as shown. The objective is to replace it by a single equivalent resistance

From Fig. 2.3(a), we have

$$v = v_1 + v_2 = ir_1 + ir_2 \qquad ...(2.3)$$

and from Fig. 2.3(*b*) we have

$$v = ir$$

Therefore, $$r = r_1 + r_2$$

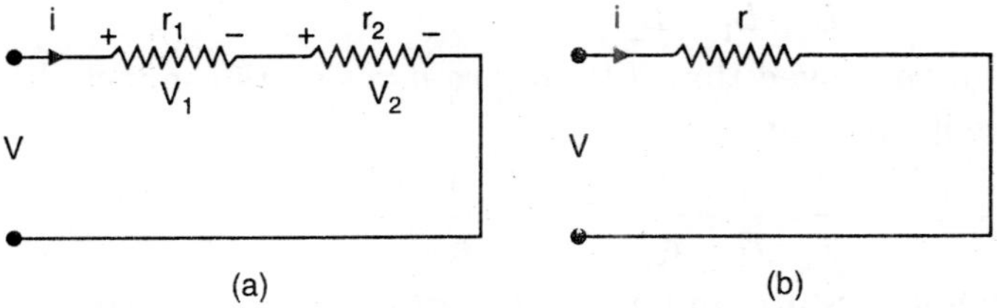

Fig. 2.3 (a) Series connection of resistances, (b) Equivalent of (a).

Similarly if there are n number of resistances in series, the equivalent single resistance would be

$$r = r_1 + r_2 + \ldots\ldots + r_n \qquad ...(2.4)$$

Similarly if the elements are inductances and v_1, v_2 are the drops across L_1, L_2 respectively, and i is the current then applying Kirchhoff's voltage law we have

$$v = v_1 + v_2 = L_1 \frac{di}{dt} + L_2 \frac{di}{dt}$$

$$= (L_1 + L_2) \frac{di}{dt} = L \frac{di}{dt}$$

Therefore, $$L = L_1 + L_2$$

If there are n inductors in series it can be shown that single equivalent inductor

$$L = L_1 + L_2 + \ldots\ldots + L_n \qquad ...(2.5)$$

If two capacitors are connected in series then

$$v = v_1 + v_2 = \frac{1}{C_1} \int i\,dt + \frac{1}{C_2} \int i\,dt$$

$$= \left(\frac{1}{C_1} + \frac{1}{C_2}\right) \int i\,dt = \frac{1}{C} \int i\,dt$$

Therefore, $$\frac{1}{C} = \frac{1}{C_1} + \frac{1}{C_2} \qquad ...(2.6)$$

Similarly by using KCL, law of parallel combination of these elements can be obtained. We will try here for inductors. Say there are two inductors connected in parallel.

From Fig. 2.4(*a*)

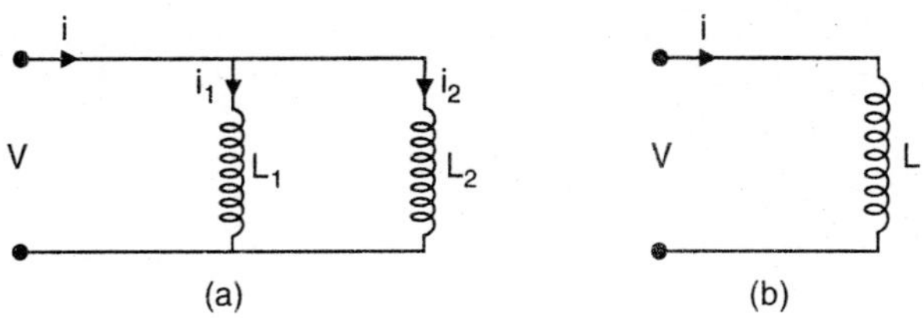

Fig. 2.4 (a) Inductor in parallel (b) Equivalent of (a).

$$i = i_1 + i_2 = \frac{1}{L_1}\int v\,dt + \frac{1}{L_2}\int v\,dt = \left(\frac{1}{L_1} + \frac{1}{L_2}\right)\int v\,dt$$

$$= \frac{1}{L}\int v\,dt$$

Therefore, $$\frac{1}{L} = \frac{1}{L_1} + \frac{1}{L_2} \qquad ...(2.7)$$

Similarly it can be proved that if there are n resistors connected in parallel its single equivalent resistance is given by

$$\frac{1}{r} = \frac{1}{r_1} + \frac{1}{r_2} + ... + \frac{1}{r_n} \qquad ...(2.8)$$

and if n number of capacitors are connected in parallel, then single equivalent capacitance is given as

$$C = C_1 + C_2 + ... + C_n \qquad ...(2.9)$$

2.3 STAR-DELTA TRANSFORMATION

By using the same definition of equivalence of two networks i.e. the two networks are equivalent between two terminals if the v-i relations are same i.e. the impedances as seen between the two terminals are same.

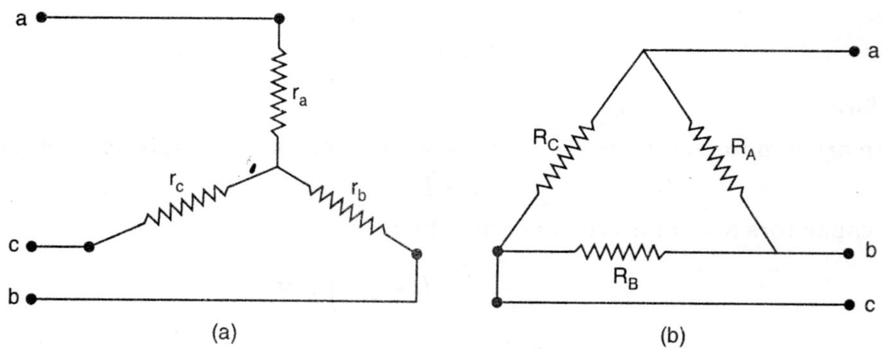

Fig. 2.5 (a) Star, (b) Delta.

Suppose we are given star connected elements as shown in Fig. 2.5(a) and we have to find it's equivalent delta connected elements then we should equal the equivalent resistance between the same two terminals.

Between a and b

$$\frac{R_A(R_B + R_C)}{R_A + R_B + R_C} = r_a + r_b \qquad ...(2.10)$$

Between b and c

$$\frac{R_B(R_A + R_C)}{R_A + R_B + R_C} = r_b + r_c \qquad ...(2.11)$$

Between c and a

$$\frac{R_C(R_A + R_B)}{R_A + R_B + R_C} = r_c + r_a \quad \text{...(2.12)}$$

Similarly shorting bc and measuring resistance between a and bc, we have

$$\frac{R_A R_C}{R_A + R_C} = r_a + \frac{r_b r_c}{r_b + r_c} \quad \text{...(2.13)}$$

Shorting ca and resistance between b and ca

$$\frac{R_A R_B}{R_A + R_B} = r_b + \frac{r_c r_a}{r_c + r_a} \quad \text{...(2.14)}$$

Shorting ab and resistance between c and ab

$$\frac{R_B R_C}{R_B + R_C} = r_c + \frac{r_a r_b}{r_a + r_b} \quad \text{...(2.15)}$$

Substracting equation (2.13) from (2.10) we have

$$r_b - \frac{r_b r_c}{r_b + r_c} = \frac{R_A(R_B + R_C)}{R_A + R_B + R_C} - \frac{R_A R_C}{R_A + R_C}$$

$$\frac{r_b^2}{r_b + r_c} = \frac{R_A^2 R_B}{(R_A + R_C)(R_A + R_B + R_C)}$$

Substituting for $(r_b + r_c)$ from equation (2.11)

$$\frac{r_b^2(R_A + R_B + R_C)}{R_B(R_A + R_C)} = \frac{R_A^2 R_B}{(R_A + R_C)(R_A + R_B + R_C)}$$

or

$$r_b^2 = \frac{R_A^2 R_B^2}{(R_A + R_B + R_C)^2}$$

or

$$r_b = \frac{R_A R_B}{R_A + R_B + R_C} \quad \text{...(2.16)}$$

r_a and r_c can be obtained using equation (2.10) and (2.11) along with the expression for r_b and it can be seen that

$$r_a = \frac{R_A R_C}{R_A + R_B + R_C} \quad \text{...(2.17)}$$

and

$$r_c = \frac{R_B R_C}{R_A + R_B + R_C} \quad \text{...(2.18)}$$

To transform a given star into delta following steps are used.

From equation (2.13) we have

$$\frac{R_A R_C}{R_A + R_C} = \frac{r_a r_b + r_b r_c + r_c r_a}{r_b + r_c}$$

using equations (2.11) and (2.16) we have

$$r_a r_b + r_b r_c + r_c r_a = R_c r_b$$

Therefore, $$R_C = \frac{r_a r_b + r_b r_c + r_c r_a}{r_b} \quad ...(2.19)$$

Similarly, $$R_B = \frac{r_a r_b + r_b r_c + r_c r_a}{r_a} \quad ...(2.20)$$

$$R_A = \frac{r_a r_b + r_b r_c + r_c r_a}{r_c} \quad ...(2.21)$$

An easier way to remember these formulae is as follows. Suppose we want to transform star into delta. The numerator is same for all the three branches and is equal to the product of impedance taken two at a time i.e. $Z_a Z_b + Z_b Z_c + Z_c Z_a$ and in the denominator it will be the star branch opposite to the equivalent delta branch i.e. if Z_A is required the denominator will have impedance Z_C. Similarly for Z_B it will be Z_a and for Z_C it will be Z_b Fig. 2.6.

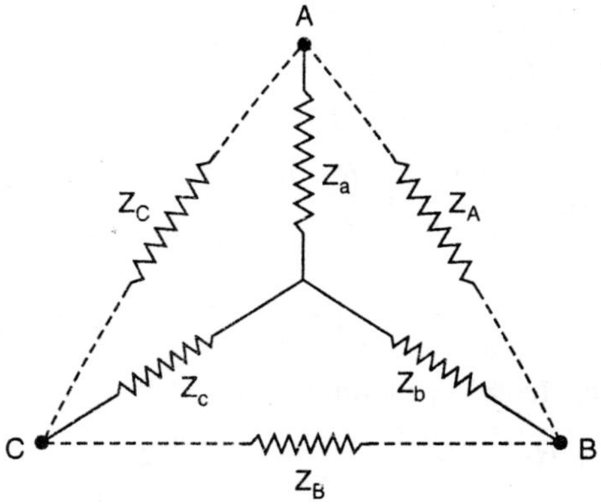

Fig. 2.6

Again for transforming delta into star consider the Fig. 2.7.

To find Z_a open the delta vertex at A. The two impedances associated at vertex A are Z_A and Z_C. The equivalent star.

$$Z_a = \frac{Z_A Z_C}{Z_A + Z_B + Z_C}$$

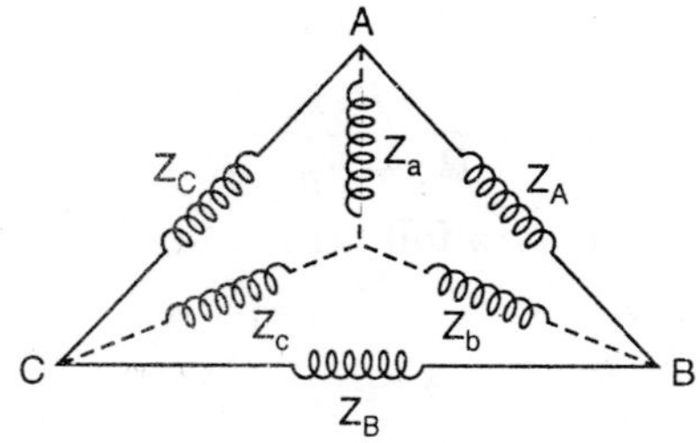

Fig. 2.7 Dalta to star transformation.

Here denominator is same for all the three legs of the equivalent star.

2.4 LOOP ANALYSIS

Let us consider a few application of KVL in terms of loop or mesh analysis.

Example 2.1: Determine the current in all the branches of the network using loop analysis.

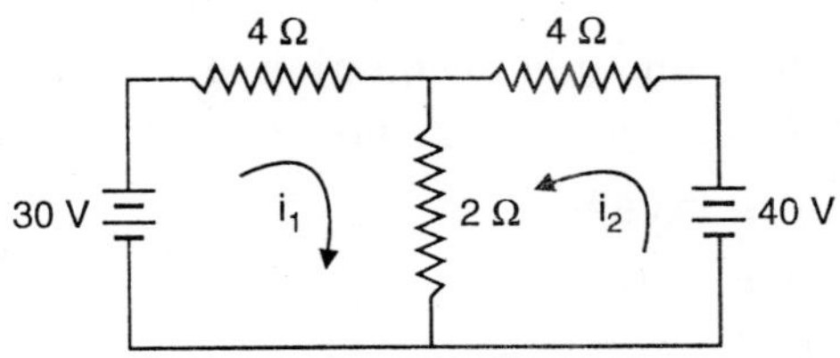

Fig. E2.1

Solution: Let the currents i_1 and i_2 be flowing in the two loops as shown. It is to be noted that the direction of the currents is arbitrary. However, depending upon the values of i_1 and i_2 thus obtained we can ascertain the direction. If they are coming out to be positive, the directions remain as shown in the figure, otherwise one with negative value, the direction of this current is reversed.

Writing loop equation for loop 1

$$4i_1 + 2(i_1 + i_2) = 30$$

or $$6i_1 + 2i_2 = 30 \quad \text{...(1)}$$

Loop 2 $$4i_2 + 2(i_1 + i_2) = 40$$

or $$2i_1 + 6i_2 = 40 \quad \text{...(2)}$$

Multiplying equation (2) by 3 and subtracting equation (1) from this we have

$$16i_2 = 90 \quad \text{or} \quad i_2 = 5.625 \text{ A}$$

Hence $$2i_1 + 6 \times 5.625 = 40$$

$$2i_1 = 40 - 33.75$$

or $$2i_1 = 6.25, i_1 = 3.125 \text{ A} \quad \textbf{Ans.}$$

Hence $i_3 = i_1 + i_2 = 5.625 + 3.125 = 8.75$ A

The current through 2 Ω branch is 8.75 A.

Since both the currents i_1 and i_2 are found to be positive hence the directions shown in Fig. E2.1 is correct.

Suppose we take the direction of current for i_2 as shown in Fig. E2.1.1 shown here

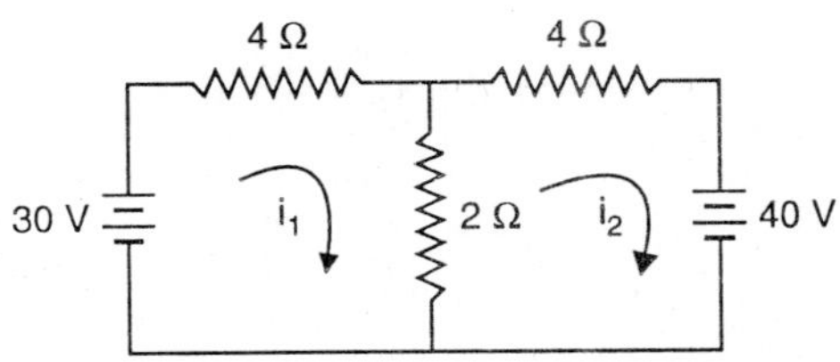

Fig. E2.1.1

For loop 1

$$4i_1 + 2(i_1 - i_2) = 30$$

or $$6i_1 - 2i_2 = 30 \quad ...(1)$$

$$4i_2 + 2(i_2 - i_1) + 40 = 0$$

or $$-2i_1 + 6i_2 = -40 \quad ...(2)$$

Multiply equation (2) by 3 and add to equation (1)

$$16i_2 = -120 + 30 = -90$$

$$i_2 = -5.625$$

and $$-2i_1 - 33.75 = -40$$

or $$-2i_1 = -6.25$$

or $$i_1 = 3.125 \text{ A}$$

We find the i_2 is coming out to be negative and hence the direction of current in loop 2 should be opposite to what is shown in Fig. E2.1.1.

Example 2.2: Determine the current in each branch of the network shown in Fig. E2.2 using loop analysis.

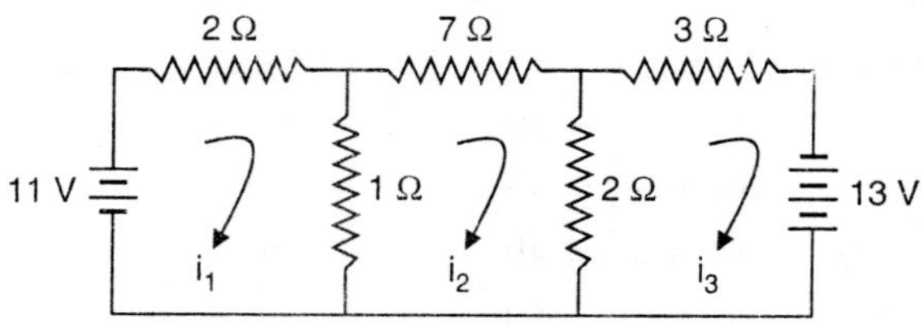

Fig. E2.2

Solution: Since there are three loops, three independent loop equations are required to solve the network. Suppose the directions of currents are as shown in Fig. E2.2. Writing equations for the three loops

Loop I $$2i_1 + 1(i_1 - i_2) = 11$$

$$3i_1 - i_2 = 11 \quad ...(1)$$

Loop II $$7i_2 + 2(i_2 - i_3) + (i_2 - i_1)1 = 0$$

or $$-i_1 + 10i_2 - 2i_3 = 0 \quad ...(2)$$

Loop III $$3i_3 + 2(i_3 - i_2) = 13$$

$$-2i_2 + 5i_3 = 13 \quad ...(3)$$

In matrix form the equations can be written as

$$\begin{vmatrix} 3 & -1 & 0 \\ -1 & 10 & -2 \\ 0 & -2 & 5 \end{vmatrix} \begin{vmatrix} i_1 \\ i_2 \\ i_3 \end{vmatrix} = \begin{vmatrix} 11 \\ 0 \\ 13 \end{vmatrix}$$

Hence det [A] $$= \begin{vmatrix} 3 & -1 & 0 \\ -1 & 10 & -2 \\ 0 & -2 & 5 \end{vmatrix}$$

$$= 3(50 - 4) + 1(-5) = 133$$

$$i_1 = \frac{\begin{vmatrix} 11 & -1 & 0 \\ 0 & 10 & -2 \\ 13 & -2 & 5 \end{vmatrix}}{133} = 11(50-4) + 26 = 11 \times 46 + 26$$

$$= \frac{532}{133} = 4\text{A}$$

$$i_1 = 4\text{A}$$

From equation 1

$$3 \times 4 - i_2 = 11$$

or $$-i_2 = -1$$

or $$i_2 = 1\text{A}$$

and from equation (3)

$$-2 + 5i_3 = 13 \quad \text{or} \quad 5i_3 = 15$$

or $$i_3 = 3\text{ A} \quad \textbf{Ans.}$$

Example 2.3: Determine the currents i_1, i_2 and i_3 and the Voltage V_1 and V_2 in the given network.

Using KVL we have

$$0.2\, i_1 + (i_1 - i_2)\, 3.75 = 120 \qquad \textit{Loop I}$$

$$3.75\,(i_2 - i_1) + 0.3\, i_2 + 5.45\,(i_2 - i_3) = 0 \qquad \textit{Loop II}$$

$$5.45\,(i_3 - i_2) + 0.1\, i_3 = 110 \qquad \textit{Loop III}$$

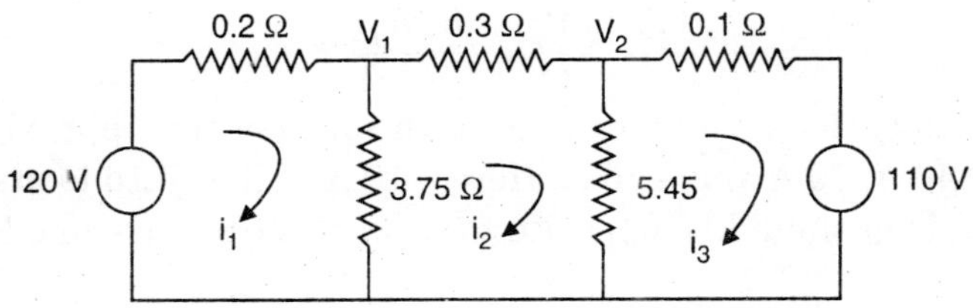

Fig. E2.3

Rearranging the equations, we have

$$(0.2 + 3.75)\, i_1 - 3.75\, i_2 = 120$$

$$-3.75\, i_1 + (3.75 + 0.3 + 5.45)\, i_2 - 5.45\, i_3 = 0$$

$$-5.45\, i_2 + (5.45 + 0.1)\, i_3 = 110$$

These equations can be written in matrix form as follows

$$\begin{bmatrix} (0.2+3.75) & -3.75 & 0.0 \\ -3.75 & (3.75+0.3+5.45) & -5.45 \\ 0.0 & -5.45 & 5.45+0.1 \end{bmatrix} \begin{bmatrix} i_1 \\ i_2 \\ i_3 \end{bmatrix} = \begin{bmatrix} 120 \\ 0.0 \\ 110 \end{bmatrix}$$

or $$ZI = V$$

Here Z matrix is the loop impedance matrix and is found to be a symmetric matrix along its leading diagonal.

It can be seen that the loop impedance matrix can be written by inspection of the loops.

(*i*) The diagonal elements are the sum of all the branch impedances in the corresponding loop.

(*ii*) The off-diagonal elements say Z_{12} is the impedance of the branches common to both the loop 1 and 2. The sign is positive if current i_1 and i_2 in the common branch flow in the same direction and is taken as negative if the two currents oppose each other.

The above set of these equations can again be written in matrix form as follows:

$$\begin{bmatrix} 3.95 & -3.75 & 0.0 \\ -3.75 & 9.50 & -5.45 \\ 0.0 & -5.45 & 5.55 \end{bmatrix} \begin{bmatrix} i_1 \\ i_2 \\ i_3 \end{bmatrix} = \begin{bmatrix} 120 \\ 0.0 \\ 110 \end{bmatrix}$$

Using Cramer's rule

$$i_1 = \frac{\begin{bmatrix} 120 & -3.75 & 0.0 \\ 0.0 & 9.50 & -5.45 \\ 110 & -5.45 & 5.55 \end{bmatrix}}{\begin{bmatrix} 3.95 & -3.75 & 0.0 \\ -3.75 & 9.50 & -5.45 \\ 0.0 & -5.45 & 5.55 \end{bmatrix}} = \Delta$$

Similarly

$$i_2 = \frac{\begin{bmatrix} 3.95 & 120 & 0.0 \\ -3.75 & 0.0 & -5.45 \\ 0.0 & 110 & 5.55 \end{bmatrix}}{\Delta}$$

Similarly expression for i_3 can be written. On solution the approximate values of i_1, i_2 and i_3 are found to be 40 A, 10 A and – 10 A. Here – 10 A in the third loop means the direction of current in the third loop should be opposite to what has been assumed at the beginning of the solution.

The voltage
$$v_1 = (i_1 - i_2)\, 3.75$$
$$= 30 * 3.75$$
$$= 112.5 \text{ Volts}$$

and
$$V_2 = 20 \times 5.45 = 109 \text{ volts}$$

Example 2.4: Determine the current in the battery, the current in each branch and the p.d. across *AB* in the network shown.

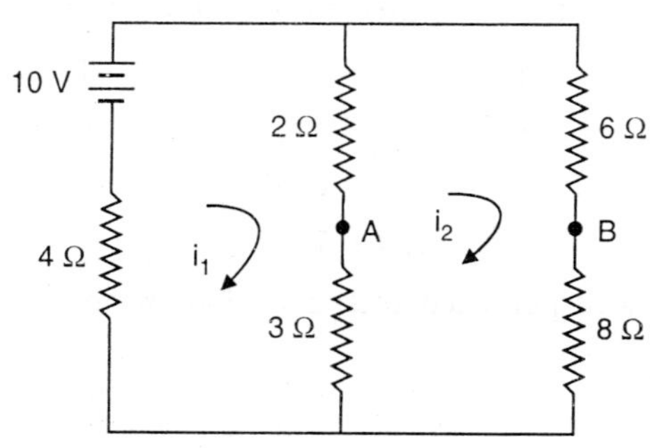

Fig. E2.4

Using KVL we have the equations in the matrix form.

$$\begin{bmatrix} 9 & -5 \\ -5 & 19 \end{bmatrix} \begin{bmatrix} i_1 \\ i_2 \end{bmatrix} = \begin{bmatrix} 10 \\ 0 \end{bmatrix}$$

From second equation

$$-5\, i_1 + 19\, i_2 = 0$$

or $$i_1 = \frac{19}{5}\, i_2$$

From first equation

$$9\, i_1 - 5\, i_2 = 10$$

$$\frac{171}{5}\, i_2 - 5\, i_2 = 10$$

or $$i_2 = \frac{10 \times 5}{146} \quad \text{or} \quad i_1 = \frac{190}{146} = 1.3 \text{ A}$$

Alternative solution:

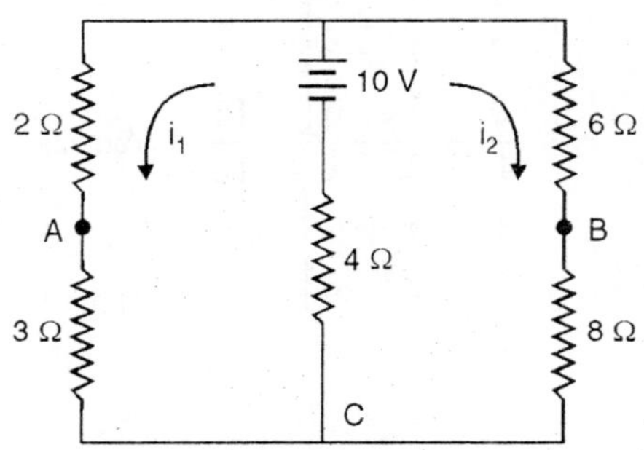

Fig. E2.4.1

Using KVL we have

$$9\, i_1 + 4\, i_2 = 10 \qquad \text{...(1)}$$

$$4\, i_1 + 18\, i_2 = 10$$

or $$9\, i_1 + 4\, i_2 = 4\, i_1 + 18\, i_2$$

or $$i_1 = \frac{14}{5}\, i_2$$

Substituting in (1) we have

$$9 \times \frac{14}{5} + 4\, i_2 = 10$$

or $$i_2 = 0.343$$

and $$i_1 = \frac{14}{5} \times \frac{50}{146} = 0.96 \text{ A}$$

Therefore current through the battery 1.303 A.

Potential of A w.r. to C $0.96 \times 3 = 2.88\ v$

and that of B w.r. to C $0.343 \times 8 = 2.744$

$V_{AB} = V_A - V_B = 2.88 - 2.74 = 0.14\ v$ **Ans.**

Example 2.5: Determine the current through the branch *AB* of the network shown

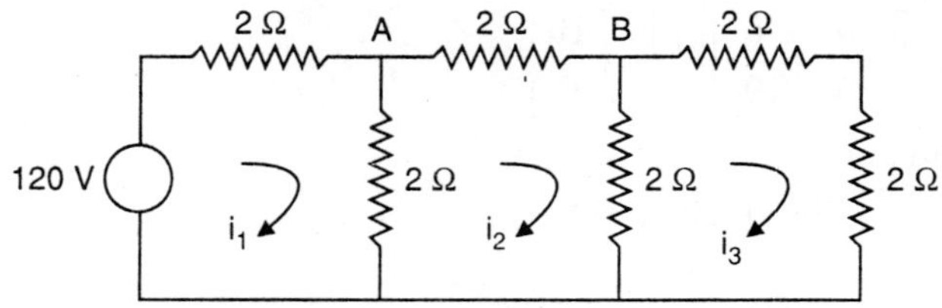

Fig. E2.5

Using KVL we have

$$\begin{bmatrix} 4 & -2 & 0 \\ -2 & 6 & -2 \\ 0 & -2 & 6 \end{bmatrix} \begin{bmatrix} i_1 \\ i_2 \\ i_3 \end{bmatrix} = \begin{bmatrix} 10 \\ 0 \\ 0 \end{bmatrix}$$

Using Cramer's rule

$$i_2 = \frac{\begin{bmatrix} 4 & 10 & 0 \\ -2 & 0 & -2 \\ 0 & 0 & 6 \end{bmatrix}}{\begin{bmatrix} 4 & -2 & 0 \\ -2 & 6 & -2 \\ 0 & -2 & 6 \end{bmatrix}} = \frac{15}{13} A \textbf{ Ans.}$$

2.5 NODAL ANALYSIS

We now illustrate the application of KCL for solving the networks. Here whenever a voltage source is given it should be converted into a current source before nodal equations using KCL are written.

Example 2.6: Determine the current in the branches of the network in Fig. E2.6 using nodal analysis.

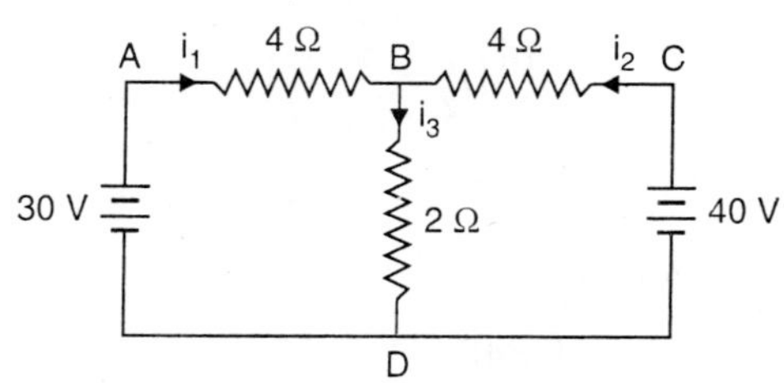

Fig. E2.6

Solution: There are four nodes *A, B, C, D* as shown in the network. Taking node *D* as reference, we are left with three nodes *A, B* and *C*. Voltages of nodes *A* and *C* with respect to reference node are known to be 30V and 40 V respectively. Hence voltage of only one node is not-known. Hence we need write only one equation at node *B*.

$$i_1 + i_2 = i_3$$

or $$\frac{30 - V_B}{4} + \frac{40 - V_B}{4} = \frac{V_B}{2}$$

or $$30 - V_B + 40 - V_B = 2V_B$$

or $$4V_B = 70 \quad \text{or} \quad V_B = 17.5 \text{ volts}$$

Hence $$i_1 = \frac{30 - 17.5}{4} = 3.125 \text{ A}$$

$$i_2 = \frac{40 - 17.5}{4} = 5.625$$

and $$i_3 = i_1 + i_2 = 8.75 \text{ A} \quad \textbf{Ans.}$$

Example 2.7. Determine the current's in all branches of the network shown in Fig. E2.7

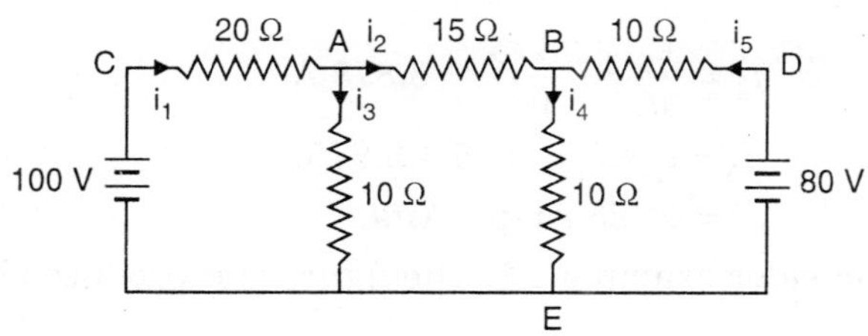

Fig. E2.7

Solution. Taking node E as reference and since the potentials of nodes *C* and *D* with respect to *E* are known hence only two independent nodes are *A* and *B* where the voltages be V_A and V_B respectively.

Writing nodal equation at node A

$$\frac{100 - V_A}{20} - \frac{V_A - V_B}{15} - \frac{V_A}{10} = 0$$

or $$5 = \frac{V_A}{20} + \frac{V_A}{15} - \frac{V_B}{15} + \frac{V_A}{10}$$

$$25 = \frac{V_A}{4} + \frac{V_A}{3} - \frac{V_B}{3} + \frac{V_A}{2}$$

$$300 = 3\,V_A + 4V_A - 4V_B + 6V_A$$

or $$13V_A - 4V_B = 300 \qquad (1)$$

At node *B*

$$\frac{V_A - V_B}{15} + \frac{80 - V_B}{10} = \frac{V_B}{10}$$

$$\frac{V_A - V_B}{3} + \frac{80 - V_B}{2} = \frac{V_B}{2}$$

$$2V_A - 2V_B + 240 - 3V_B = 3V_B$$

$$2V_A - 8V_B = -240$$

or $$-V_A + 4V_B = 120 \qquad ...(2)$$

Adding equation (1) and (2) we have

$$12V_A = 420$$
$$V_A = 35 \text{ volts}$$

Hence $\quad -35 + 4V_B = 120$

or $\quad 4V_B = 155$

or $\quad V_B = 38.75$ volts

Hence current $\quad i_1 = \dfrac{100-35}{20} = \dfrac{65}{20} = 3.25$ A

$$i_2 = \frac{35-38.75}{15} = -\frac{3.75}{15} = -0.25$$

The value of i_2 is coming out to be negative which means in the actual circuit the direction of current should be opposite to what is shown in Fig. E2.7

Hence $i_3 = i_1 + i_2 = 3.25 + 0.25 = 3.5$

$$i_4 = \frac{V_B}{10} = \frac{38.75}{10} = 3.875 \text{ A}$$

and hence $\quad i_5 = i_2 + i_4 = 0.25 + 3.875$

$= 4.125$ Amp **Ans.**

Example 2.8: Consider example 2.5. The figure is reproduced here

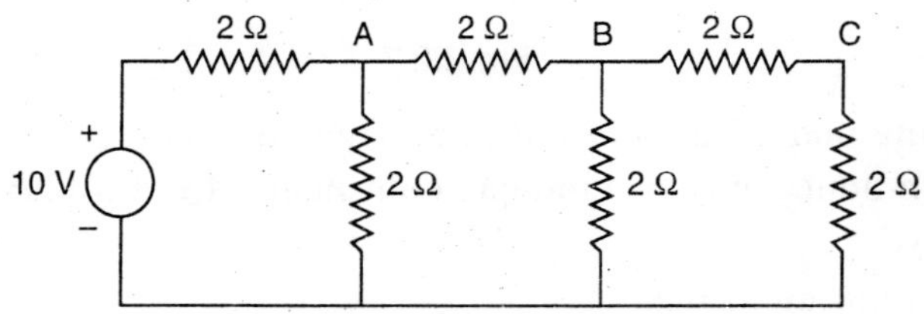

Fig. E2.8

First we transform the voltage source into a current source and the diagram becomes

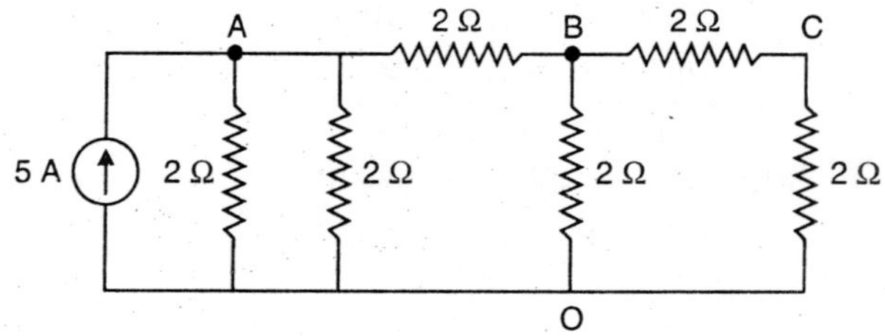

Fig. E2.8.1

Taking node O as reference, we write down KCL equations

At node A

$$\frac{V_A}{2} + \frac{V_A}{2} + \frac{V_A - V_B}{2} = 5$$

$$V_A\left[\frac{1}{2}+\frac{1}{2}+\frac{1}{2}\right] - V_B/2 + 0 \,.\, V_c = 5$$

At node B $\quad -V_A/2 + \left(\frac{1}{2}+\frac{1}{2}+\frac{1}{2}\right) V_B - V_C/2 = 0$

At node C $\quad 0.V_A - \frac{V_B}{2} + V_C \left(\frac{1}{2}+\frac{1}{2}\right) = 0$

These equations can be written in matrix form as follows:

$$\begin{bmatrix} \frac{1}{2}+\frac{1}{2}+\frac{1}{2} & -\frac{1}{2} & 0 \\ -\frac{1}{2} & \frac{1}{2}+\frac{1}{2}+\frac{1}{2} & -\frac{1}{2} \\ 0 & -\frac{1}{2} & \frac{1}{2}+\frac{1}{2} \end{bmatrix} \begin{bmatrix} V_A \\ V_B \\ V_C \end{bmatrix} = \begin{bmatrix} 5 \\ 0 \\ 0 \end{bmatrix}$$

Here the first matrix is known as nodal admittance matrix and is denoted by Y and the two column matrices are known as nodal voltage and nodal current matrices denoted by V and I respectively.

Hence, the above matrices can be written in compact form as

$$YV = I \qquad \text{...(2.22)}$$

It can be seen that the nodal admittance matrix can be written simply by looking at the interconnections of the network using the following procedure:

(*i*) The diagonal element of each node is the sum of the admittances connected to it.

(*ii*) The off diagonal element is the negative admittance between the nodes. If there is no connection between the two nodes, the entry in the off-diagonal for both the nodes is zero i.e. if node i is not connected to the node j, then $Y_{ij} = Y_{ji} = 0$. In general every node will not be connected to all other nodes in the network and hence normally nodal admittance matrix is a highly sparse matrix i.e. it will have very small number of non-zero elements. In fact, in a large network consisting of say 100 nodes, these non zero elements would be as small as 2% of the total elements.

We come back to original problem of finding out V_{AB}

Using Cramer's rule

$$V_A = \frac{\begin{bmatrix} 5 & -\frac{1}{2} & 0 \\ 0 & \frac{3}{2} & -\frac{1}{2} \\ 0 & -\frac{1}{2} & 1 \end{bmatrix}}{\begin{bmatrix} \frac{3}{2} & -\frac{1}{2} & 0 \\ -\frac{1}{2} & \frac{3}{2} & -\frac{1}{2} \\ 0 & -\frac{1}{2} & 1 \end{bmatrix}} = \frac{5 \times \frac{5}{4}}{\frac{13}{8}}$$

Therefore, $\quad V_A = \frac{50}{13}$ volts

Similarly,

$$V_B = \frac{\begin{bmatrix} \frac{3}{2} & 5 & 0 \\ -\frac{1}{2} & 0 & -\frac{1}{2} \\ 0 & 0 & 1 \end{bmatrix}}{\Delta}$$

Therefore, $V_B = \frac{5}{2} \times \frac{8}{13} = \frac{20}{13}$ and $V_{AB} = \frac{50}{13} - \frac{20}{13} = \frac{30}{13}$ volts

2.6 SUPERPOSITION THEOREM

This theorem is applicable to linear networks. A network is said to be linear where the response is proportional to the excitation. If X_1 is the excitation and Y_1 the response of a given network related by $Y_1 = CX_1$ and for an excitation X_2 the response Y_2 is related by $Y_2 = CX_2$ then for excitation $(X_1 + X_2)$ the response will be $Y_1 + Y_2$ for a linear network. It is to be noted that if the excitation and response are related by the linear equation $Y_1 = CX_1 + d_1$ the network is not linear as for double the excitation; response will not be doubled. Thus a circuit with R, L, C, elements with constant values and initially relaxed (for L $i(o) = O$ and for C $V(o) = O$) is a linear circuit. However, the initial conditions could be considered as sources or excitation and the overall response of the circuit can be obtained.

Superposition theorem can be stated as follows: In a linear network, if more than one number of sources (excitation) are acting, the response in a certain branch can be obtained by superimposing algebraically the responses due to various sources taking one excitation at a time and considering the remaining voltage sources (ideal) short circuited or replaced by their equivalent internal impedances and current sources (ideal) open circuited or replaced by their equivalent admittances.

The theorem is further illustrated with a few examples.

Example 2.9: Using superposition principle determine the current through 2 Ω resistor connected between A and B shown in the circuit here.

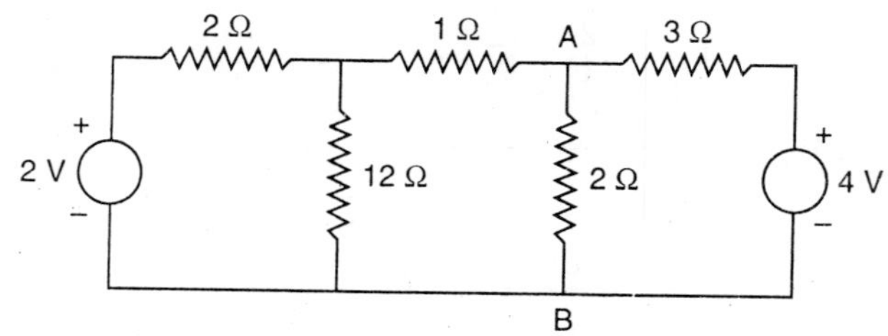

Fig. E2.9

Solution: Since there are two sources in the circuit, there will be two circuits for analysis purposes where in one source will be present and the other short circuited, being the voltage source.

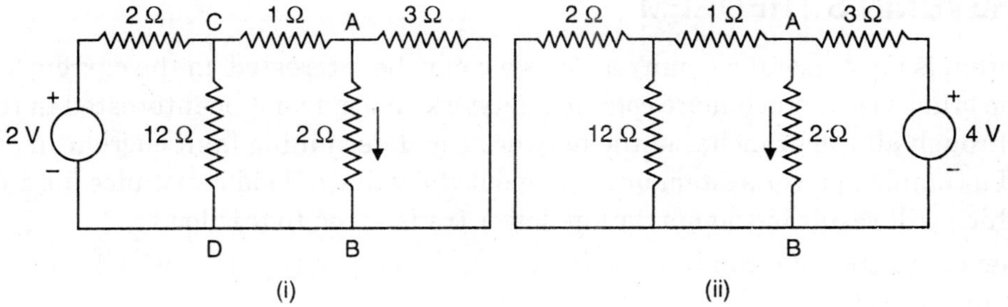

Fig. E2.9.1

The resistance between CD excluding 12 Ω resistance

$$1 + \frac{6}{5} = 1 + 1.2 = 2.2.\ \Omega, \text{ including } 12\ \Omega \text{ resistance}$$

$$\frac{1}{R} = \frac{1}{12} + \frac{10}{22} = \frac{(11 + 60)}{132} = \frac{71}{132}$$

Total resistance $2 + \frac{132}{71} = \frac{(142 + 132)}{71} = \frac{274}{71}$

Therefore, current through the source is $\frac{2 \times 71}{274}$

The current through 1 Ω resistor is

$$\frac{2 \times 71}{274} \times \frac{12 \times 10}{142} = \frac{120}{274}$$

Therefore current through 2 Ω resistor

$$\frac{120}{274} \times \frac{3}{5} = \frac{72}{274}\ \text{A}$$

(ii)

$$1 + \frac{24}{14} = \frac{38}{14}\ \Omega$$

$$\frac{1}{R} = \frac{1}{2} + \frac{14}{38} = \frac{(19 + 14)}{38} = \frac{33}{38}$$

or

$$R = \frac{38}{33}\ \Omega$$

Total resistane $3 + \frac{38}{33} = \frac{137}{33}\ \Omega$

Current through the source $\frac{4 \times 33}{137}$ A

Current through 2 Ω resistor

$$\frac{4 \times 33}{137} \times \frac{38}{14} \times \frac{14}{66} = \frac{76}{137}$$

Therefore, total current is

$$\frac{72}{274} + \frac{76}{137} = \frac{224}{274} = 0.8\ \text{A}$$

2.7 THEVENIN'S THEOREM

This theorem is most useful as many a times we may be interested in the current/voltage in a part of the otherwise a much more complex network. We may not be interested in the current/voltage through all the branches of the network e.g. determining fault current in a particular branch of a complex power system or finding out the value of load resistance for an electronic circuit which will result in the maximum power transferred to this load.

Thevenin's theorem can be explained with the help of two networks A and B connected as shown in Fig. 2.8.

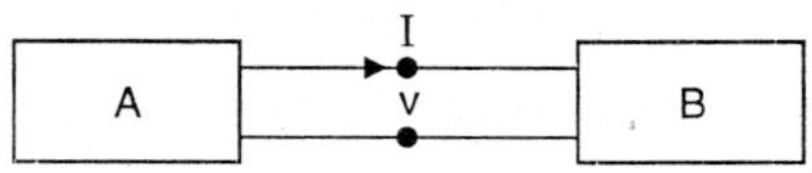

Fig. 2.8 Two networks A and B connected.

The original network can be considered as a combination and connection of networks A and B. Network A contains linear elements which may have initial conditions. It may also have an independent and dependent sources. However, there is no mutual magnetic coupling between the networks A and B nor does a controlled source in A couple to B.

For network B, the elements need not necessarily be linear. These could be non-linear or time varying or both. However, in our analysis we will assume that the network elements are linear and many a times this will be a single branch (load) of the network.

Network A is to be replaced by an equivalent network under the condition that the current I and voltage V identified in the Fig. remain invariant when the replacement is made. This amounts to replacing network A by an equivalent Thevenin's voltage source and an equivalent Thevenin's series impedance as shown in Fig. 2.8 (*a*).

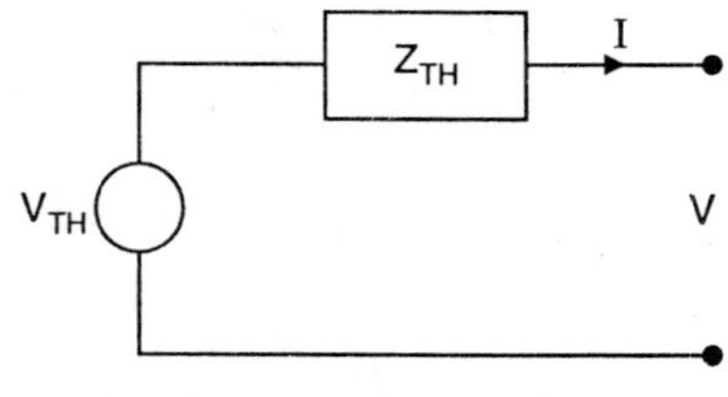

Fig. 2.8 (a) Thevenin equivalent of Fig. 2.8.

The following procedure is followed :

To determine V_{TH} we remove the network B (could be a load resistor or the element through which current is required) and solve the network A to find out voltage across the open circuited terminal. This voltage is the Thevenin's equivalent voltage source V_{TH}. Also Z_{eq} is the impedance as seen in the network between the two open circuited terminals replacing the voltage sources by their equivalent series impedances and the current sources by their equivalent shunt admittances. The impedance as seen is known as Thevenin's equivalent series impedance. While finding Thevenin's equivalent, dependent sources continue to operate in the network.

If Z_L is the load impedance and Z_{eq} the Thevenin's equivalent series impedance (Fig. 2.9) the current through the load impedance is given by

$$I = \frac{V_{TH}}{Z_{eq} + Z_L} \quad \text{...(2.14)}$$

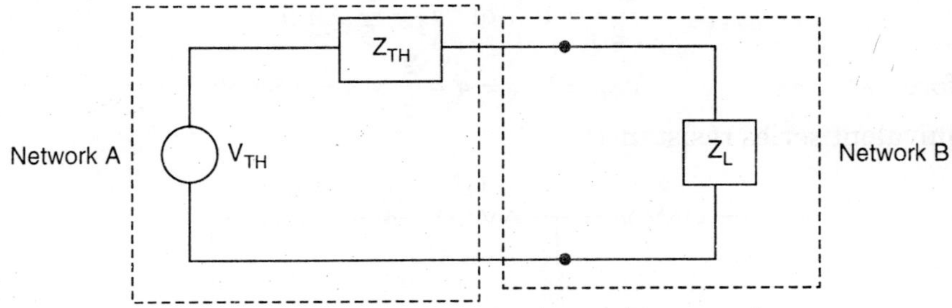

Fig. 2.9 Thevenin's equivalent with load.

A few examples will further illustrate the application of Thevenin's theorem.

Example 2.10: Determine the current through 2 Ω resistor connected between A and B in the circuit shown using Thevenin theorem.

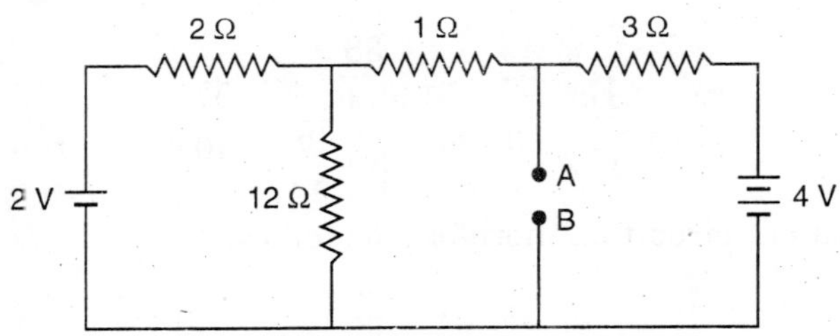

Fig. E2.10

Solution: To obtain V_{TH} refer Fig. below:

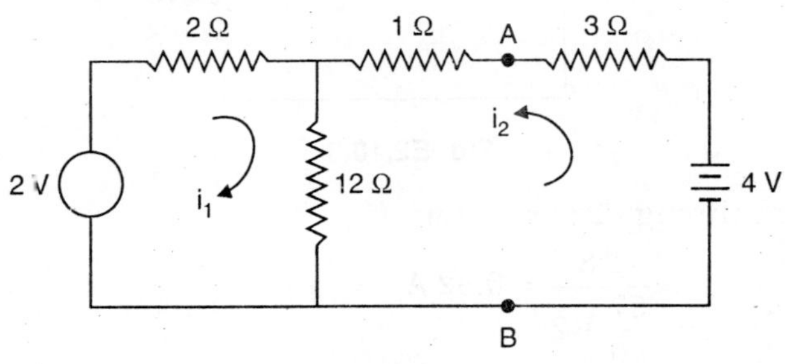

Fig. E2.10.1

$$2i_1 + 12(i_1 + i_2) = 2,\ 4i_2 + 12i_1 + 12i_2 = 4$$

$$14i_1 + 12i_2 = 2 \quad \text{or} \quad 7i_1 + 6i_2 = 1$$

$$12i_1 + 16i_2 = 4 \quad \text{or} \quad 3i_1 + 4i_2 = 1 \quad \text{...(1)}$$

or $$7i_1 + 6i_2 = 3i_1 + 4i_2$$

or $$4i_1 = -2i_2$$

or $$i_1 = \frac{-i_2}{2}$$

Substituting in (1) we have

$$\frac{-3i_2}{2} + 4i_2 = 1$$

or $$\frac{5i_2}{2} = 1 \quad \text{or} \quad i_2 = \frac{2}{5} \text{ Amp.}$$

Therefore, $$V_{TH} = V_{AB} = 4 - \frac{2}{5} \times 3 = 2.8 \text{ volt}$$

The equivalent series resistance

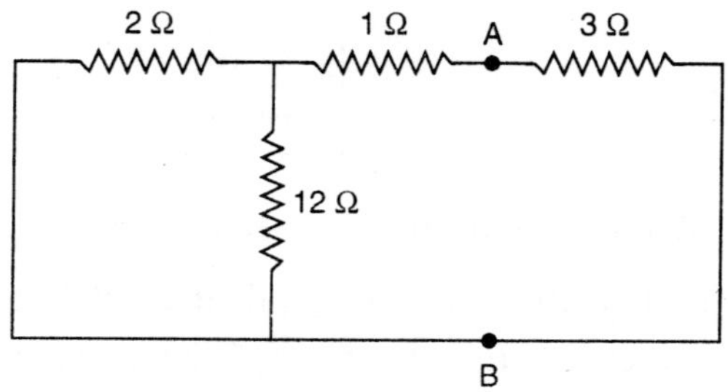

Fig. E2.10.2

$$\left(1 + \frac{24}{14}\right) \parallel 3 \quad \text{or} \quad \frac{\frac{38}{14} \times 3}{3 + \frac{38}{14}} = \frac{57}{40}\ \Omega$$

Thevenin's equivalent circuit is, therefore, as follows :

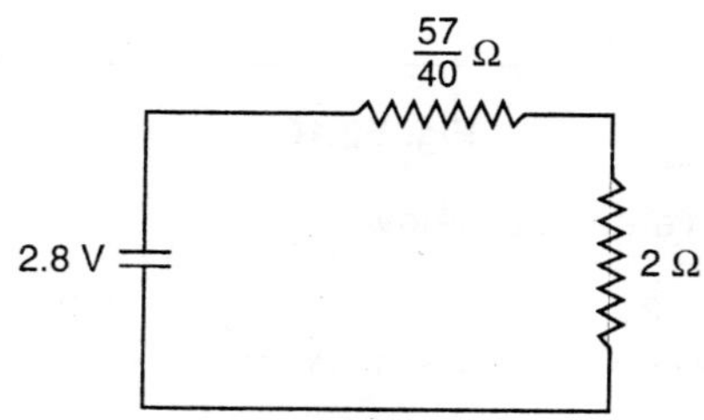

Fig. E2.10.3

Therefore, current through 2 Ω resistor

$$\frac{2.8}{\frac{57}{40} + 2} = 0.82 \text{ A}$$

Example 2.11: Determine the current through the branch *AB* of the network shown here using Thevenins equivalent.

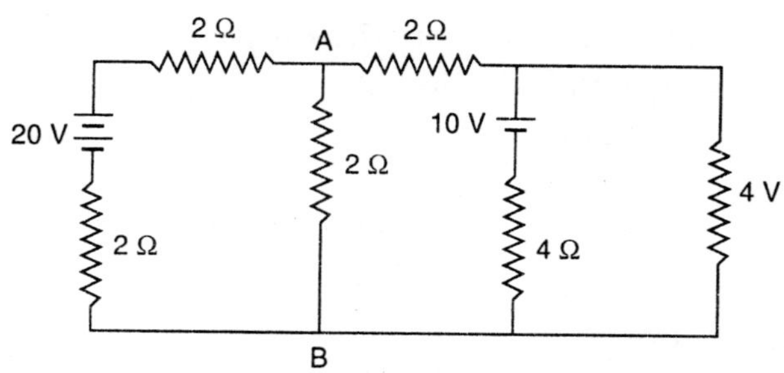

Fig. E2.11

Solution: To obtain V_{TH} the equivalent circuit is given as

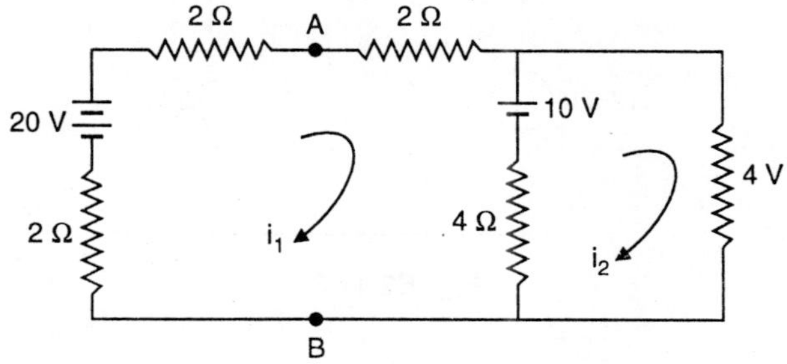

Fig. E 2.11.1

$$2i_1 - 20 + 4i_1 + 10 + 4(i_1 - i_2) = 0$$

$$10i_1 - 4i_2 = 10$$

or $$5i_1 - 2i_2 = 5$$

$$4(i_2 - i_1) - 10 + 4i_2 = 0$$

$$-4i_1 + 8i_2 = 10$$

$$-2i_1 + 4i_2 = 5$$

or $$-2i_1 + 4i_2 = 5i_1 - 2i_2$$

or $$6i_2 = 7i_1, \quad -2i_1 + \frac{14}{3} i_1 = 5$$

or $$i_2 = \frac{7}{6} i_1, \quad \frac{8}{3} i_1 = 5$$

$$V_{AB} + \frac{15}{2} = 20 \qquad i_1 = \frac{15}{8},$$

$$V_{AB} = 12.5 \text{ V}$$

Therefore, $$V_{AB} = V_{TH} = 12.5 \text{ volts}$$

To calculate R_{TH} or R_{eq}

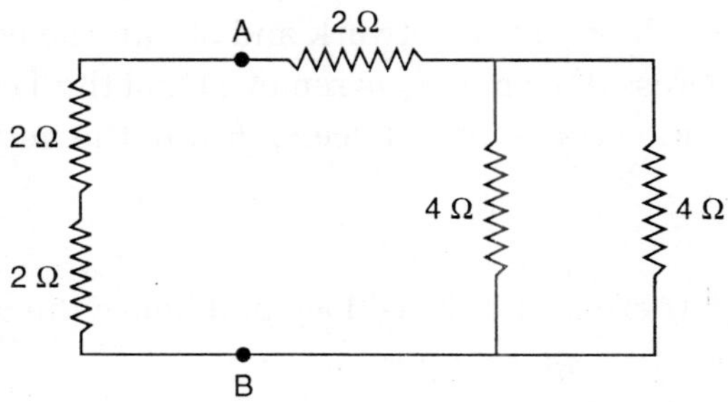

Fig. E2.11.2

Therefore current through the AB branch

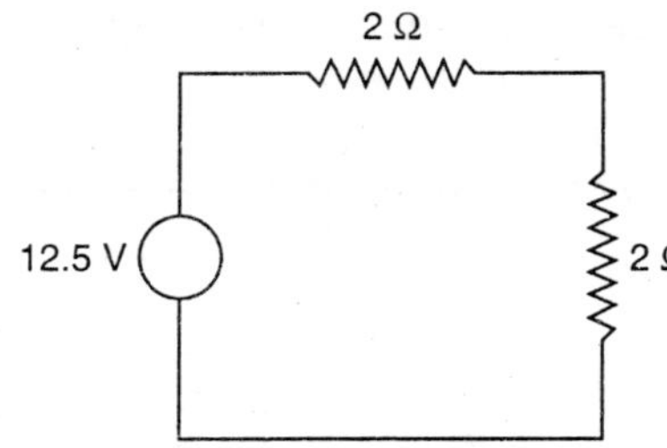

Fig. E2.11.3

$$I_{AB} = \frac{12.5}{4} = 3.125 \text{ A}$$

2.8 NORTON'S THEOREM

Norton's theorem is dual of Thevenins theorem and can be stated as follows:

It states that any two terminal linear network can be replaced by an equivalent network consisting of a current source I_0 in parallel with a network of impedance Z_0. The impedance is same as that found for Thevenins equivalent and I_0 is the current between the two terminals when these terminals are short circuited. The following example will further clarify the concepts associated with the theorem.

Example 2.12: Determine the current through the resistance R_L = 1, 2, 5 Ω for the network shown in Fig. E 2.12. Using Thevenin's theorem.

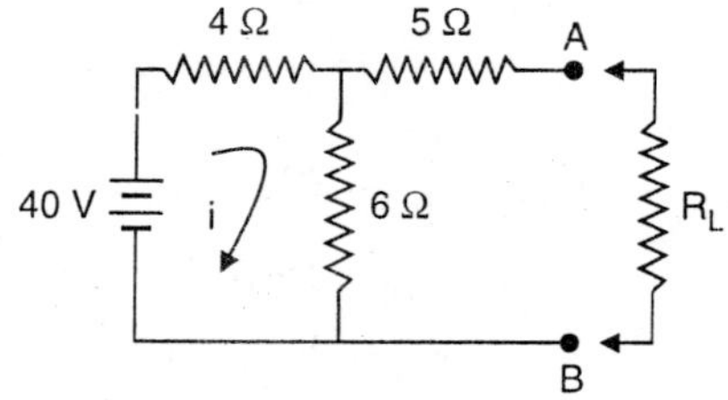

Fig. E2.12

Solution: We remove R_L from the network and obtain the equivalent Thevenin's voltage source i.e. the voltage across AB when R_L is removed and the Thevenin's equivalent series resistance which is the resistance as seen between *AB* into the network shorting the voltage sources.

To find out V_{TH}

When R_L is removed, there is one closed loop and hence the current in the loop is

$$i = \frac{40}{4+6} = 4\text{A}$$

Now when R_L is removed, voltage across terminals *AB* is the voltage across 6 Ω resistor

i.e. $V_{TH} = 6 \times 4 = 24\text{V}$

To find out the equivalent Thevenin series resistance the resulting circuit is

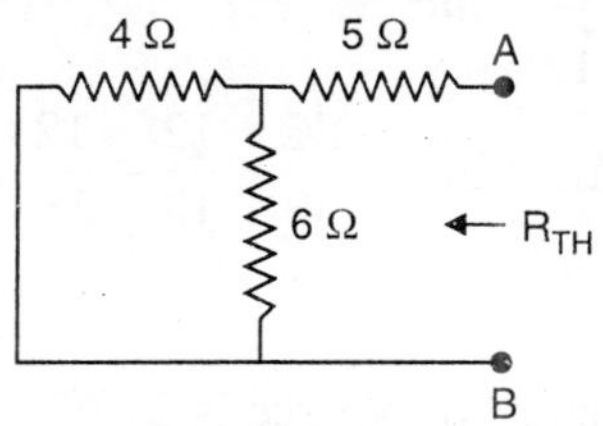

Fig. E2.12.1

Hence $R_{TH} = 5 + \frac{6 \times 4}{10} = 7.4\ \Omega$

So the equivalent Thevenin's circuit is

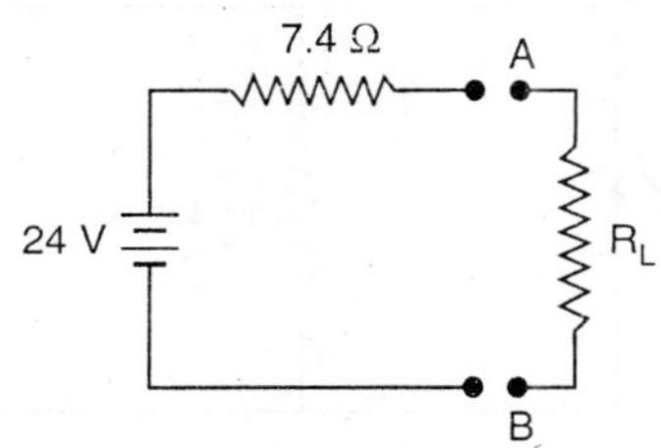

Fig. E2.12.2

When $R_L = 0$ Current in $R_L = \frac{24}{7.4} = 3.24$ A

When $R_L = 2\ \Omega$ Current in $R_L = \frac{24}{7.4 + 2} = \frac{24}{9.4} = 2.55$ A

When $R_2 = 5\ \Omega$ Current in $R_L = \frac{24}{7.5 + 5} = \frac{24}{12.5} = 1.935$ A **Ans.**

Example 2.13: Solve the network of example E 2.10 using Norton's theorem.

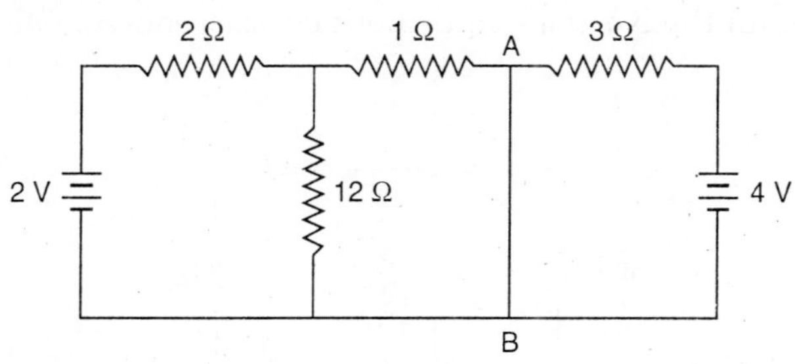

Fig. E2.13

Norton's equivalent current source will be the current through the short circuited branch *AB*.

Using superposition principle to find current through branch *AB* we first short circuit 2 *V* battery, the current through *AB* due to 4 volt will be $\frac{4}{3}$ A. When 4 volt battery is shorted, the current supplied by 2 V battery is

$$\frac{2}{2+\frac{12}{13}} = \frac{26}{38}$$

Therefore current through AB will be $\frac{26}{38} \times \frac{12}{13} = \frac{12}{19}$ A

so total current is $\frac{4}{3} \times \frac{12}{19} = \frac{112}{57}$ A

So that Norton's equivalent current source is 112/57 A. The Norton's equivalent impedance is same as in case of Thevenins equivalent impedance except that it is to be connected across the current source. Along with this the resistance through which current is required should be connected across the current source. This forms the complete equivalent Norton's circuit and is drawn here.

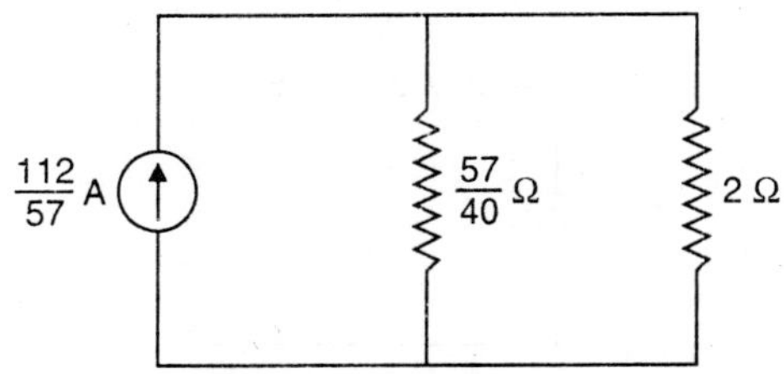

Fig. E2.13.1

Current through 2 Ω resistor is

$$\frac{112}{57} \times \frac{57}{40} \times \frac{1}{\frac{57}{40}+2}$$

$$= \frac{112}{137} = 0.82 \text{ A}$$

Example 2.14: For the network shown in Example 2.12, determine the current in R_L = 0, 2, 5 Ω using Norton's theorem.

Solution: To obtain the Norton's equivalent current source, following circuit would be used

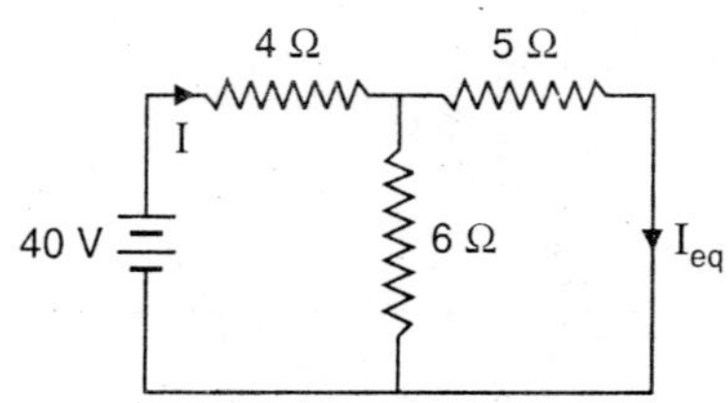

Fig. E2.14

Current $\quad I = \frac{40}{4+\frac{30}{11}} = \frac{440}{74} = 5.946$ A

Hence $\quad I_{eq} = \frac{440}{74} \times \frac{6}{11} = 3.243$

The equivalent Norton's resistance is same as in Thevenin's equivalent i.e. 7.4 Ω. The Norton's equivalent circuit is

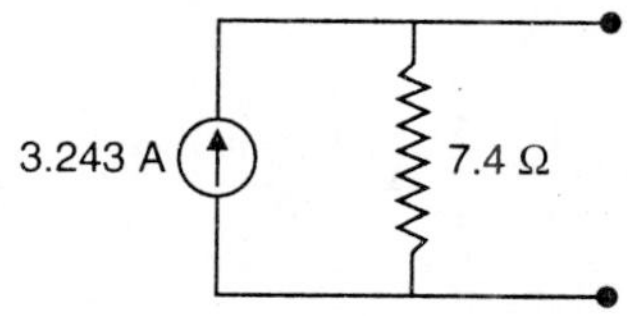

Fig. E2.14.1

Now current through R_L when $R_L = 0$. Obviously the total current will flow through R_L and no current through resistance 7.4 Ω

$$R_L = 2\Omega\text{, current through } R_L = 3.243 \times \frac{7.4}{9.4} = 2.55 \text{ A}$$

$$R_L = 5\ \Omega \text{ current through } R_L = 3.243 \times \frac{7.4}{12.4} = 1.935 \text{ A}$$

Example 2.15: Solve example 2.11 using Norton's theorem

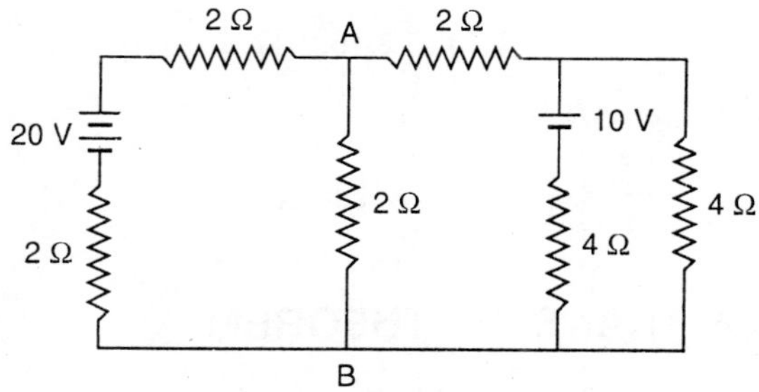

Fig. E2.15

Solution: Shorting branch *AB* and using superposition principle to find current through *AB*.

20 Volts in the circuit and shorting 10 volt source

The current is $\frac{20}{4}$ A = 5 A

10 volts in the circuit and shorting 20 V

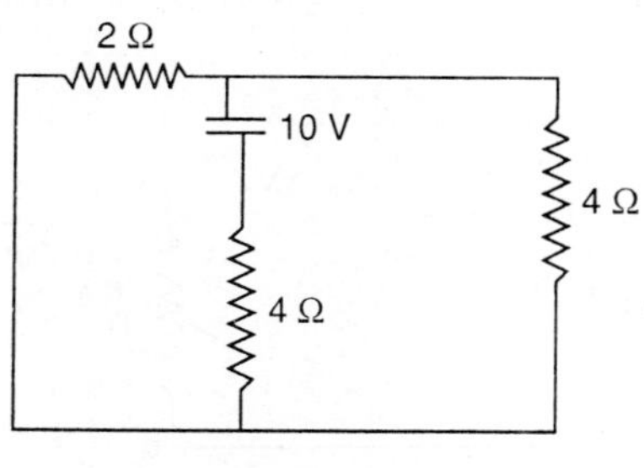

Fig. E2.15.1

Current supplied by the battery

$$\frac{10}{4+4/3} = \frac{3\times 10}{16} = \frac{15}{8}\text{ A}$$

Current through 2 Ω resistor

$$\frac{15}{8}\times\frac{4}{6} = \frac{5}{4}\text{ A}$$

Total current through AB is

$$5 + \frac{5}{4} = \frac{25}{4}\text{ A}$$

The Norton's equivalent circuit, therefore is

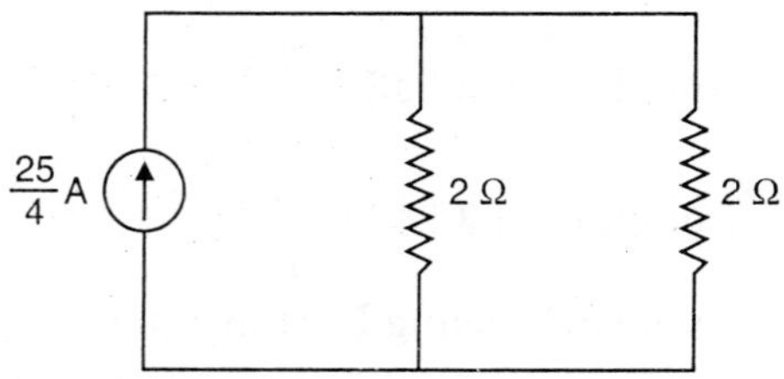

Fig. E 2.15.2

Therefore, current through AB branch is

$$\frac{25}{4}\times\frac{1}{2} = \frac{25}{8} = 3.125\text{ A}$$

2.9 MAXIMUM POWER TRANSFER THEOREM

Sometimes in electronics circuits it is desirable to transfer maximum power through a source to a load e.g. impedance matching is carried out in a loud speaker so that maximum sound could be heard. However in power system maximum power transfer is never tried as huge amount of power in terms of MW are transferred from source to the load and the efficiency of transfer is 50% as will be seen shortly if power transferred is based on maximum power transfer criterion. Also under this, the terminal voltage becomes half of the source voltage and hence this results in poor voltage regulation, therefore, maximum power transfer criterion is never considered in power system.

We will consider the theorem both for d.c. and a.c. circuits.

d.c. circuits :

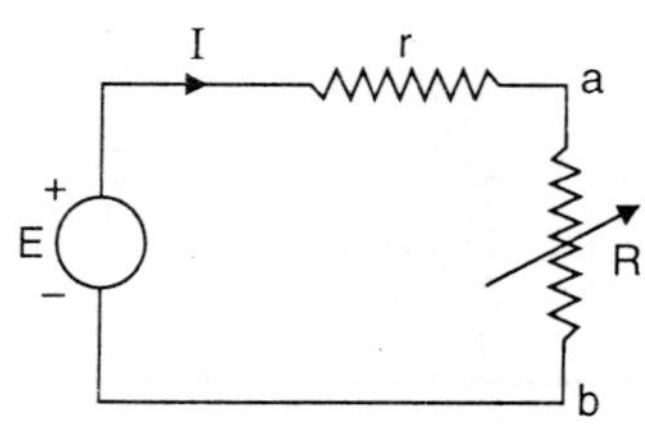

Fig. 2.10 Maximum power transfer-d.c. circuit.

Here r is the internal resistance of the source or the equivalent resistance as seen between a b or the load resistance R. Usually r in a circuit or in a system is fixed, the load resistance R is variable.

The problem is : what should be R expressed in terms of r so that power delivered to the load resistance is maximum.

The current in the circuit $I = E/(r + R)$ and power transferred is

$$P = \frac{E^2}{(r+R)^2} \cdot R$$

For this power to be maximum $\frac{dP}{dR} = 0$

$$= \frac{E^2[(r+R)^2 - R \cdot 2(R+r)]}{(r+R)^4} = 0$$

or $\quad r + R - 2R = 0$

or $\quad R = r \quad$ (2.22)

This shows that for maximum power transfer the load resistance R should equal the internal resistance r or the equivalent resistance r. Under this condition

Power $\quad P_{max} = \frac{E^2}{(2r)^2} \cdot r = \frac{E^2}{4r}$

Same amount of power is dissipated in the internal resistance r and hence the efficiency is 50%. Also it can be seen that voltage across the load is $\frac{E}{2}$ and voltage regulation is 50% which is very poor in power system operation.

a.c. Circuits:

(*i*) Load is $R + jX$ and internal impedance $(r + jx)$

Therefore, $\quad I = \frac{E}{(r+R) + j(x+X)}$

Power transfer $\quad P = I^2R = \frac{E^2 R}{(r+R)^2 + (x+X)^2} \quad$ ...(2.23)

For power to be maximum $\frac{\partial P}{\partial X} = 0$

$$= \frac{E^2 R \cdot \{-2(x+X)\}}{D^2} = 0$$

or $\quad x + X = 0 \quad$ or $\quad X = -x \quad$ ...(2.24)

which means the load reactance must be equal to the negated value of the source internal reactance. The maximum power becomes

$$P_{max} = \frac{E^2 R}{(r+R)^2} \quad ...(2.25)$$

Again taking $\frac{dP}{dR} = 0$ for maximum power

We have just now seen the condition is $R = r$ and, therefore, in a.c. circuits the load impedance should be complex conjugate of the internal impedance that is $Z_L = Z_I^*$ where Z_L is the load impedance and Z_I is source impedance. Usually source impedance is inductive $r + jx$ and , therefore, the load impedance should be capacitive for maximum power transfer condition. Here again the efficiency is 50%.

(*ii*) Load resistance R and internal impedance $(r + jx)$

Here current $$I = \frac{E}{(r + R) + jx}$$

And power $$P = \frac{E^2 R}{(r + R)^2 + x^2} \quad ...(2.26)$$

Therefore, $$\frac{dP}{dR} = 0 \text{ for maximum power transfer}$$

$$E^2 [\{r + R)^2 + x^2\} - R \,.\, 2(R + r)] = 0$$

or $$r^2 + x^2 = R^2$$

or $$R = \sqrt{r^2 + x^2} = Z_I \quad ...(2.27)$$

This means when load resistance R equals the impedance of the source the power transfer is maximum. Let us find out the magnitude of power under this condition

$$P = \frac{E^2 Z_I}{(r + Z_I)^2 + x^2} = \frac{EI_{sc} Z_I^2}{r^2 + Z_I^2 + 2rZ_I + x^2}$$

$$= \frac{EI_{sc} Z_I^2}{2Z_I^2 + 2rZ_I} = \frac{EI_{sc} Z_I}{2(Z_I + r)} \quad ...(2.28)$$

This power is certainly less than the power under the previous two conditions.

Example 2.16: Determine the current in the 2 Ω resistor branch of the network using superposition theorem.

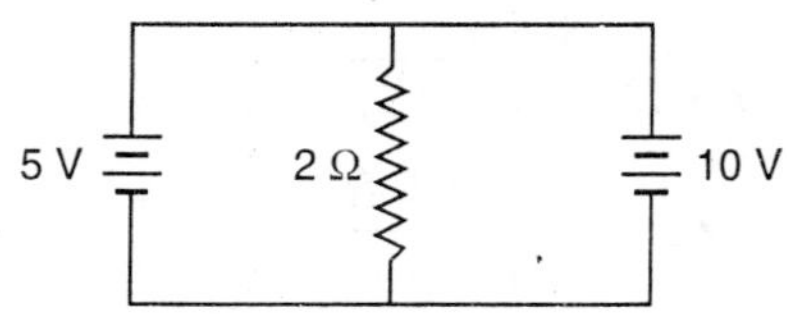

Fig. E2.16

Solution: By simple logic it can be said that the current through 2 Ω resistor is zero as two voltage sources of unequal magnitude (dc) should not be connected in parallel as they act short circuit for each other.

The equivalent circuits using superposition principle are

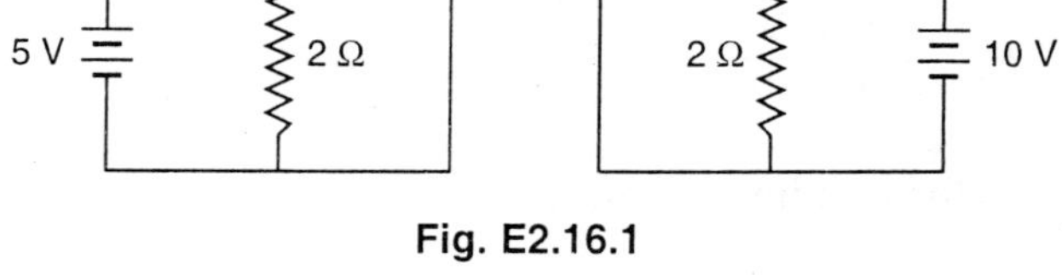

Fig. E2.16.1

Hence current through 2 Ω resistor is zero.

Example 2.17: Determine the value of R_L in the network shown for maximum power transfer and calculate the value of power.

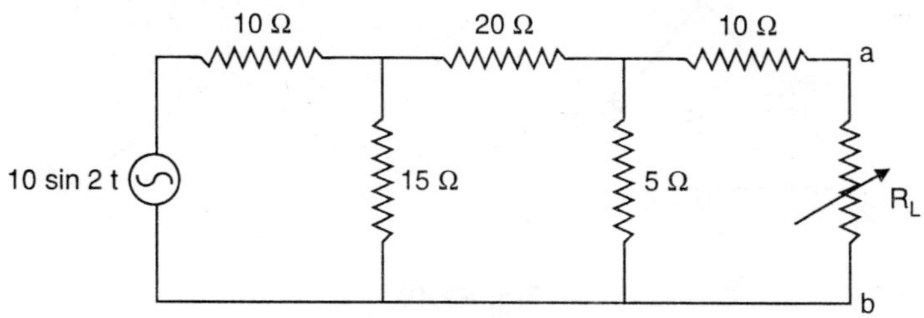

Fig. E2.17

Solution: For maximum power transfer, the resistance should equal the resistance between *a b* when looked into the network replacing the voltage source by a short circuit.

$$\frac{10 \times 15}{25} + 20 = 26\ \Omega$$

$$r_{ab} = \frac{26 \times 5}{31} + 10 = 14.2\ \Omega$$

$$V_{TH} = \frac{10}{\sqrt{2}} \times \frac{8}{155} \times \frac{15}{40} \times 5 = 0.684 \text{ volt}$$

$$\text{Power transfer} = \frac{V^2}{4\, r_{ab}} = \frac{(0.684)^2}{4 \times 14.2} = 8.24 \text{ mW} \quad \textbf{Ans.}$$

Example 2.18: Determine delta equivalent of the star circuit shown here in Fig. E 2.18.

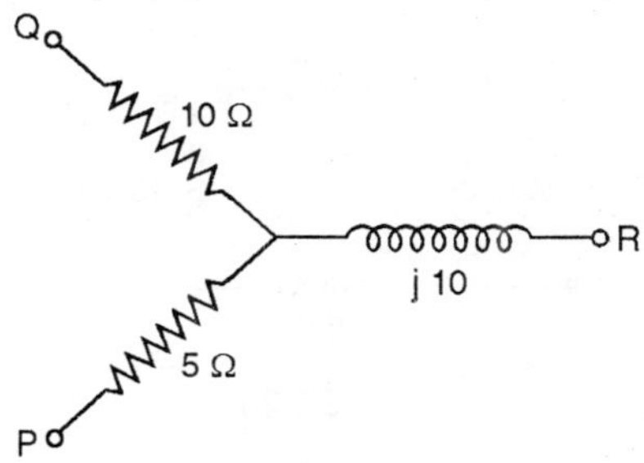

Fig. E2.18

Solution:

$$Z_{PQ} = \frac{Z_p Z_Q + Z_Q Z_R + Z_R Z_P}{Z_R}$$

$$= \frac{5 \times 10 + 10(j\,10) + 5(j\,10)}{j\,10}$$

$$= \frac{50 + j100 + j\,50}{j\,10} = \frac{j(5 + j\,15)}{j\,j}$$

$$= -j5 + 15\ \Omega = 15 - j5$$

$$Z_{QR} = \frac{50 + j\,150}{5} = 18 + j30\ \Omega$$

$$Z_{PR} = \frac{50 + j\,150}{10} = (5 + j15)\ \Omega \quad \textbf{Ans.}$$

Example 2.19: Determine the star equivalent of the delta circuit shown in E 2.19.

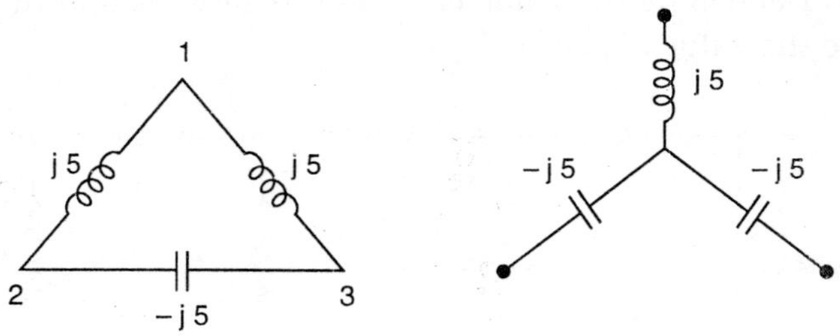

Fig. E2.19

Solution:

$$Z_1 = \frac{(j\,5)\,(j\,5)}{j\,5 + j\,5 - j\,5} = j5$$

$$Z_2 = \frac{(j\,5)(-\,j\,5)}{j\,5} = -\,j5$$

$$Z_3 = \frac{(j5)\,(-\,j5)}{j\,5} = -\,j5 \quad \textbf{Ans.}$$

Example 2.20: Determine the equivalent resistance between the terminals X and Y of the circuit shown using star-delta transformation.

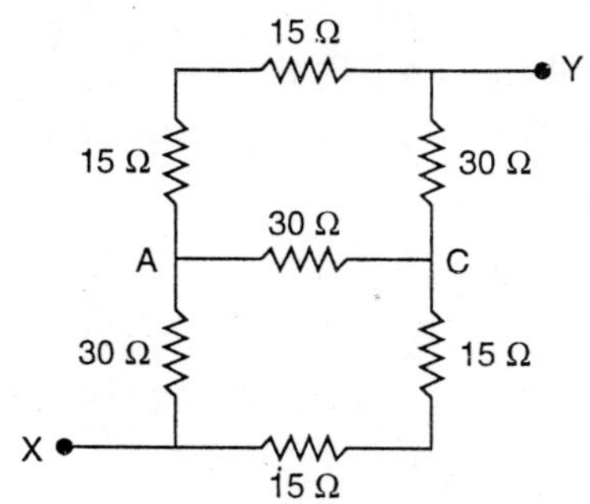

Fig. E2.20

Solution:

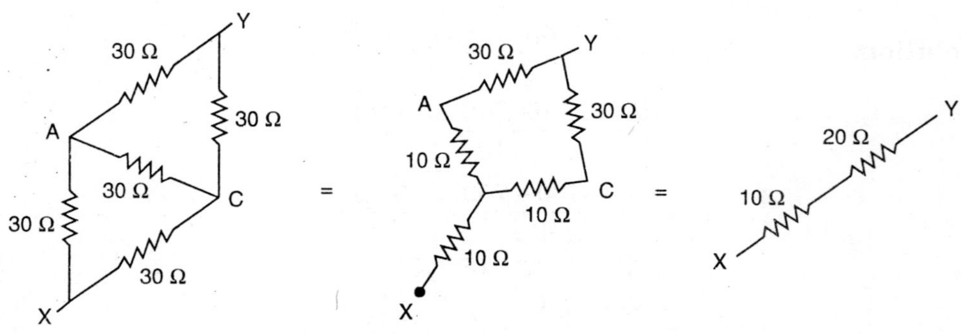

Fig. E2.20.1

Hence the total resistance between X and Y is 30 Ω. **Ans.**

PROBLEMS

2.1. Determine the current in the battery the current in each branch and the potential difference across *AB* in the network shown using loop analysis.

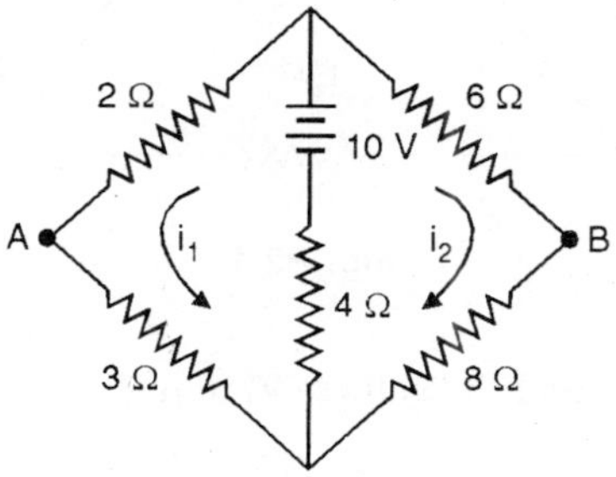

Fig. P2.1

[**Ans.** $i_1 = 0.96$A, $i_2 = 0.343$A, $V_{AB} = 0.14$ V]

2.2. In the network shown determine the direction and magnitude of current flow in the ammeter *A* having a resistance of 10 Ω.

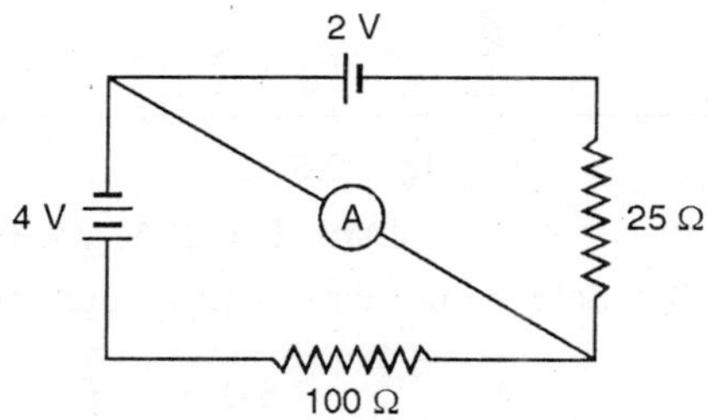

Fig. P2.2

[**Ans.** 26.7 mA]

2.3. Determine the current in all the branches of the network shown in Fig. P. 2.3 using loop analysis.

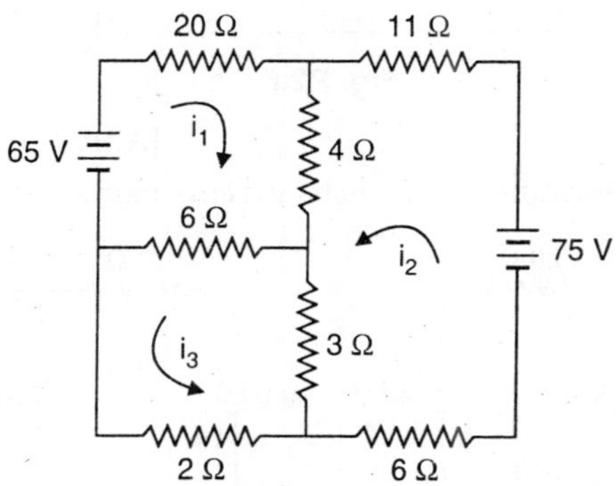

Fig. P2.3

[**Ans.** 1.844 A, 2.787 A, – 0.246A]

2.4 In the network shown determine the current using loop analysis.

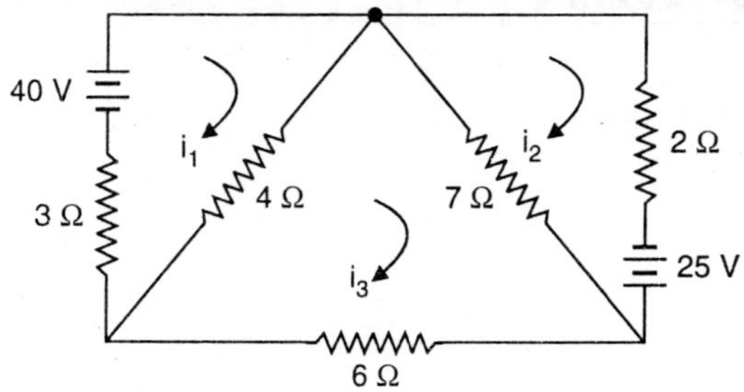

Fig. P2.4

[**Ans.** 8.32 A, 6.32 A, 4.56 A]

2.5. Find the currents I_1, I_2 and I_3 and the voltage V_1 and V_2 in the given network using nodal analysis.

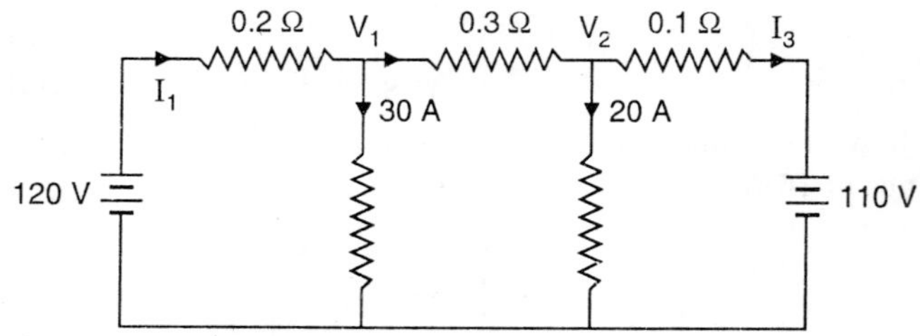

Fig. P2.5

[**Ans.** I_1 = 40A, $I_2 = I_3$ = 10A, V_1 = 112 V, V_2 = 109 V]

2.6. In the network shown in P. 2.6 determine currents in various branches using nodal analysis.

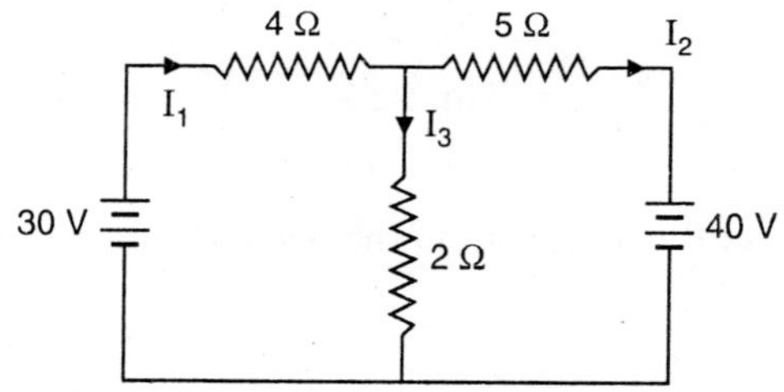

Fig. P2.6

[**Ans.** I_1 = 3.42A, I_2 = – 4.74, I_3 = 8.16 A]

2.7. In the networks shown determine the branch currents using nodal analysis.

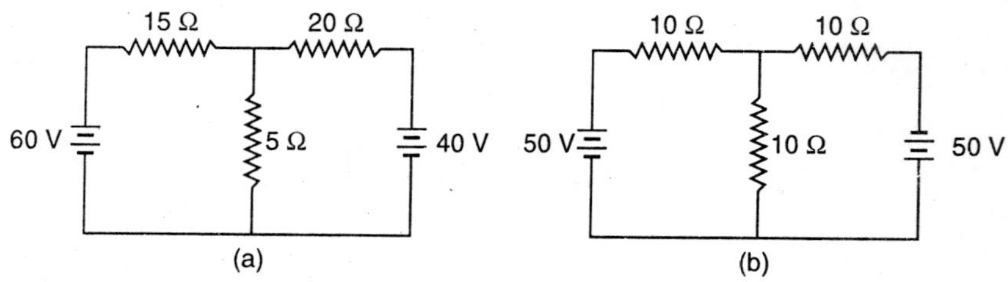

Fig. P2.7 (a) and (b)

[**Ans.** (*a*) 2.74, 1.05, 3.79 A, (*b*) 5A, 0A, 5A]

2.8. Determine the branch current in the networks shown using nodal analysis.

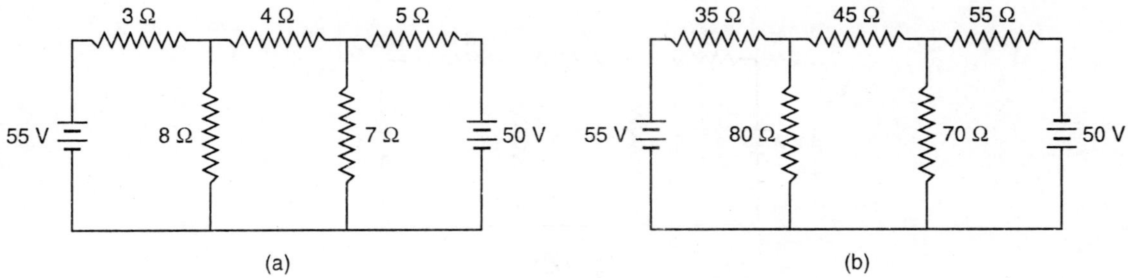

Fig. P2.8 (a) and (b)

[**Ans.** 5.866, 1.19, 4.675, 3.472, 4.662, (*b*) 0.938, 0.661, 0.277, 0.1089, 0.7705]

2.9. Determine the difference of potential between the points *X* and *Y* in the network shown.

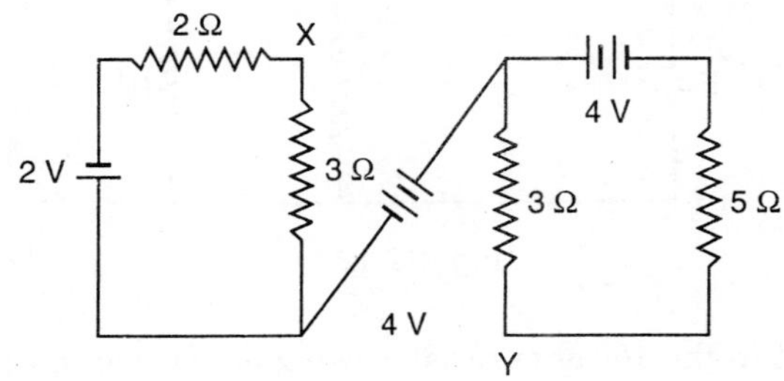

Fig. P2.9

[**Ans.** *Y* is 3.7 V + ve w.r.t. *X*]

2.10. Determine the equivalent (*a*) Thevenin's and (*b*) Norton's circuits which may be used to represent the given network at the terminals *AB*.

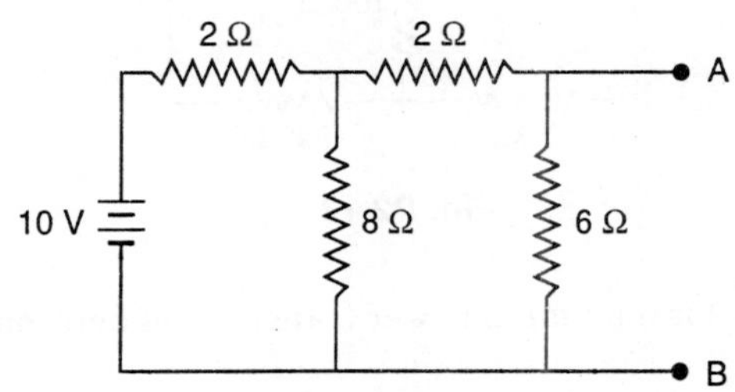

Fig. P2.10

[**Ans.** (*a*) 5V 2.25 Ω (*b*) 2.22 A, 2.25 Ω]

2.11. Determine the current through 3 Ω resistor branch as shown in Fig. P 2.11 using (*a*) Thevenin's theorem (*b*) Norton's theorem.

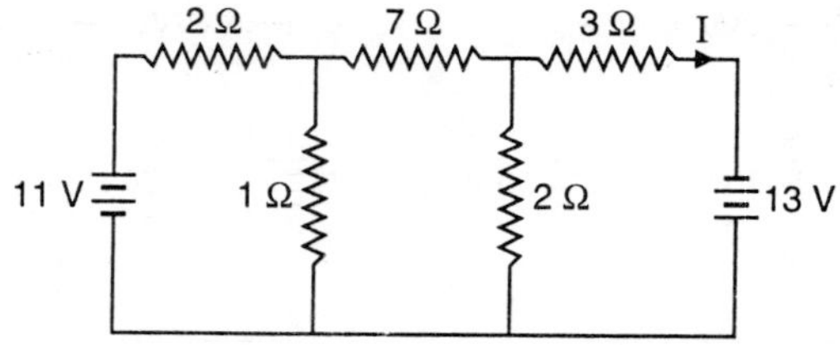

Fig. P2.11

[**Ans.** 3.03 A]

2.12. Determine the equivalent circuit and the current in R_L by (*a*) Thevenin's theorem (*b*) Norton's theorem, E is 100 V and all resistors in the circuit have the value of 20 Ω.

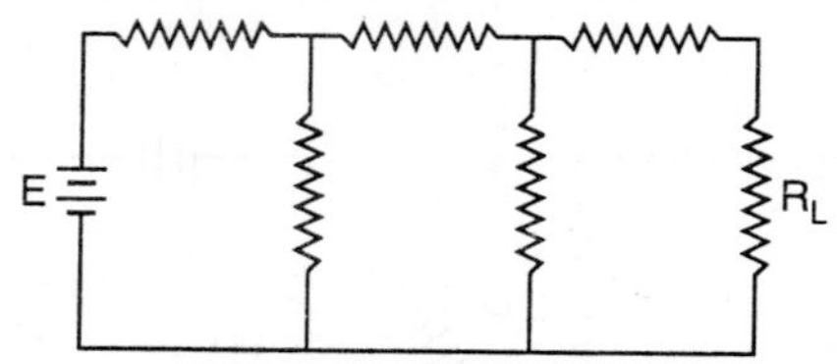

Fig. P2.12

[**Ans.** 0.385 A]

2.13. Determine the current in the 100 ohm resistor using (*a*) Thevenin's theorem (*b*) Norton's theorem.

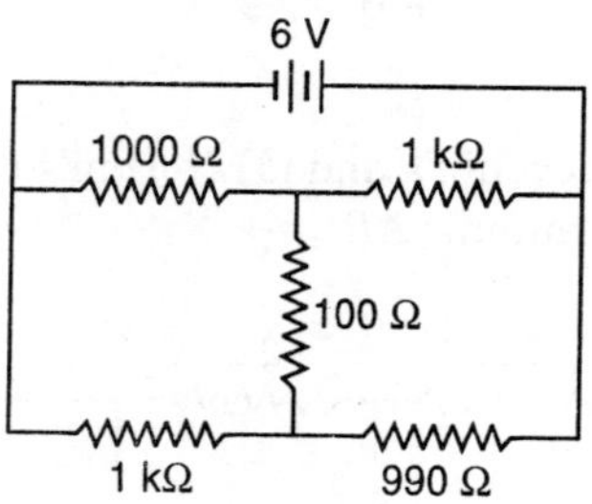

Fig. P2.13

[**Ans.** 13.7 μA]

2.14. Determine the value of R_L for maximum power transfer and determine this maximum power for the circuit given.

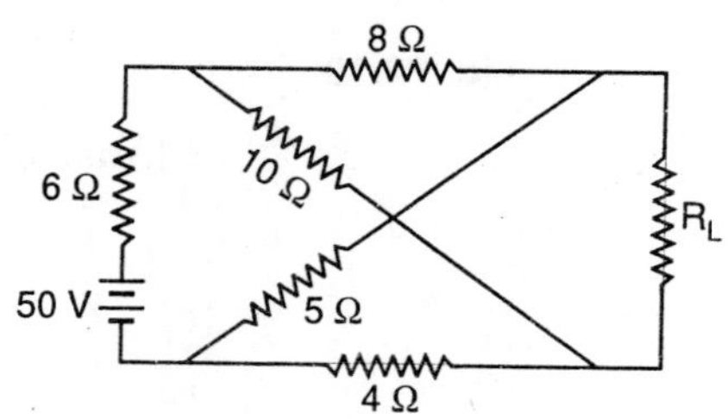

Fig. P2.14

[**Ans.** 6.67 Ω, 0.257 watt.]

2.15. Determine the resistance between the terminals *AB* of the network shown.

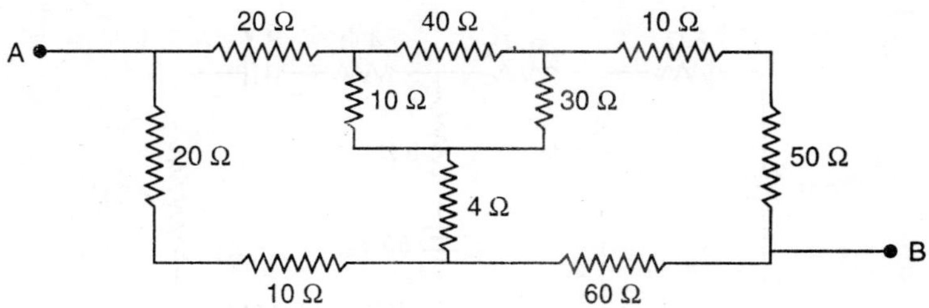

Fig. P2.15

[Ans. 47.2 Ω**]**

2.16. Determine the resistance between the terminals *A* and *B* if the resistance of each branch is 6 Ω in the given circuit.

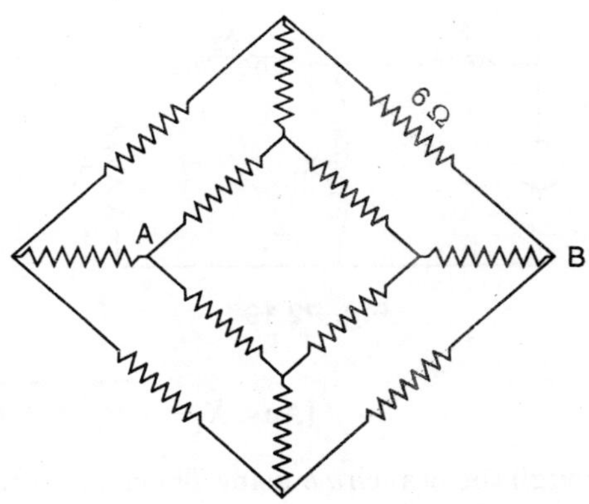

Fig. P2.16

[Ans. 5 Ω**]**

2.17. Determine the current through 10 Ω resistor in the network by (*i*) star-delta transformation (*ii*) Thevenin's theorem.

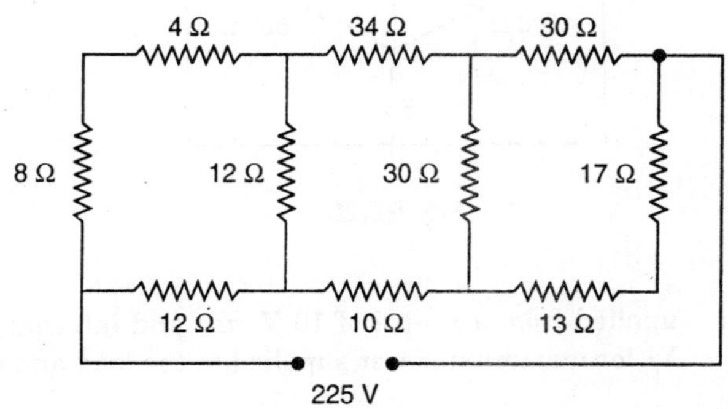

Fig. P2.17

[Ans. 5 A**]**

2.18. Determine the current through 2.5 Ω resistor using Thevenin's and Norton's theorem in the network shown.

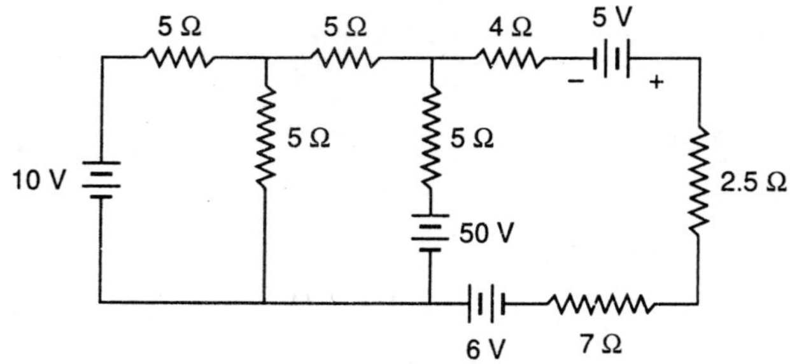

Fig. P2.18

[**Ans.** 2.44 A]

2.19. Determine X_1 and X_2 in terms of R_1 and R_2 to give maximum power dissipation in R_2.

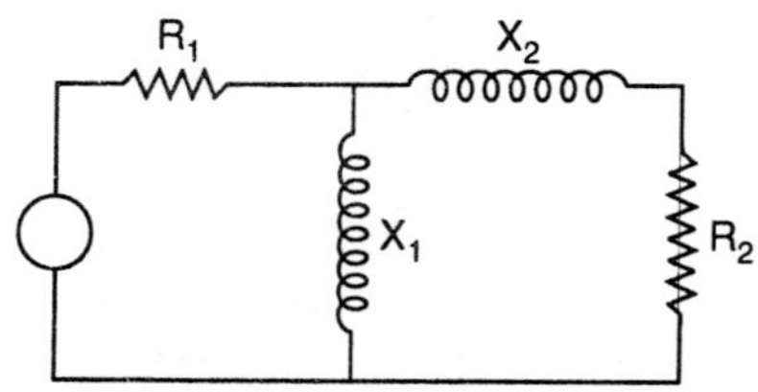

Fig. P2.19

[**Ans.** $X_1 = \sqrt{R_1^2 R_2/(R_1 - R_2)}$, $X_2 = \sqrt{R_2(R_1 - R_2)}$]

2.20. Determine the current through the branch $a\ b$ using Thevenin's theorem and superposition theorem.

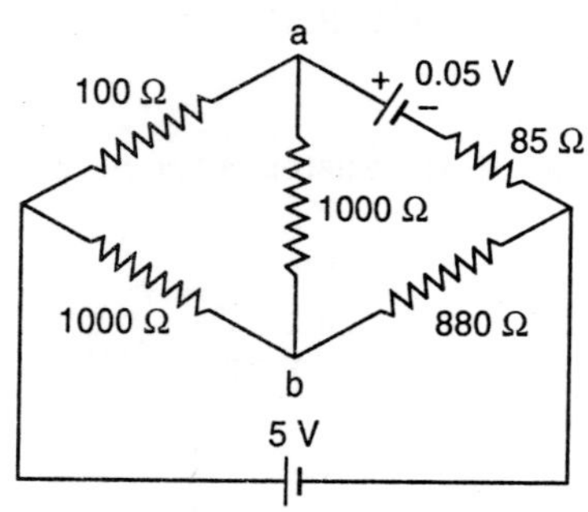

Fig. P2.20

[**Ans.** 10.62 A]

2.21. A load of $(20 - jX_c)$ is supplied from a source of 10 V rms and internal impedance $(10 + j20)$. Determine the value of X_C for maximum power supplied to the load and determine this power.

[**Ans.** 20 Ω, 2.22 W]

2.22. Two reactors each of 50 Ω reactance and negligible resistance are connected in series across a 200 V 50 Hz supply. Determine the value of resistance to be connected in parallel with one of the coils for maximum power dissipation.

[**Ans.** 25 Ω]

2.23. A source of voltage $2\sqrt{2}\sin 2t$ has an internal impedance $R = 1\ \Omega$ and $L = 1/3$ H. It is connected to a load having parallel combination of C and resistance of 1Ω. Determine C for maximum power consumption in $1\ \Omega$ load resistance.

$$\left[\textbf{Ans. } \frac{1}{13} F\right]$$

2.24. In the circuit shown R_g is variable between $2\ \Omega$ and $55\ \Omega$. What value of R_g results in maximum power transfer to the load R_L ?

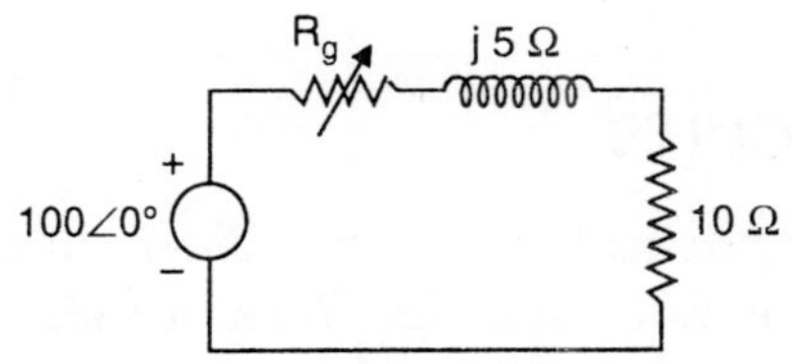

Fig. P2.24

[**Ans.** 8.66 Ω, 268 W]

CHAPTER

3 Three Phase Supply

3.1 THREE PHASE CIRCUITS

The use of d.c. for day to day application is much older than that of a.c. The first Central Electric Station was installed by Edison in New York in 1882 which operated at 110 V d.c.

The use of transformers for transmitting power over longer distances and higher voltages justified the use of 3-phase a.c. especially when the source of energy is at a far off place from the place of load centers and the power to be transmitted is in terms of thousands of MW.

A 3-phase circuit is more efficient, reliable and cost effective as compared to 1-phase system. A 3-phase circuit would obviously require a 3-phase supply. Hence a brief review of how 3-phase voltages are generated will not be out of place even though the book is only on analysis aspects of 3-phase circuits.

Generation of 3-phase voltages is also based on Faraday's laws of electromagnetic induction. Here the coil where the voltage is induced, is stationary whereas the magnetic field of the constant magnitude is made to rotate. This arrangement is preferred as the voltages induced are in terms of kilovolt and the currents to be handled by the winding are of large magnitude depending upon the size of the generator. The largest size of the generator (3-phase) in this country is of 500 MW capacity. Therefore, this winding is made stationary whereas the field winding which normally operates at relatively much lower voltage and currents is made rotating.

Fig. 3.1 shows an elementary 3-phase generator or alternator. Fig. 3.1 shows a two-pole rotor, the field winding of which is fed from a d.c. source to create magnetic flux for the poles. The poles are so shaped that these produce sinusoidal flux in space. The stator has a balanced three-phase winding with the axis of each phase displaced by 120°. The stator winding is balanced if the machine is mechanically symmetrical and if the electrical characteristics of the three coils are identical. The complete winding is represented in Fig. 3.1 by a single coil for simplicity.

Now, if the rotor is driven by a prime mover in the anticlockwise direction as shown, at synchronous speed, the voltage induced in the coil will be $v \times Bl$ and the direction of emf. induced will be given as per Flemmings right hand rule or by the cross product $v \times B$ rule.

With this rule the induced emf. in the coil sides $a\ b'\ c'$ is directed outside the paper whereas in coil sides $a'bc$ it is directed into the paper. Also since coil side a and a' are located along the pole axis where the flux density is maximum, the induced emf. in phase 'a' is maximum.

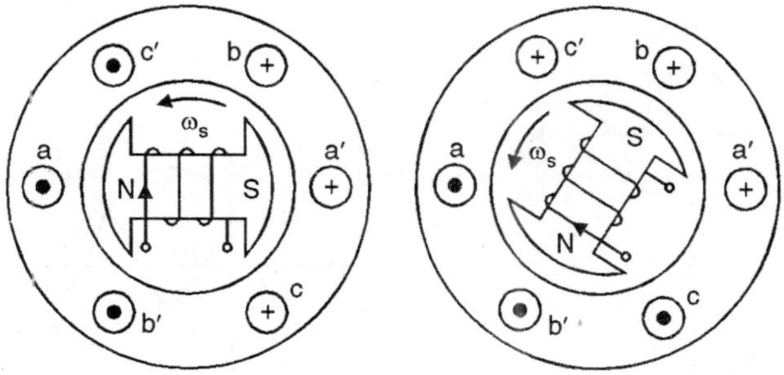

Fig. 3.1. (a) Voltage induced when pole axis along aa′
(b) Voltage induced when pole axis along bb′.

The corresponding emf. induced in phase *b* as identified by coil side *b* is found to be of opposite sign and of smaller magnitude. The magnitude is smaller as the flux density at the same instant is smaller as compared to what it is along the axis of the pole. In fact the distribution of *B* is cosinusoidal when it is maximum along the axis *aa*′, it will have a value *B*/2 along *bb*′ as the angle between *a* and *b* is 60° and cos 60° = ½. A similar reasoning yields the same results for phase *c*. It is to be noted that the angles given here are electrical degrees. However, for a two pole rotor, the electrical angles are equal to mechanical angles.

Next consider that the rotor has advanced 60° in counterclockwise direction. Now, North Pole coincides with *b*′ and south pole with *b* indicating that the induced emf. is now a negative maximum in phase *b*. In the first position it was – *Vm*/2 now it becomes – *Vm*. Also coil *cc*′ now comes under the influence of flux of reversed polarity as shown in Fig. 3.1(*b*). Hence the direction of the induced emf. in *c* is now out of the paper and into the paper for *c*′. The instantaneous values of the three phase voltages for the first and second time instants are shown in Fig. 3.2(*d*) and are identified as t_1 and t_2 respectively. By repeating the procedure as outlined above for various instants and plotting the results for each instant of time we obtain the curves for the three phases as shown in Fig. 3.2 (a)–(c). It is to be noted that the differences between the voltages of the three coils is not in the amplitude or waveform or the frequency but a time difference between the three waveforms. One complete rotation of the rotor gives one cycle of the voltage wave as shown by dotted vertical line.

Physical connection of the ends of each phase (points *a*′, *b*′, *c*′) gives a Y-connection or Wye connection of the stator winding.

Fig. 3.2(*d*) shows the typical pattern of three phase system of voltages when the three waveforms are drawn on the same time axis. These waveforms can also be developed as the projections of the rotating phasor as shown in Fig. 3.2(*e*). In our analysis we will always assume that we have a 3-phase supply i.e. the three voltages *Ea′a*, *Eb′b* and *Ec′c* are equal in magnitude and displaced in time phase by 120°.

In a basic alternator the three coil ends *a*′, *b*′, *c*′ are tied together and brought externally to a common lead called the neutral *N*. Leads *a*, *b* and *c* are brought externally to line terminals called phase *a*, phase *b* and phase *c* as shown in Fig. 3.3. The three phase voltages are specifically 120° apart. First phase *a* has positive maximum value, next phase *b* has its maximum positive value and another 120° time apart phase *c* has a positive maximum value. If the

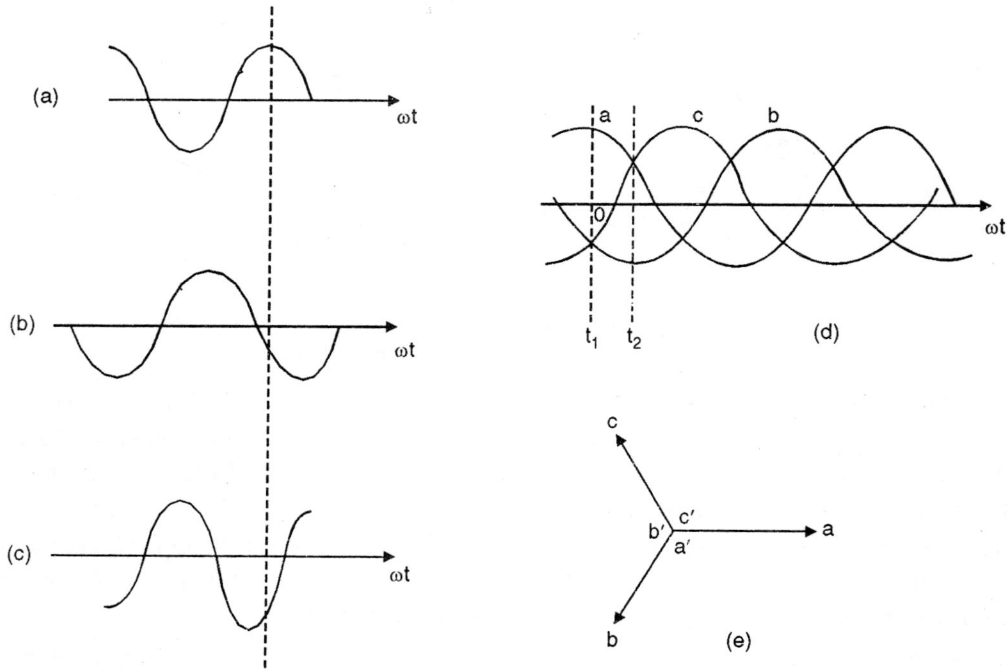

Fig. 3.2 (a) Voltage in phase *a*, (*b*) and (*c*) simultaneous voltages in phases *b* and *c* (*d*) all three voltages on a common time axis (*e*) phasor diagram.

variations of voltages of the three phases follow this pattern, the voltages are said to be of positive sequence and the phase ordering is positive phase sequence and it is denoted by *a b c* sequence. However, if the rotation of the rotor is reversed i.e. it is made to rotate clockwise the phase sequence would be *acb*. Here at first phase *a* attains its positive maximum value, next phase *c* and finally phase *b* attains its positive maximum and this is known as negative phase sequence.

Let us redraw the phasor diagram of the three phase voltages for the alternator of Fig. 3.3 when the phase voltages are expressed in their rms values. Let this diagram be as shown in Fig. 3.4(a).

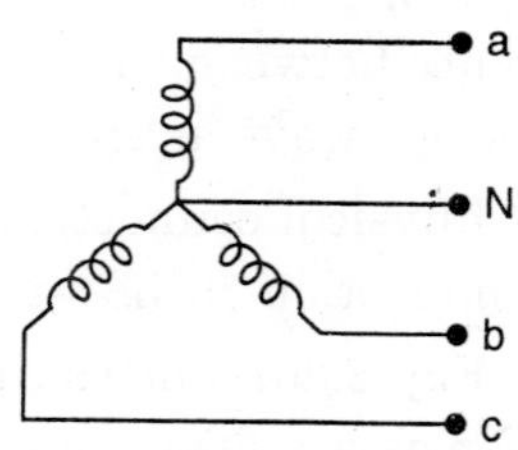

Fig. 3.3 Wye connection in a 3-phase alternator.

Here the phasor voltages for the phases *a b* and *c* are V_{Na}, V_{Nb} and V_{Nc} respectively and these are known as phase voltages and usually denoted by V_{ph} or V_{ϕ}. If we take phase *a* voltage as the reference, the three phase voltages are expressed mathematically as

$$V_{Na} = V_{ph} \angle 0° \quad V_{Nb} = V_{ph} \angle -120°$$

and $$V_{Nc} = V_{ph} \angle 120°$$

The voltage between phases *a* and *b* is denoted as V_{ab}. Similarly, V_{bc} and V_{ca} are known as voltages between phases *b* and *c* & *c* and *a* respectively. These are known as line to line voltages or simply line voltage. Voltage V_{ab} is expressed as $V_{ab} = V_a - V_b$. Here $-V_b$ means the phasor Voltage V_b is reversed. This is illustrated in Fig. 3.4(*b*). Using law of parallelogram of forces where two sides are $|V_a| = |V_b| = |V_{ph}|$ and the angle between the two phasors is 60°.

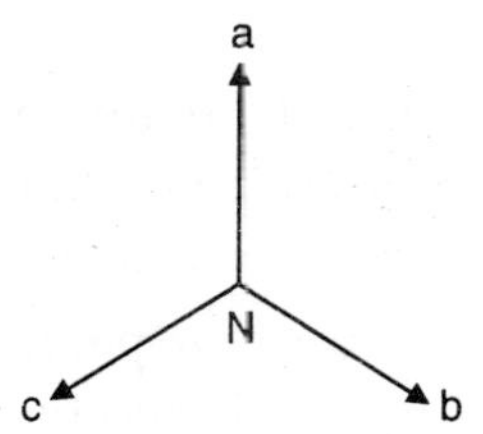

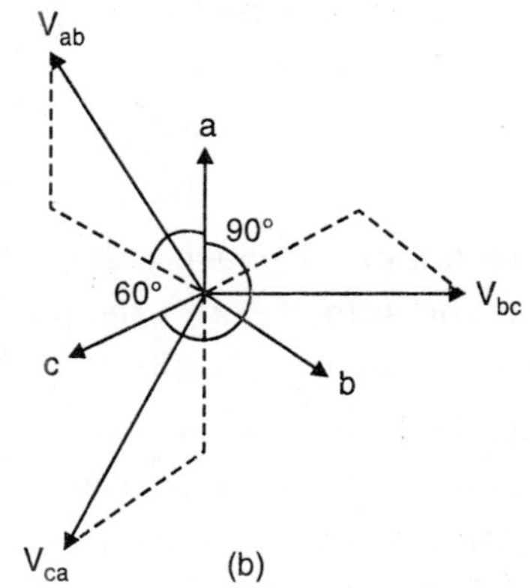

Fig. 3.4 (a) Three phase phasor diagram (b) Line voltage phasor diagram.

Hence $|V_{ab}|^2 = V_{ph}^2 + V_{pn}^2 + 2\,|V_{ph}|\,.\,|V_{ph}|\cos 60°$

$= 3\,|V_{ph}|^2$

or $|V_{ab}| = \sqrt{3}\,|V_{ph}|$ ...(3.1)

This means in a star connected system the line voltages equal $\sqrt{3}$ times the phase voltages.

Let us consider a star connected 3-phase supply system connected to a 3-phase star connected load as shown in Fig. 3.5. It can be seen that for star connected supply the line currents I_a, I_b and I_c are same as phase currents.

Therefore, we conclude that for a star connected system

$$V_L = \sqrt{3}\,V_{ph}$$

and $$I_L = I_{ph}$$

For a balanced 3-phase supply the neutral N is always held at zero or ground potential. The common terminal of star connected load is known as star point rather than neutral.

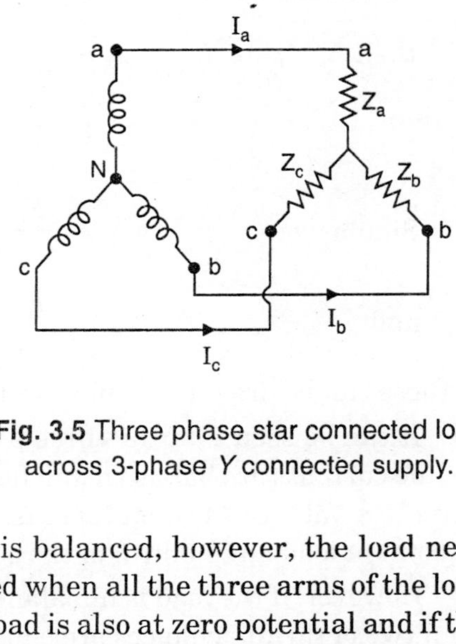

Fig. 3.5 Three phase star connected load across 3-phase *Y* connected supply.

The three-phase supply under normal conditions is balanced, however, the load need not necessarily be balanced. The load is said to be balanced when all the three arms of the load are identical. If the load is balanced the star point of the load is also at zero potential and if the star point and the neutral point of the supply are connected by a wire the current in this wire will be zero. However, if the load is unbalanced, the star point of the load is no more at zero potential and there will be a current, I_n in the wire connecting the star point and the neutral of the supply. This system is known as three-phase 4-wire system. Every effort is made by the electric utilities to maintain balanced load on the system, as unbalanced load on the system gives rise to what is known as negative sequence current which adversely affect the operation of the rotating machines in the system.

The sum of the three phase voltages of a 3-phase balanced supply is zero.

$$V_a + V_b + V_c = V_{ph} \angle 0 + V_{ph} \angle -120 + V_{ph} \angle 120°$$
$$= V_{ph} + (-0.5 - j\,0.866)\,V_{ph} + (-0.5 + j\,0.866)\,V_{ph}$$
$$= 0$$

The three balanced phase voltages form a closed equilateral triangle and hence, this also shows that the sum of the three phase voltages is zero. Similarly, if a balanced load is connected across a balanced supply, the sum of three phase or line currents is again zero.

However, if the three phase load is unbalanced even though the star-point is not at zero potential, the sum of the three currents I_a, I_b and I_c is zero if it is a three-phase three wire system. For 3-phase 4 wire system for unbalanced load the sum of three phase currents equals neutral current $I_a + I_b + I_c = I_N$.

3.2 ANALYSIS OF STAR CIRCUITS

We will assume that the supply is star connected and balanced. However, the star connected load could be balanced or unbalanced.

If the load is balanced i.e. all the three star legs have identical impedances, in such a case, the circuit can be analysed based on 1-phase basis.

Consider Fig. 3.5. Here if $Z_a = Z_b = Z_c = Z$

then $$I_a = \frac{E_a}{Z}, \quad \text{If} \quad Z = |Z| \angle\phi$$

$$I_a = \frac{E_a \angle 0}{Z \angle\phi} = \frac{|E_a|}{|Z|} \angle -\phi$$

Similarly $$I_b = \frac{E_b}{Z \angle\phi} = \frac{|E_b| \angle -120°}{Z \angle\phi} = \frac{|E_b|}{|Z|} \angle -120 - \phi$$

and $$I_c = \frac{|E_c| \angle 120}{Z \angle\phi} = \frac{|E_c|}{|Z|} \angle 120 - \phi$$

and these can be drawn on a phasor diagram as shown in Fig. 3.6.

It can be seen that the three phase current or in this case line currents are balanced and hence the resultant sum is zero. In a balanced system there is no need to connect the star point of load and neutral of the source.

However, if the load is unbalanced if $Z_a \neq Z_b \neq Z_c$, and even if two of the impedances are equal and the third one is different, it is still an unbalanced load and the analysis can be carried out using loop equations or by using Millman Theorem. Fig. 3.5 is reproduced here for convenience. Taking loop consisting of voltage sources E_a and E_b and the impedances Z_a and Z_b we have

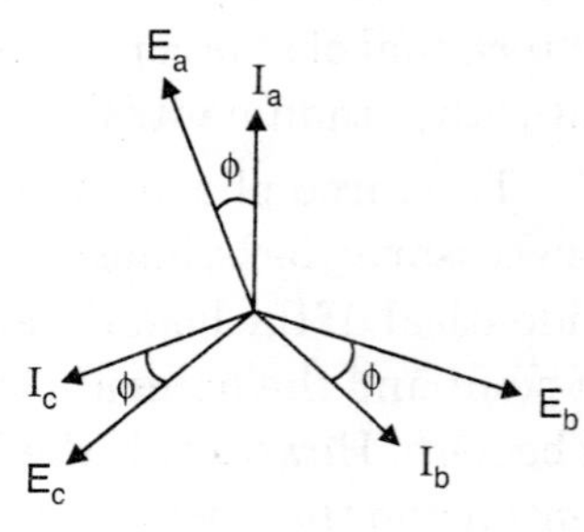

Fig. 3.6 Phasor diagram for a balanced lagging *p.f.* load.

$$E_a - I_a Z_a + I_b Z_b - E_b = 0 \quad ...(3.2)$$

Similarly for the other loop this equation is

$$E_a - I_a Z_a + I_c Z_c - E_c = 0 \quad ...(3.3)$$

and $$I_a + I_b + I_c = 0 \quad ...(3.4)$$

It is to be noted that all the quantities in these three equations viz. E_a, E_b, E_c, Z_a, Z_b and Z_c must be used in phasor form to obtain the solution for the currents I_a, I_b and I_c.

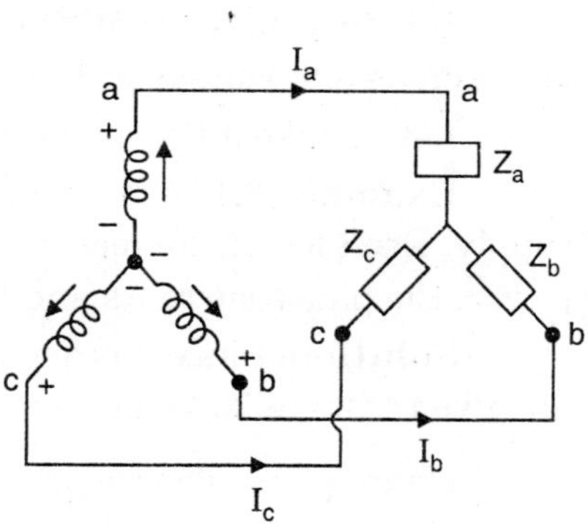

Fig. 3.6(a) Unbalanced loading.

Millman Theorem: Since the load is unbalanced the star point of the load will not be at zero potential. Let the potential be V_s. This is given by Millman's Theorem as

$$V_s = \frac{\dfrac{E_a}{Z_a} + \dfrac{E_b}{Z_b} + \dfrac{E_c}{Z_c}}{\dfrac{1}{Z_a} + \dfrac{1}{Z_b} + \dfrac{1}{Z_c}} \quad ...(3.5)$$

Again, all the quantities on the *rhs* are to be used in phasor form. Once V_s is known

$$I_a = \frac{E_a - V_s}{Z_a}, \quad I_b = \frac{E_b - V_s}{Z_b}$$

and $$I_c = \frac{E_c - V_s}{Z_c}$$

Even though it is an unbalanced system, since the star point is not connected to the neutral of the supply, the sum of the three currents $I_a + I_b + I_c = 0$.

The potential of the star point is an indication of the degree of unbalancing of the load. The higher the value of V_s, the larger is the degree of unbalancing and hence it is an undesirable operation of the system.

Consider 3-phase 4-wire system as shown in Fig. 3.7.

Whenever 1-phase loads are present in the system, 3-phase 4-wire system is used. The distribution is, therefore, invariably a 3-phase 4-wire system.

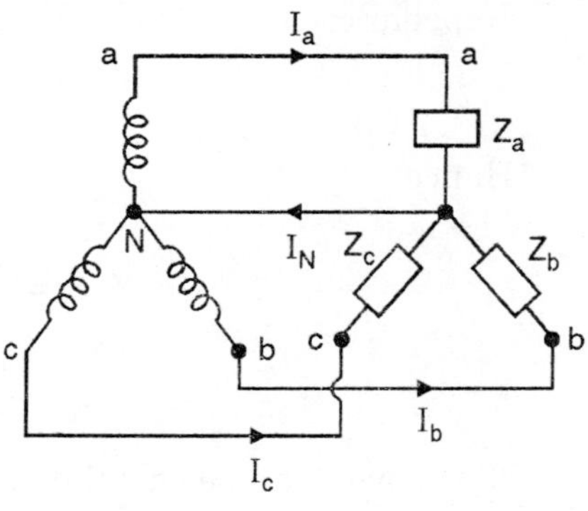

Fig. 3.7 3-phase 4-wire system.

The analysis here again is on single phase basis

$$I_a = \frac{E_a \angle 0}{Z_a}, \quad I_b = \frac{E_b \angle -120°}{Z_b}$$

$$I_c = \frac{E_c \angle 120°}{Z_c}$$

and $$I_a + I_b + I_c = I_n$$

Here Z_a, Z_b and Z_c may include the impedance of the leads between the alternator and the load and between the star point and the neutral of the alternator if these are present.

It is to be noted that whenever, in a problem, system voltage is specified, it is always to be taken as line voltage or line to line voltage, unless it is specified as phase voltage.

Let us take a few examples to illustrate what we have learnt so far.

Example 3.1: A balanced 3-phase star connected load is fed from a 208 volt 3-phase supply. Each leg of the load has a resistance of 35 Ω. Determine, the power factor, the total power, the phase currents and line currents of the system.

Solution: Since it is a balanced circuit, the analysis is to be made on 1-phase basis. Also the given voltage 208 since not specified, it is to be taken as line voltage.

Hence phase voltage is $208/\sqrt{3} = 120$ volt.

$$I_a = \frac{120\angle 0}{35} = 3.428\angle 0°$$

$$I_b = \frac{120\angle -120}{35} = 3.428\angle -120°$$

$$I_c = \frac{120\angle 120}{35} = 3.428\angle 120°$$

Since the three currents are balanced, the phasor sum of the three currents is zero. Since the load is resistive power p per phase is

$$I^2R = 3.428^2 \times 35 = 411.3 \text{ watts}$$

Hence total power $3 \times 411.3 \simeq 1234$ watts and the load is resistive, the p.f. is unity. Since the load is star connected $I_L = I_{ph} = 3.428$ A.

Example 3.2: Repeat example 3.1 if each leg of the load has an impedance of $(15 + j20)\Omega$.

Solution: Again as it is balanced load we analyse it on 1-phase basis.

Impedance $$z = |z|\angle\phi = \sqrt{15^2 + 20^2}\tan^{-1}\frac{20}{15}$$

$$= 25\angle 53°$$

Hence $$I_a = \frac{120\angle -0}{25\angle 53°} = 4.8\angle -53°$$

$$I_b = \frac{120\angle -120}{25\angle 53°} = 4.8\angle -173°$$

$$I_c = \frac{120\angle 120}{25\angle 53°} = 4.8\angle 67°$$

Now power can be calculated either using the formula, power loss = $3I^2R$

or $$P = 3V_{ph}I_{ph}\cos\phi$$

$$= \sqrt{3}V_LI_L\cos\phi$$

Using the ohmic loss formula, since resistance in each phase is 15Ω, the power loss is $3 \times 4.8^2 \times 15 = 1036.8$ watts.

In the formulae $P = \sqrt{3}\,V_LI_L\cos\phi$, the angle ϕ is the angle between the phase voltage and the phase current corresponding to the same phase. It is to be noted with caution that the angle ϕ is not the angle between the line voltage and line current or phase current but it is the angle between phase voltage and phase current. Since the impedance angle is 53° it is an inductive load and the current in each phase lags by its voltage by 53°.

Hence $P = \sqrt{3} \times 208 \times 4.8 \cos 53°$

$= 1037$ watts. **Ans.**

$I_L = I_{ph} = 4.8$ A

Example 3.3: A symmetrical 3-phase 440 volt system supplies a star connected load. The branch impedances are $Z_R = 10\ \angle 30°$, $Z_Y = 12\ \angle 45°$ and $Z_B = 15\ \angle 40°$. Assuming the neutral of the supply to be earthed, calculate the potential of the star point and the line currents, phase sequence $R\ Y\ B$.

Solution: Using Millman Theorem

$$V_s = \frac{\frac{254\angle 0}{10\angle 30} + \frac{254\angle -120}{12\angle 45°} + \frac{254\angle 120°}{15\angle 40}}{0.1\angle -30 + \frac{1}{12}\angle -45° + \frac{1}{15}\angle -40}$$

$$= \frac{25.4\angle -30° + 21.17\angle -165 + 16.9\angle 80}{0.0866 - j0.05 + 0.0589 - j0.0589 + 0.05 - j0.0428}$$

$$= \frac{21.996 - j\,12.702 - 20.45 - j\,5.48 + 2.94 + j\,16.68}{0.1966 - j\,0.152}$$

$$= \frac{4.73\angle -18.5}{0.2485\angle -37.71}$$

$$= 19.03 < 19.21$$

$$I_a = \frac{254\angle 0 - 19.03\angle 19.21}{10\angle 30}$$

$$= \frac{236.1\angle -15}{10\angle 30}$$

$$I_a = 23.61\angle -31.5$$

$$I_b = \frac{254\angle -120 - 19.03\angle 19.21}{12\angle 45°}$$

$$= \frac{-127 - j\,219.96 - 17.97 - j\,6.26}{12\angle 45°}$$

$$= \frac{268.68\angle -122.7}{12\angle 45°}$$

$$= 22.39\angle -167.7$$

$$I_c = \frac{254\angle 120 - 19.03\angle 19.21}{15\angle 40}$$

$$= \frac{-144.97 + j\,213.70}{15\angle 40} = \frac{258.2\angle +124.2}{15\angle 40}$$

$$= 17.21\angle 84.2$$

$$I_a + I_b + I_c = 23.61\angle -31.5 + 22.39\angle -167.7 + 17.21\angle 84.2$$

$$= 20.13 - j\,12.336 - 21.87 - j\,4.77 + 1.74 + j\,17.12$$

$$= 21.87 - 21.87 - j\,17.10 + j\,17.12 = 0$$

Hence it is found that for a 3-phase 3 wire system with unbalanced loading the sum of the three currents is zero (KCL at star point).

Example 3.4: Three equal star-connected inductors take 8 KW at p.f. 0.8 when connected to a 460 volt 3 – phase 3-wire supply. Determine line currents if one inductor is short circuited.

Solution. Refer to Fig. E 3.4 (a) and (b).

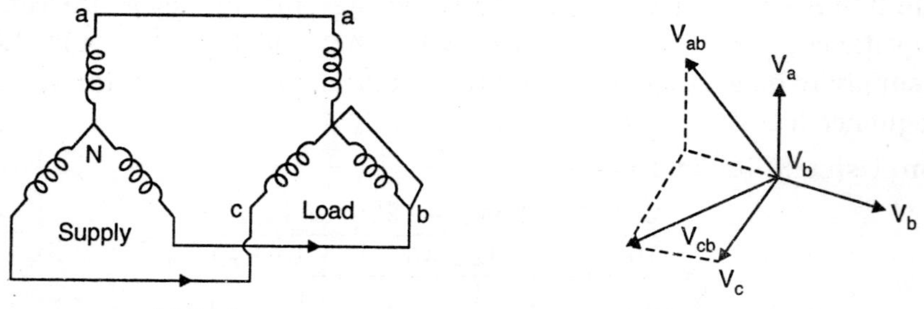

Fig. E 3.4

Suppose inductor in phase *b* is shorted. Therefore, the voltage across inductor in phase a is $V_a - V_b$ as the potential of the star point equals V_b. Similarly voltage across the inductor in phase *c* is

$$V_{cb} = V_c - V_b$$

These voltages have been shown in the phasor diagram. We take here V_{ab} as the reference.

Now the phase current

$$= \frac{8 \times 1000}{\sqrt{3} \times 460 \times 0.8} = 12.55 \text{ A}$$

Hence impedance per phase

$$= \frac{V_{ph}}{I_{ph}} = \frac{460}{\sqrt{3} \times 12.55} = 21.16\ \Omega$$

Hence $\quad Z = 21.16\,(0.8 + j\,0.6)$

$$= 16.93 + j\,12.7 = 21.16\ \angle 36.86$$

Current in phase a $\quad = \dfrac{460\ \angle 0}{21.16\ \angle 36.86} = 21.7\ \angle -36.8$

Current in phase c $\quad = \dfrac{460\ \angle 60}{21.16\ \angle 36.86} = 21.7\ \angle 23.14$

I_a and I_c are at an angle of 60° = (23.14° + 36.86°)

Hence $\quad I_b^2 = I_a^2 + I_c^2 + 2I_a I_c \cos 60°$

$$= 21.7^2 + 21.7^2 + 2 \times 21.7 \times 21.7 \times \frac{1}{2}$$

$$= 3 \times 21.7 = 37.6 \text{ Amp.}$$

3.3 DELTA CIRCUITS

Each coil has two terminals, start and finish terminal. Fig. 3.8 shows that winding terminals $a'\ b'\ c'$ are called the start terminals whereas abc the finish terminals.

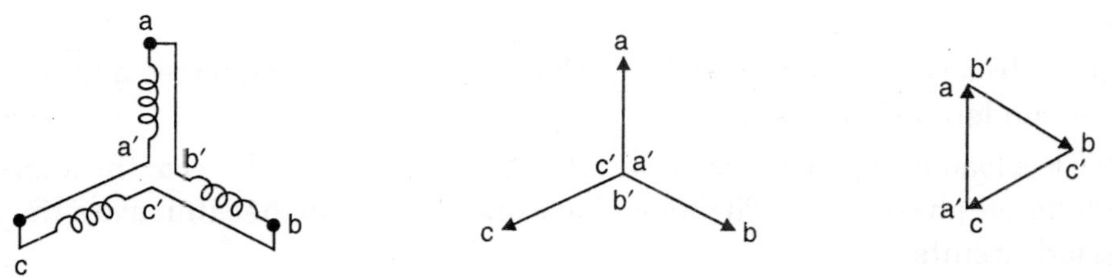

Fig. 3.8 (a) Delta (b) Star phasor (c) delta phasor.

If the start point represents the tail of the phasor and the finish point the arrow head of the phasor, when $a'\ b'\ c'$ are joined together forming the neutral of the system, a, b, c form the phase terminals of the system.

However, if the finish of phase a is connected to start of phase $b(b')$ and finish of phase b is connected to the start of phase $c(c')$ and finally the finish of phase c is connected to start of phase a, a closed circuit is formed and if the phasors are also drawn on these lines, a Δ like shape is obtained. This is known as delta circuit. In the phasors diagram of Fig. 3.8(c), the finish of phasor a (arrow head of phase a) is connected to start of phase b (tail of arrow of phase b).

From Fig. 3.8 (a) and (c) it is seen that there is no common terminal for all the three windings and hence delta winding does not have a neutral.

To find out relation between line quantities and phase quantities for a delta connected system refer to Fig. 3.9

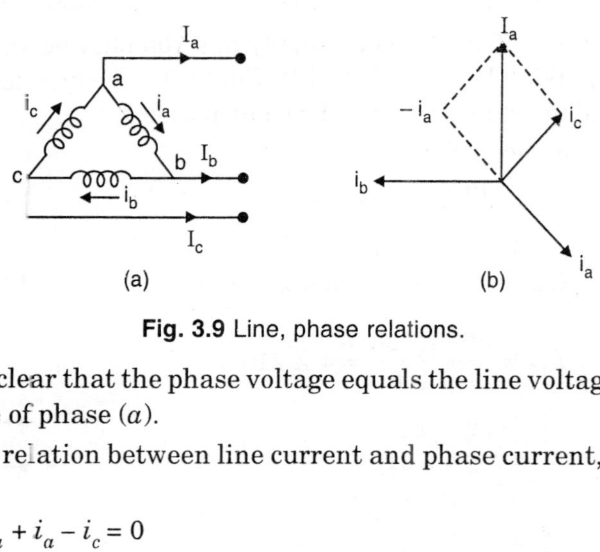

Fig. 3.9 Line, phase relations.

From the Fig. 3.9 it is clear that the phase voltage equals the line voltage *i.e.* line voltage V_{ab} is also the phase voltage of phase (a).

Now to find out relation between line current and phase current, we apply *KCL* at node a of Fig. 3.9(a)

$$I_a + i_a - i_c = 0$$

or $$I_a = i_c - i_a$$

Assuming of course balanced system $|\, i_a \,| = |\, i_b \,| = |\, i_c \,| = i_{ph}$ and referring to phasor diagram of Fig. 3.9(b) it is seen that $I_a = i_c - i_a = \sqrt{3}\, i_{ph}$ i.e. the line current is $\sqrt{3}$ times the phase current.

Normally the supply is star connected as neutral is available, hence single phase loads can be supplied. Also, for the same line voltage, the star winding requires less insulation as the insulation is to be provided for $\frac{V_L}{\sqrt{3}}$ voltage as against delta connection, where the insulation per phase is V_L.

We have already studied the analysis when supply is star connected and the load is also star connected under various options.

In case the load is delta connected whether balanced or unbalanced, the analysis is to be made on individual phase basis. The following example will further illustrate the solution of delta connected circuits.

Example 3.5. A balanced delta connected load of $(8 + j6)\ \Omega$ per phase is connected to a 3-phase 230 V supply. Find the line current, p.f., power, reactive power and total volt amperes.

Solution. Since it is a balanced system the phase current

$$i_R = \frac{230\angle 0}{8 + j6} = 23 \angle -36.8 \text{ i.e. the p.f. is 0.8}$$

Line current $\quad 23\sqrt{3} = 39.8$ A

Power $\quad \sqrt{3}\ V_L I_L \cos\phi = \sqrt{3} \times 230 \times 39.8 \times 0.8$

$= 12.696$ kw

Reactive power $\sqrt{3}\ V_L\, I_l \sin\phi$

$= \sqrt{3} \times 230 \times 39.8 \times 0.6$

$= 9.513$ KVAr

and $\quad$ Total KVA $= \sqrt{3} \times 230 \times 39.8 = 15.855$ KVA **Ans.**

Example 3.6. A 400 v, 50 Hz 3-phase supply has 100 ohm between R and Y, $j\ 100\ \Omega$ between Y and B and $-j\ 100\ \Omega$ between B and R. Find (a) line currents for phase sequence RYB(b) star connected balanced resistors for the same power.

Solution: Phase sequence RYB

$$I_{RY} = \frac{400\angle 0}{100} = 4\angle 0$$

$$I_{YB} = \frac{400\angle -120}{j\,100} = 4\angle -210$$

$$I_{BR} = \frac{400\angle 120}{-j\,100} = 4\angle 210$$

$$I_R = I_{RY} - I_{BR}$$
$$= 4 - 4\angle 210$$
$$= 4 + 3.46 + j2$$
$$= 7.46 + j2$$
$$= 7.72\angle 14.5$$

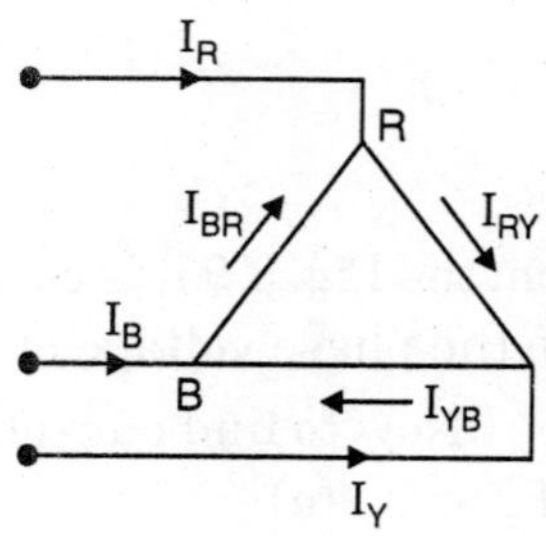

Fig. E 3.6

$$I_Y = I_{YB} - I_{RY}$$
$$= 4\angle -210 - 4$$

$$= -3.46 + j2 - 4$$
$$= -7.46 + j2$$
$$= 7.72 \angle 165$$

Similarly

$$I_B = I_{BR} - I_{YB}$$
$$= 4 \angle 210 - 4 \angle -210$$
$$= -3.46 - j2 + 3.46 - j2$$
$$= -j2 = 2 \angle -90°$$

(*b*) Since power consumed is only in the arm *RY* where the current is 4 A. Total power $4^2 \times 100 = 1600$ watts.

If the resistors are connected in star, voltage across each star leg is $400/\sqrt{3}$ and if *R* is the resistance per star leg, total power

$$3 \times \left(\frac{400}{\sqrt{3}}\right)^2 \cdot \frac{1}{R} = 1600$$

$$\frac{16 \times 10^4}{R} = 1600 \quad \text{or} \quad R = 100\ \Omega.$$

Example 3.7. A 3-phase 3-wire supply feeds a load consisting of 3 equal resistors. By how much is the load reduced if one of the resistors be removed (a) when the load is star connected (b) when the load is mesh connected ?

Solution: (a) Star connected load.

Let *R* be the resistance per phase and V_{ph} is the phase voltage.

The power 3 phase $= 3\dfrac{V_{ph}^2}{R}$

However, if one of the resistors is removed the voltage across the remaining two resistors is $V_L = \sqrt{3}\,V_{ph}$. Hence power drawn by the load $= \dfrac{(\sqrt{3}\,V_{ph})^2}{2R}$

$$= \frac{3V_{ph}^2}{2R}$$

Comparing the two results, the power under the second condition is just ½ of the original power. Hence there is reduction of load by 50%.

(*b*) Mesh Connected load

Since it is mesh connected load, the voltage across each resistor is $V_L = \sqrt{3}\,V_{ph}$. Hence 3-phase power $= 3(\sqrt{3}\,V_{ph})^2 \cdot \dfrac{1}{R} = 3 \times 3\dfrac{V_{ph}^2}{R}$. Comparing this result with the star connected load, it is found that for the same system voltage (Normally star connected), and same value of resistance if it is desired to have more power, the resistances should be connected in delta as then the power is three times what we have when these are connected in star.

Now if one of the resistors from delta connected load is removed, the voltage across each resistor is still line voltage hence, power drawn.

$$2.(\sqrt{3}V_{ph})^2 \cdot \frac{1}{R} = \frac{2\times 3V_{ph}^2}{R}$$

The difference in power is $\frac{3V_{ph}^2}{R}$ which is 33.3% of the original power.

Similarly, if capacitors are to be connected in the circuit for supplying reactive power to the system, for the same capacitance, it is desirable to connect them in delta for higher reactive power supply.

3.4 MEASUREMENT OF POWER IN 3-PHASE CIRCUITS

We first study measurement of active power

The following are the methods.

1. Three wattmeter method → Both balanced and unbalanced loads.
2. Two wattmeter method → Both balanced and unbalanced loads.
3. One wattmeter method → for balanced loads only.

3.4.1 Three Wattmeter Method

Fig. 3.10 shows the connection of the three wattmeters to a star connected load. These three wattmeters could be three single phase wattmeters or a single unit consisting of three wattmeteric elements.

The arrows in the diagram show the direction of current and voltage which are conventionally considered positive. If the letters representing currents and voltages denote instantaneous values then, instantaneous power in the load,

$$= e_1 i_1 + e_2 i_2 + e_3 i_3$$

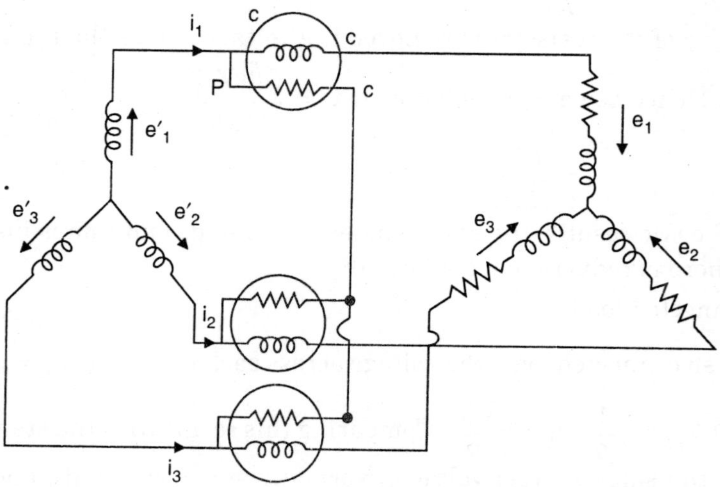

Fig. 3.10 Three wattmeter method for power measurement.

If it is an unbalanced load the potential of the star point is v, therefore,

$$e_1 = e_1' + v$$

$$e_2 = e_2' + v$$

and $$e_3 = e_3' + v$$

Therefore, instantaneous power of the load,

$$= (e_1' + v)i_1 + (e_2' + v)i_2 + (e_3' + v)i_3$$

$$= e_1' i_1 + e_2' i_2 + e_3' i_3 + v(i_1 + i_2 + i_3)$$

$$= e_1' i_1 + e_2' i_2 + e_3' i_3$$

as $i_1 + i_2 + i_3 = 0$ for a star connected load whether it is balanced or unbalanced.

Now $e_1' i_1 + e_2' i_2 - e_3' i_3$ is the total instantaneous power measured by the three wattmeters and thus the sum of the readings of the wattmeter will give the mean value of the total power.

3.4.2 Two Wattmeter Method

Fig 3.11 shows two wattmeter method of measurement of power. This is the most commonly used method for measurement of power as this can be used both for balanced and unbalanced load and requires two wattmeters.

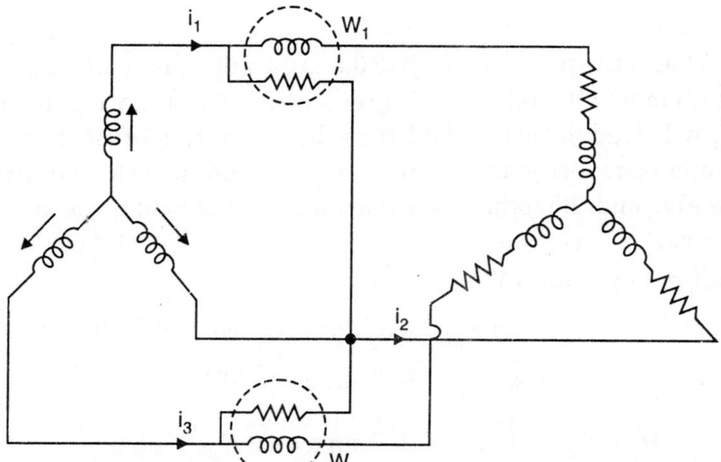

Fig. 3.11 Two-wattmeter method of measuring 3-phase power.

The voltage across the pressure coil of W_1 is $(e_1 - e_2)$ and the current i_1 whereas voltage across the pressure coil of W_2, the voltage is $(e_3 - e_2)$ and the current i_3. Hence, instantaneous power recorded by the two wattmeters.

$$= (e_1 - e_2)\, i_1 + (e_3 - e_2)\, i_3$$
$$= e_1 i_1 + e_3 i_3 - e_2(i_1 + i_3)$$
$$= e_1 i_1 + e_3 i_3 + e_2 i_2$$

as $\quad i_1 + i_3 = -i_2$

Hence the two wattmeters connected in this fashion i.e. current coil in phase 1 and 3 and potential coils between phase 1 and 2 & 3 and 2, measures instantaneous 3-phase power.

However, if the load is balanced, the following analysis can be used to calculate the p.f. of the balanced load.

Fig. 3.12 shows the phasor diagram for the connections of Fig. 3.11 for balanced loads. Phasors have been drawn taking rms values into considerations.

Power $W = W_1 + W_2 = E_{12} I_1 \cos(30 + \phi) + E_{32} I_3 \cos(30 - \phi)$

Since it is a balanced circuit,

$$|E_{12}| = |E_{32}| = \sqrt{3}\ |E_{ph}|$$

and $\quad |I_1| = |I_3| = |I| \quad$ (say)

Therefore,

$$W = \sqrt{3}\, E_{ph} I \cos(30 + \phi) + \sqrt{3}\, E_{ph} I \cos(30 - \phi)$$
$$= \sqrt{3}\, E_{ph} I\, [\cos(30 + \phi) + \cos(30 - \phi)]$$
$$= 2\sqrt{3}\, E_{ph} I . \cos 30 \,.\, \cos\phi$$
$$= 3\, E_{ph} I \cos\phi$$

Which is nothing but the three phase power in the balanced load.

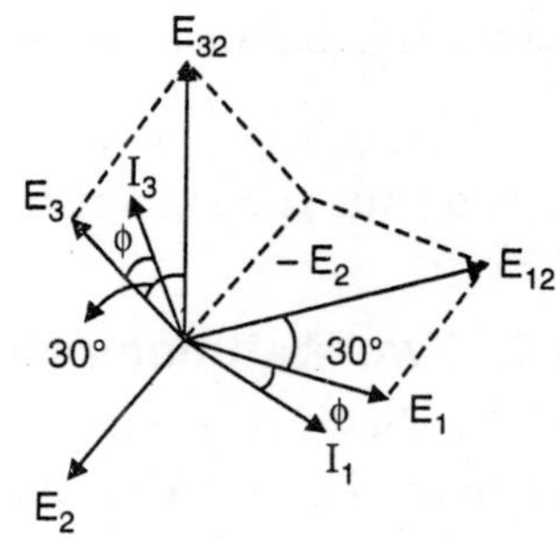

Fig. 3.12 Phasor diagram for Fig. 3.11.

It can be seen that if the p.f. of the load is 0.5 i.e. $\phi = 60°$ the reading of W_1 becomes zero and if the p.f. is less them 0.5, the p.f. angle is greater than 60° and with the usual connection of the wattmeter W_1 will read negative and hence the connection of either the current coil or the pressure coil should be reversed to obtain its reading and in such a situation this reading is to be reckoned negative and the total power consumed by the load is the algebraic sum of the two wattmeter readings.

Now to find out p.f. we obtain $W_1 - W_2$

$$= \sqrt{3}\, E_{ph} I\, [\cos(30 + \phi) - \cos(30 - \phi)]$$
$$= \sqrt{3}\, E_{ph} I\, [-2 \sin\phi \sin 30°]$$

or
$$W_2 - W_1 = \sqrt{3}\, E_{ph} I \cdot 2 \cdot \frac{1}{2} \sin\phi = \sqrt{3}\, E_{ph}\, I \sin\phi$$

Hence
$$\frac{W_2 - W_1}{W_2 + W_1} = \frac{\sin\phi}{\sqrt{3}\cos\phi} = \frac{\tan\phi}{\sqrt{3}}$$

or
$$\tan\phi = \frac{\sqrt{3}\,(W_2 - W_1)}{W_2 + W_1}$$

Hence $\cos\phi$, the p.f. of the load can be calculated.

3.4.3 One Wattmeter Method

This method can be used when the load is balanced. Fig. 3.13 shows connection for a star connected balanced load. Current coil of the wattmeter is connected in one of the lines (say '1') and one terminal of the pressure coil is connected to line 1 and the other terminal to line 2 and 3 alternatively for the two readings of the wattmeter using the switch as shown in the Fig. 3.13.

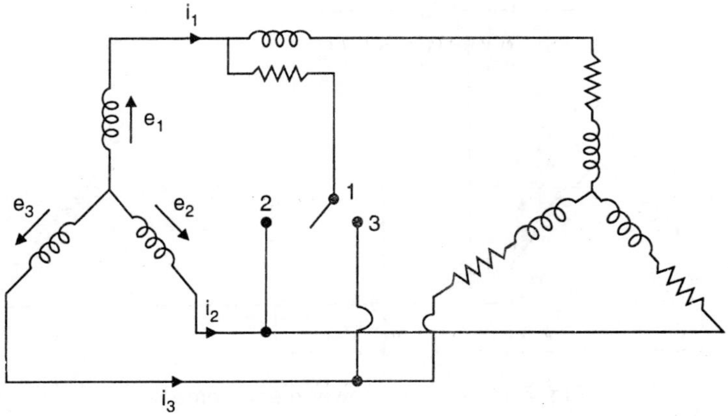

Fig. 3.13 One wattmeter method.

The two wattmeter readings are obtained first by connecting to terminal 2 and then to terminal 3. The two readings are

$$= (e_1 - e_2)\, i_1 + (e_1 - e_3)\, i_1$$
$$= e_1 i_1 + e_1 i_1 - (e_2 + e_3)\, i_1$$
$$= 3e_1 i_1$$

as $e_1 + e_2 + e_3 = 0$

Since it is a balanced load, this gives the average instantaneous power measured by the wattmeter.

To obtain p.f. of the load, refer to phasor diagram Fig. 3.14.

The wattmeter readings are

$$W = \sqrt{3}\, E_{ph}\, I \cos (30 - \phi) + \sqrt{3}\, E_{ph}\, I \cos (30 + \phi)$$
$$= W_2 + W_1$$

First term is when switch is connected to terminal 3 and the second term when it is connected to terminal 2.

This is similar to the two wattmeter method and

$$\tan \phi = \frac{\sqrt{3}\,(W_2 - W_1)}{W_2 + W_1}$$

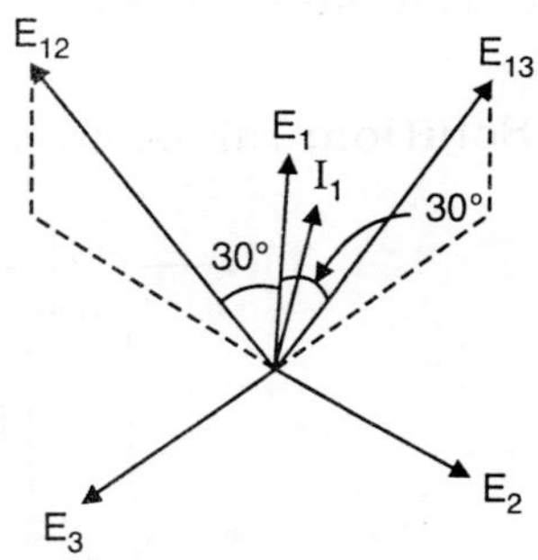

Fig. 3.14. Phasor diagram to Fig. 3.13.

3.5 MEASUREMENT OF REACTIVE POWER

If three phase power and the p.f. of the circuit are known reactive power = $W \tan \phi$. However, Fig. 3.15 gives connection for reactive power measurement using one wattmeter when the load is balanced. Here current coil of the wattmeter is connected in one line and the pressure coil is connected across the other two lines.

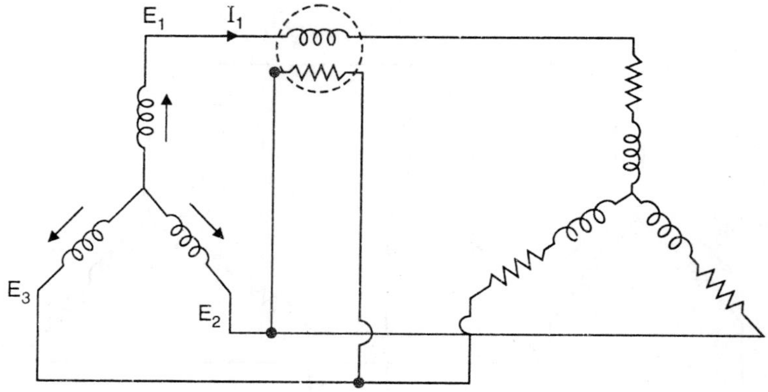

Fig. 3.15 Reactive power measurement.

Fig. 3.16 shows the phasor diagram for the system in Fig. 3.15.

The wattmeter reading is

$$\sqrt{3}\, E_{ph}\, I \cos (90 - \phi)$$

$$= \sqrt{3}\, E_{ph}\, I \sin \phi$$

which is the reactive power $\sqrt{3}$ times per phase, the wattmeter can be so calibrated that it indicates three phase reactive power by having a multiplication factor of $\sqrt{3}$.

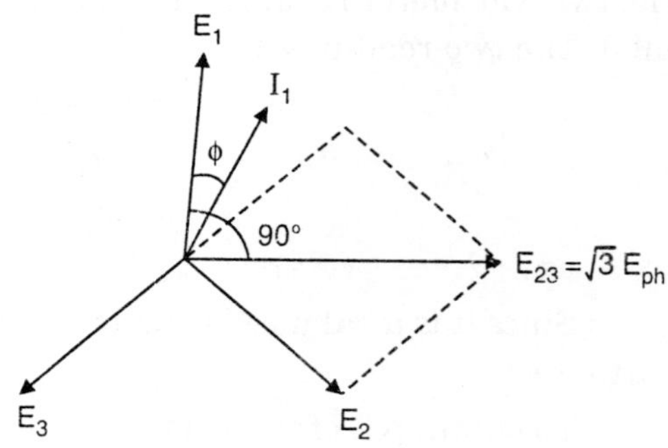

Fig. 3.16 Phasor diagram for Fig. 3.15.

Example 3.7: Find the reading of the wattmeter when the network shown is connected to symmetrical 440 V, 3-ϕ supply. Phase sequence $R\ Y\ B$.

Solution: Taking E_{RY} as reference

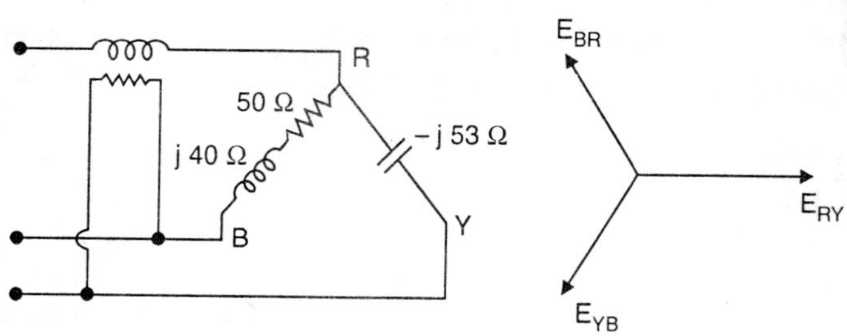

Fig. E3.7

$$I_{RY} = \frac{440\angle 0}{-j53} = 8.3\angle 90$$

$$I_{BR} = \frac{440\angle 120}{50 + j40} = \frac{440\angle 120}{64\angle 38.6}$$

$$= 6.875\angle 81.4$$

$$I_R = I_{RY} - I_{BR} = 8.3\angle 90 - 6.875\angle 81.4$$

$$= j8.3 - 1.03 - j6.8 = -1.03 + j1.5$$

Power $P = Re[VI^*] = Re[E_{YB} I_R^*]$

$$= Re\,[(-220 - j381)(-1.03 - j1.5)]$$

$$= 226.6 - 571 = -344.4 \text{ W} \quad \textbf{Ans.}$$

Example 3.8 : A 500 volt 3-phase motor has an output of 37.3 KW and operates at a p.f. of 0.85 with an efficiency of 90%. Calculate the reading on each of the two wattmeters connected to measure the input.

Solution:

The input $\dfrac{37.3}{0.9} = 41.44$ KW

Now $W_1 + W_2 = 41.44$ KW

$$\tan\phi = \frac{\sqrt{3}\,(W_2 - W_1)}{W_2 + W_1} = 0.6197$$

$$= \frac{\sqrt{3}\,(W_2 - W_1)}{41.44}$$

or $W_2 - W_1 = 14.82$

$$W_1 + W_2 = 41.44$$

$$2W_2 = 56.26 \quad \text{or} \quad W_2 = 28.13 \text{ KW}$$

and $W_1 = 13.31$ KW **Ans.**

Example 3.9: Three nonreactive resistances are connected across a symmetrical 3-phase 400 volt system as shown in Fig. E 3.9. Determine current in each line, the total power input to circuit and the reading on each of the wattmeters. Phase sequence *RYB*.

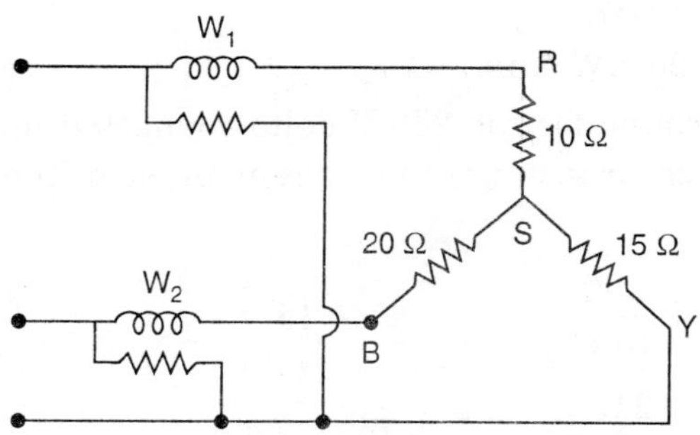

Fig. E3.9.

Solution: Taking V_R as reference, the potential of the star point.

$$V_S = \frac{\frac{231}{10} + \frac{231\angle-120}{15} + \frac{231\angle120}{20}}{\frac{1}{10} + \frac{1}{15} + \frac{1}{20}}$$

$$= 44.39 - j15.36$$

$$I_R = \frac{V_R - V_S}{10} = \frac{231 - 44.39 + j15.36}{10}$$

$$= 18.66 + j1.5 \simeq 18.7\text{A}$$

$$I_Y = \frac{231\angle-120 - V_S}{15}$$

$$= \frac{115.5 - j200 - 44.39 + j15.36}{15}$$

$$= \frac{71.1 - j184.64}{15}$$

$$= 4.7 - j12.3 = 13.2 \text{ A}$$

$$I_B = \frac{231\angle120 - V_S}{20} = \frac{-115.5 + j200 - 44.39 + j15.36}{20}$$

$$= \frac{-159.89 + j215.36}{20}$$

$$= -7.99 + j10.77 = 13.4 \text{ Amp.}$$

Voltage across pressure coil

$$V_{RY} = V_R - V_Y = \sqrt{3}\,V_{ph}\,\angle30°$$

Hence power in wattmeter W_1 $Re\,[V I^*]$

$$= Re\,[\sqrt{3}\,(200 + j115.5)(18.66 - j1.5)]$$

$$= 6.76 \text{ KW}$$

$$V_{BY} = \sqrt{3}\,V_{ph}\,\angle90°$$

$$W_2 = Re\,[j400\,(-7.99 - j10.77)]$$

$$= 4.3 \text{ KW}$$

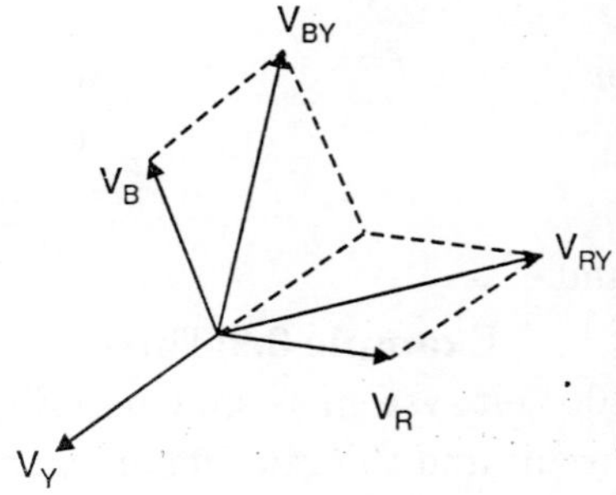

Fig. E3.9.1

Hence total power 11.06 KW **Ans.**

Example 3.10: Determine for the 250 V delta connected unbalanced 3-phase system shown (*i*) the current in each branch (*ii*) the current in each line and the reading of each wattmeter.

$$I_{RY} = \frac{250\angle0}{10 + j10} = 17.6\angle-45°$$

$$I_{YB} = \frac{250\angle-120}{20 - j15} = 10\angle-83.2$$

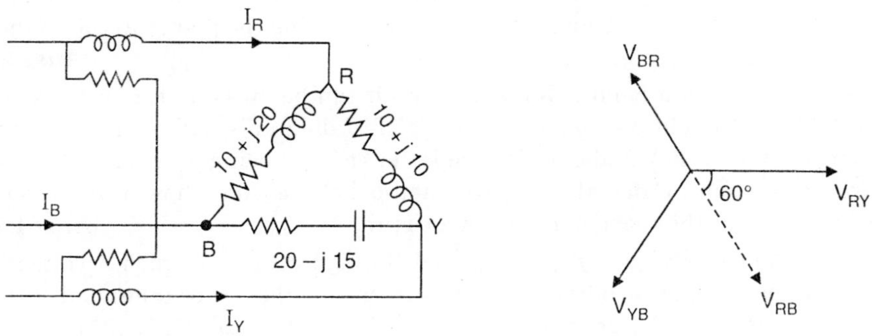

Fig. E3.10. (a) and (b)

$$I_{BR} = \frac{250\angle 120}{10+j20} = \frac{250\angle 120}{22.36\angle 63.4} = 11.18\angle 56.6$$

$$I_R = I_{RY} - I_{BR} = 17.6\angle -45° - 11.18\angle 56.6$$
$$= 12.44 - j12.44 - 6.15 - j9.33$$
$$= 6.29 - j21.77$$
$$= 22.66 \text{ A}$$

$$I_Y = I_{YB} - I_{RY}$$
$$= 10\angle -83.2 - 17.6\angle -45°$$
$$= 1.18 - j9.92 - 12.44 + j12.44$$
$$= -11.26 + j2.52 = 11.53 \text{ Amp}$$

$$W_1 = Re[VI^*] = Re[250\angle -60\, I_R^*]$$
$$= Re\,[(125 - j216.5)(6.29 + j21.77)]$$
$$= 787 + 4713 = 5500 \text{ Watts}$$

$$W_2 = Re\,[250\angle -120\,(-11.26 - j2.52)]$$
$$= 862 \text{ Watts.}$$

PROBLEMS

3.1. The power input to a 2000 V, 50 Hz 3-phase induction motor, running on full load at an efficiency of 90% is measured by two-wattmeters which indicate 300 KW and 100 KW respectively. Calculate (a) the input (b) the p.f. (c) the line current, (d) KW output.

[**Ans.** 400 kW, 0.76, 152, 360 kW]

3.2. Three non-reactive resistors of 3,4 and 5Ω respectively are star connected to a 3-phase 400 V symmetrical system, phase sequence *RYB*. Determine (a) current in each resistor (b) the power dissipated in each resistor (c) the phase angle between the currents and the corresponding line voltages (d) the star point voltage.

[**Ans.** (*a*) 66.5, 59.5, 51.8 A (*b*) 13.2, 14.15, 13.4 kW (*c*) 26°24′, 38°10′, 25° 20′ (*d*) 34 V]

3.3. The currents in the *RY*, *YB* and *BR* branches of a delta connected system with symmetrical voltage are 25A at p.f. 0.8 lag 30 A at p.f. 0.7 lead and 20 A at unity p.f. respectively. Determine the currents in each line, phase sequence *RYB*. [**Ans.** $I_R = 44.1$, $I_Y = 18.3$ A, $I_B = 49.5$ A]

3.4. The power input to a load was measured by two wattmeter method. The readings were 800 W and 400 W, the latter reading being obtained after reversing the potential coil connections. Find power and p.f. of the load. [**Ans.** 400 W, 0.189]

3.5. Prove that the power in a balanced 3-phase circuit can be measured from the readings of two wattmeters. Draw the relevant connection and phasor diagrams. If the readings on two wattmeters so connected are 5 KW and 0.5 KW, the latter reading being obtained after reversal of the current coil connections. Calculate the power and p.f. of the load. Show how the wattmeters are connected to measure the reactive power in a circuit. [**Ans.** 4.5 kW, 0.426]

3.6. A load of impedances each $9 + j21$ is supplied through a line to which a voltage of 415 V is applied. The impedance of each line is $2 + j4\Omega$. Determine the power input and output when the load is (a) star connected (b) mesh connected. [**Ans.** (a) 2556, 2091W (b) 5925, 3555.]

CHAPTER

4 Basic Instruments

4.1 INTRODUCTION

Ammeters and voltmeters are classified together as there is no difference in the principle for torque production. A voltmeter carries a current proportional to the voltage which produces operating torque whereas in case of an ammeter, the torque is produced by the current to be measured or by a definite fraction of it.

The basic philosophy of measurement is that whenever we insert an instrument for measurement of certain quantity, its insertion should not disturb the corresponding already existing quantity in the network.

An ammeter is used to measure current in a circuit and hence it is to be connected in series with the circuit. Now if there is certain current in the circuit, the insertion of ammeter in the circuit should not disturb the current value. This requires that the ammeter must have low resistance. Similarly, a voltmeter is connected across the voltage to be measured and must, therefore, have a high resistance so that the current taken by it should be small. A low range ammeter *i.e.* the one which can measure small values of current can be used as a voltmeter by connecting a high resistance in series with it. Of course, the current through the converted voltmeter should not exceed it's range when used as an ammeter.

It is to be noted that torque in an a.c. instrument can be produced only when there are two fluxes displaced in time and space. If an ammeter or a voltmeter whose operating torque is proportional to the current passing through it, will produce a pulsating torque if used in ac circuits and hence the inertia of the moving system of the instrument will not be able to respond to the pulsating torque and hence no deflection would be observed.

It is, therefore, essential that while measuring a.c. quantities, the torque should be proportional to the square of the current passing through the instrument. The deflection would then be proportional to the mean value of the square of current (assuming spring control is used) and the instrument can thus be used for the measurement of rms values of current or voltage.

4.2 TYPES OF INSTRUMENTS

The following types of ammeters and voltmeters are in common use :

(*a*) Moving coil

(*i*) Permanent magnet form

(*ii*) Dynamometer form

(*b*) Moving iron

(*c*) Hot wire

(*d*) Electrostatic (voltmeters only)

(*e*) Induction.

Here we will concentrate mainly on moving coil and moving iron types whereas a brief introduction for the others will be given.

Of the above mentioned, the permanent magnet moving coil is used for dc only whereas induction type are used for ac measurement. The other types can be used both for dc as well as ac circuits.

4.2.1 Permanent Magnet Moving Coil Type

The basic principle of operation is that when a current carrying conductor is brought in a magnetic field (they should not be parallel to each other) a torque on the conductor is produced. The instrument consists of a rectangular coil pivoted so that its sides lie in the air gap between the two poles of a permanent magnet and a soft-iron cylinder. The air gap between the magnet poles and iron core is small and the flux density is uniform and is in a radial direction, so that the flux lines are always at right angle to the current carrying conductor and hence when current passes through the coil, a deflecting torque is produced owing to the interaction between the two fluxes, one due to permanent magnet and the other due to the magnetic field of the coil. This is shown in Fig. 4.1.

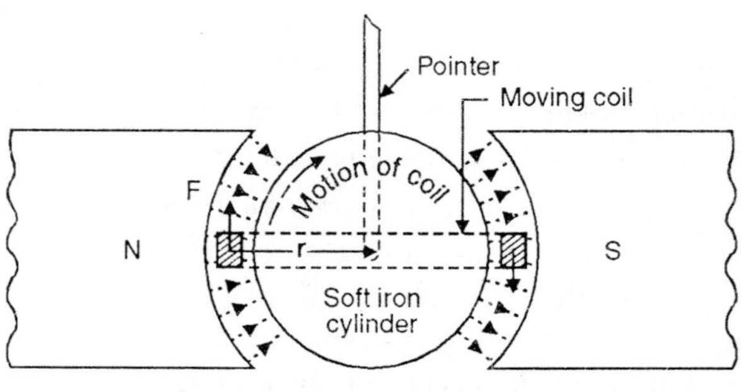

Fig. 4.1. Action of permanent-magnet moving-coil instrument.

If I is the current flowing in the moving coil in the direction shown, forces F, F will act on the two sides of this coil, the direction of force being determined by Fleming's left-hand rule and is in the direction for this system as shown in Fig. 4.1. The torque causing the coil to rotate is given as F. $2r$ where r is the mean distance of the wires forming the sides of the coil, from the axis of rotation.

Now, if N is the no. of turns in the coil, B the magnetic flux density due to permanent magnet, I the current in the coil, l the effective length, the force acting on the coil is given by

$$F = NBIl \quad \text{Newton} \qquad ...(4.1)$$

Hence the torque on the coil is

$$T = NBIl.\ 2r \quad \text{N-m} \qquad ...(4.2)$$

In any measuring (electromagnetic) instrument there are three torques acting on the moving mechanism to which a pointer is connected which moves on the dial of the instrument and indicates the reading of the quantity being measured. One of the three torques is the deflecting torque which we have just studied. The other torques are (*i*) control torque (*ii*) Damping torque. In case of PMMC instruments spring made of phosphor bronze provides control torque. The spring also serves as leads to the moving coil. When deflecting torque acts on the coil, both the control torque and damping torques come into action. The control torque restrains the rotation of the coil whereas the deflecting torque tries to rotate the coil. At balance, if the coil has moved through an angle θ and if k is the spring constant, we have

$$\text{NBI}l \,.\, 2r = k\theta \qquad ...(4.3)$$

or

$$\text{I} = \frac{k\theta}{\text{NB}l \,.\, 2r} \qquad ...(4.4)$$

$$= k'\theta \qquad ...(4.5)$$

as NBl . 2r are constant of a particular meter. Hence equation (4.5) suggests that the current is proportional to θ. Thus the scale is uniformly divided i.e. it is a linear scale i.e. if for I amp the deflection is through an angle θ, for 2I amps the deflection would be 2θ.

Now if damping torque were absent, the pointer will keep on oscillating around the mean value and an exact reading cannot be recorded. These oscillations are damped out by the damping torque. The moving coil is wound on an aluminium former which is placed in the magnetic field. As the coil rotates eddy currents are induced in the aluminium former and these eddy currents would try to oppose the cause i.e. the deflecting force and finally when the pointer comes to its actual value being measured, damping is provided by eddy currents and the coil (the pointer) remains stationary at its actual value.

Following are the advantages of PMMC instruments :

(*a*) Low power consumption

(*b*) High Torque/weight-ratio

(*c*) Uniformity of the scale and the possibility of a very long scale

(*d*) Perfect damping provided by eddy currents induced in the metal former of the moving coil. The metal used is aluminium as it is light in weight

(*e*) The possibility of a single instrument being used with shunt and resistance to cover a large range of both currents and voltages

(*f*) Freedom from errors due to stray magnetic fields.

4.2.2 Dynamometer Type Moving Coil Instruments

In this instrument the permanent magnet is replaced by one or two fixed coils which carry current to be measured or a current proportional to the voltage to be measured and which are connected either in series or parallel with the moving coil. The coils are usually air cored. The torque of the instrument is dependent upon the magnetic field strengths of the fixed and moving coil i.e. the torque is proportional to square of the current in an ammeter and square of the voltage in a voltmeter. Dynamometer instruments can thus be used in alternating current circuits for which square law is essential. These can be as well used for dc circuits also.

When the meter is used as an ammeter Fig. 4.1 (*a*) or as a voltmeter Fig. 4.1 (*b*), the torque of the meter is proportional to the product if the flux of the fixed coil (FC) and the flux of the moving coil. The scale of the instrument is thus a square one as shown in Fig. 4.1(*e*).

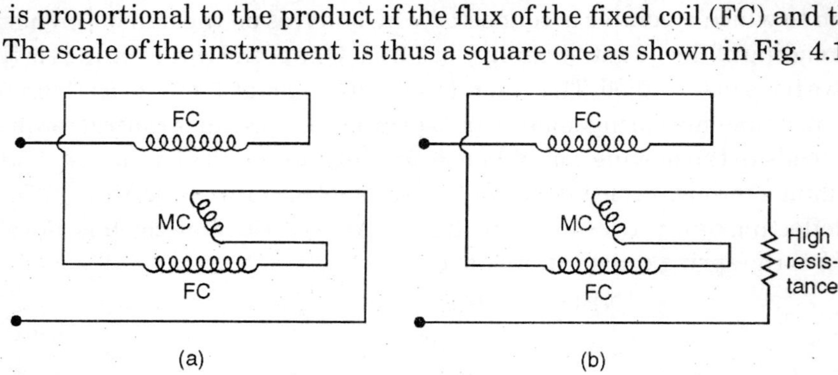

Fig. 4.1. (a) An ammeter (b) A voltmeter.

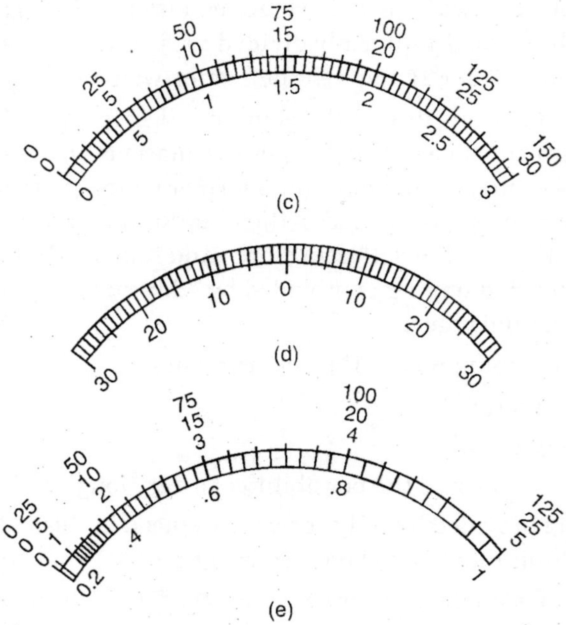

Fig. 4.1. Typical meter scales : (*c*) linear, (*d*) zero center, (*e*) squared.

Since energy must be used to create two magnetic fields, the electrodynamometer movement is less sensitive as compared to that in PMMC instruments.

The control torque is provided by spring torque. The damping is provided by either eddy currents or air piston attached to the pointer.

These instruments are not in common use as

(*a*) They have low torque/weight ratio

(*b*) Owing to $T_D \propto I^2$, the scale is non uniform

(*c*) These are expensive.

4.3 MOVING IRON TYPE

There are mainly two types of moving iron instruments (*a*) Attraction type Fig. 4.2 and (*b*) Repulsion type Fig. 4.3. In these instruments the current to be measured or current proportional to the voltage to be measured is passed through a coil of wire, the no. of turns on which depends upon the current passing through it. A certain number of ampere turns is required to operate the instrument.

When current flows through the coil, a small piece of iron is drawn into the core of the coil in case of attraction type of instruments. In the repulsion type, there are two rods or pieces of iron inside the coil one are fixed and are movable. These are similarly magnetised when current flows through the coil and repulsion of the moving iron from the fixed one takes place. The force of repulsion is roughly proportional to square of the rms value of the current as both magnets are magnetised by the same current.

Whatever be the direction of current through the coil, these are so designed that attraction takes place in case of attraction type of instruments, similarly repulsion takes place in the repulsion type of instruments. These are therefore known as unpolarised instruments in contrast to moving coil permanent magnet type of instruments which are polarised. If the direction of current through the PMMC instrument is changed, the deflection goes in the opposite direction.

Earlier the control torque was provided by attaching a weight to the moving iron and hence due to gravity the control torque is obtained. But now almost universally spring control is used. The damping torque in the moving iron instrument is provided by air friction. Two different types of damping chambers have been shown in Fig. 4.2 and 4.3. The control torque is proportional to the angle θ through which the pointer moves and the deflecting torque is proportional to square of the current being measured i.e.

$$T_C = k\theta$$

and $$T_D = k'I^2$$

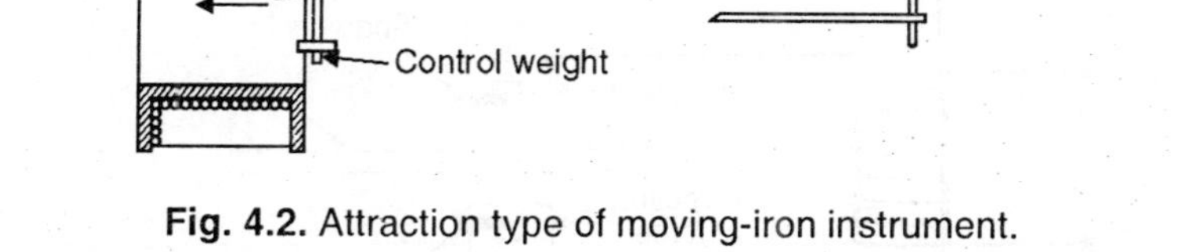

Fig. 4.2. Attraction type of moving-iron instrument.

Hence at balance

$$T_D = T_C$$

$$k'I^2 = k\theta$$

or $$\theta \propto I^2$$

and, therefore, the scale in case of MI instruments is non uniform. In fact if two meters one MI and another PMMC is placed before us and we are asked to identify looking at the scale. The one with uniform (linear) scale is PMMC and one with non-uniform (non linear) scale is MI instrument. The scale is normally cramped at the lower end and sparse at the upper end, Fig. 4.1 (*c-e*).

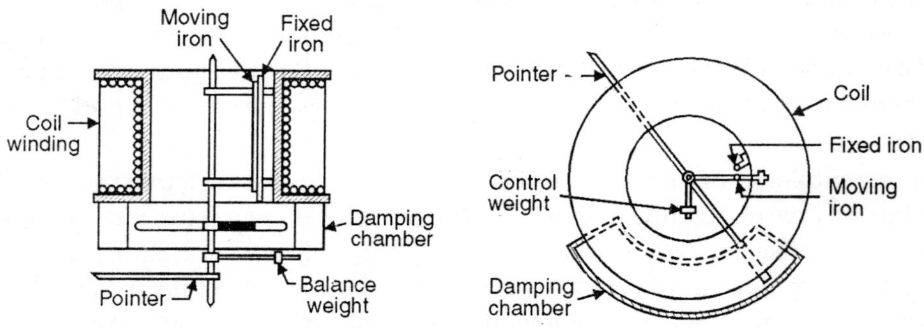

Fig. 4.3. Repulsion type of moving-iron instrument.

Hot wire instruments are normally not used and hence are not discussed.

Electrostatic instruments are used for measurements of high voltage and hence are not in the scope of the book.

4.4 INDUCTION INSTRUMENTS

As the name suggests these instruments can be used in alternating current circuits only. Their main advantage is that a full scale deflection of around 300° can be obtained giving a long and open scale.

These instruments are of two types (*i*) Induction disc and (*ii*) Induction cylinder type.

4.4.1 Induction Disc Type or Shaded Pole Type

Consider Fig. 4.4. As shown a thin aluminium disc is mounted on a spindle which is supported by jewelled bearings. The spindle carries a pointer and a control spring is attached to it. The edge of the disc moves in the air gap of a laminated electromagnet which is energized by the current to be measured or by a current proportional to the voltage to be measured. A damping magnet is placed at the opposite side of the disc from the electromagnet so that the damping torque is provided by the damping magnet and deflecting torque is provided by the electromagnet. The poles of the electromagnet are split into two halves with a thin copper band round one-half

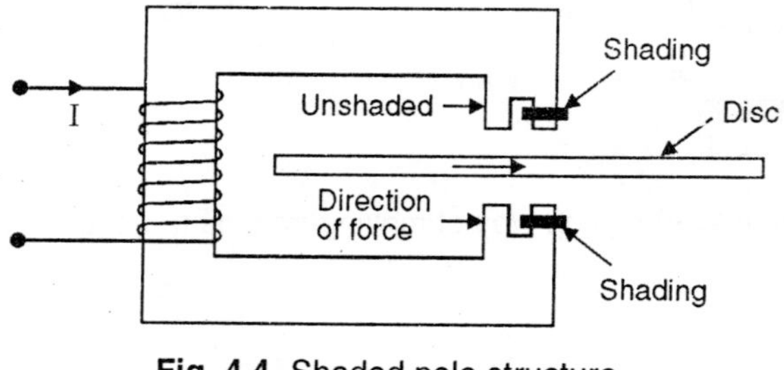

Fig. 4.4. Shaded pole structure.

of each pole (on the same side) as shown. The unshaded and shaded poles produce fluxes where the unshaded pole flux leads the shaded pole flux by an angle α. This is done so that we have two fluxes displaced in time and space acting on the moving system, here the aluminium disc. These are displaced in time because of the copper band and are displaced in space as these are from two different parts of the electromagnet. In the induction cylinder type of instruments the time displacement is obtained by inserting resistance in one circuit and inductance/capacitance in another circuit and space displacement is obtained by having windings on different pair of poles. In order to obtain uniform torque in the disc it is usually not circular but is spiral shaped. This is due to non-uniform torque provided by the spring torque which increases as the disc moves due to tightening action of the spring. The scale is usually cramped at the lower end since the deflecting torque is proportional to the square of the current or voltage to be measured and spring control is used.

Let the two fluxes be

$$\phi_1 = \phi_m \sin \omega t$$

and

$$\phi_2 = \phi_m' \sin (\omega t + \alpha)$$

Flux ϕ_1 is due to shaded pole whereas ϕ_2 due to unshaded pole. The two fluxes ϕ_1 and ϕ_2 will induce voltages e_1 and e_2 respectively in the disc due to induction. These voltages will circulate eddy current in the disc. Assuming the disc to be non-inductive, these currents will be in phase with their respective voltages. The phasor diagram in Fig. 4.5 shows the phase relation between various quantities.

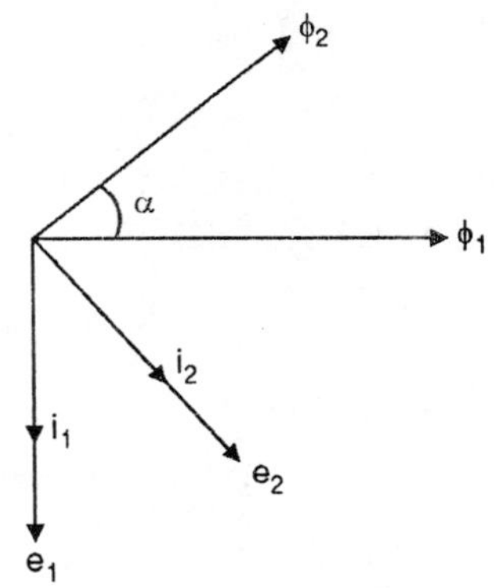

Fig. 4.5. Phasor diagram for induction instrument.

$$e_1 \propto \frac{d\phi_1}{dt}$$

$$\propto \phi_m \, \omega \cos \omega t$$

and

$$e_2 \propto \phi_m' \omega \cos (\omega t + \alpha)$$

The eddy current $i_1 \propto e_1$.

Assuming same resistance to flow of eddy currents

$$i_2 \propto e_2$$

and

$$i_1 \propto \phi_m \, \omega \cos \omega t$$

$$i_2 \propto \phi_m' \, \omega \cos (\omega t + \alpha)$$

The flux ϕ_1, will interact with eddy current i_2 and ϕ_2 will interact with i_1 and since ϕ_2 is leading ϕ_1 the torque due to ϕ_2 and i_1 will be reckoned as positive where as that due to ϕ_1 and i_2 as negative. The resultant torque is

$$T_D \propto \phi_2 \, i_1 - \phi_1 \, i_2$$

$$\therefore \quad \propto \phi_m' \sin(\omega t + \alpha)\, \phi_m\, \omega \cos \omega t - \phi_m \sin \omega t\, \phi_m'\, \omega \cos(\omega t + \alpha)$$

$$\propto \phi_m \phi_m' \sin(\omega t + \alpha) \cos \omega t - \phi_m \phi_m' \sin \omega t \cos(\omega t + \alpha)$$

$$\propto \phi_m \phi_m' \sin \alpha$$

Thus the torque is maximum when the two fluxes are displaced by 90° and since ϕ_2 leads ϕ_1 the rotation of the disc under the poles will be from unshaded pole to the shaded pole.

Now At balance $\quad T_B = T_C$

and since $\quad T_C = k\theta$

Hence $\quad k'I^2 = k\theta$

or $\quad \theta \propto I^2$

The damping torque is provided by a permanent magnet of high retentivity steel. The motion of the disc can be controlled by adjusting the position of this magnet.

4.4.2 Induction Cylinder Type Instrument

For this refer Fig. 4.6.

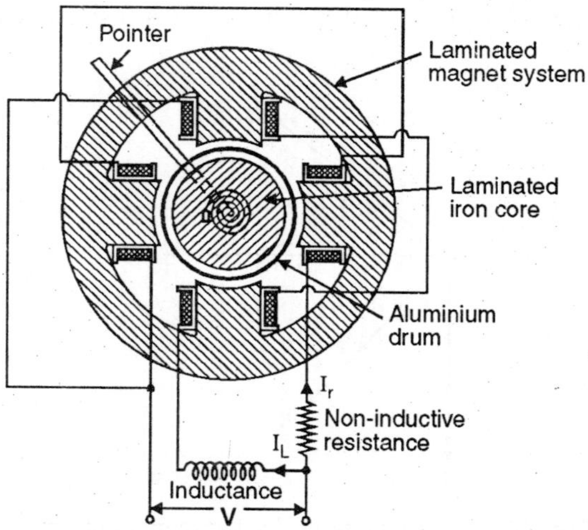

Fig. 4.6. Cylinder type induction instrument.

As mentioned earlier phase displacement between the fluxes is obtained by placing the windings on two different pairs of poles (space displacement) and by placing a high non-inductive resistance in series with one pair of winding and inductance in the other. The flux due to the winding having non-inductive resistance will lead the flux due to the other winding. The principle of operation of the instrument is exactly identical to the induction disc type of instrument.

4.4.3 The Shunt

Meters are available to read current in the ranges from a few microamperes to several hundred amperes full scale. However, the coil structure of the galvanometer movement is not capable of handling current greater than a few mA. Therefore when current to be measured exceeds this limiting value of coil current, the excess current must be diverted or by passed around the meter movement by means of a shunt. A shunt is a small resistance much smaller than the meter resistance, and is connected across the meter terminals as shown in Fig. 4.7. Let R_m be the meter resistance and R_{sh} the shunt resistance and if

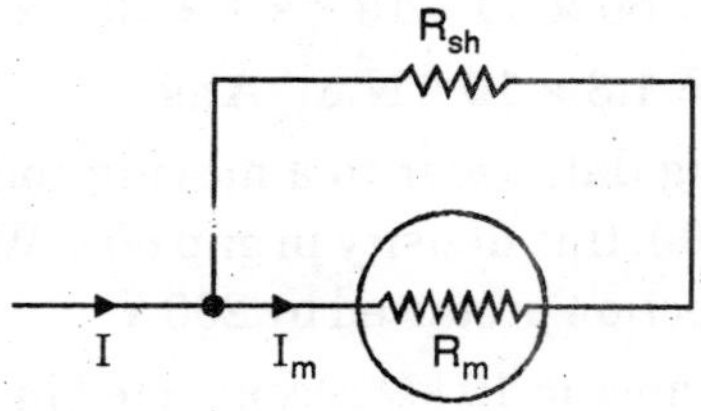

Fig. 4.7. Shunt.

I is the current to be measured the current through the meter I_m is given as

$$I_m = I \cdot \frac{R_{sh}}{R_{sh} + R_m}$$

This current I_m should not exceed the full scale deflection of the meter. The current through the shunt is much larger than I_m.

4.4.4 High Series Resistance (Voltage Indicating Instrument)

Galvanometer instrument is a current indicating device as the coil current produces the coil flux that in turn develops torque for a scale deflection. The coil has a finite value of resistance and since the current through this is also a few miliamperes, the net drop across the instrument is in mV and therefore if a voltage in hundred of volts is to be measured a high resistance in series with the galvanometer should be connected as shown in Fig. 4.8.

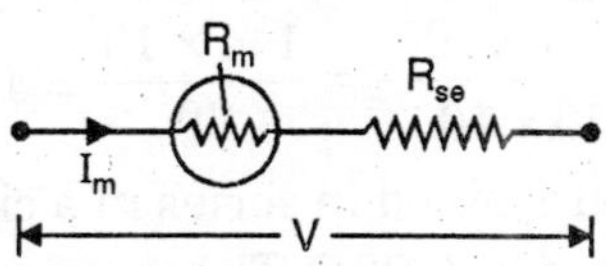

Fig. 4.8. High resistance in series.

When the meter is used to measure a voltage V greater than $I_m R_m$, the difference must be the voltage drop across the series resistance R_{se}. The equation for the voltmeter circuit is

$$V = I_m(R_m + R_{se})$$

Hence to measure V, we require a resistance R_{se} in series

$$R_{se} = \frac{V}{I_m} - R_m$$

However, it is to be calibrated for a voltage V rather than $I_m R_m$.

Example 4.1. The moving coil of a galvanometer has 60 turns, a width of 2 cm and a depth of 3 cm. It hangs in a uniform radial field of 50 mwb/m^2. Determine the torque on the coil when it is carrying a current of 1 mA.

Solution. It is to be noted that the instruments we have studied so far are galvanometers. When we connect a shunt (low resistance) across the galvanometer it works as an ammeter, of course with suitable calibration and it works as a voltmeter when a high resistance in series with the operating coil is connected.

Now $\qquad T_D = NBIl.2r$

Here $\qquad N = 60, B = 50 \text{ mWb/m}^2, I = 1 \text{ mA}$

$$l = 3 \times 10^{-2} \text{ m} \quad \text{and} \quad 2r = 2 \times 10^{-3} \text{ m}$$

Substituting the values we have

$$T_D = 60 \times 50 \times 10^{-3} \times 1 \times 10^{-3} \times 3 \times 10^{-2} \times 2 \times 10^{-3} \text{ Nm}$$
$$= 1.8 \times 10^{-6} \text{ Nm} \quad \textbf{Ans.}$$

Example 4.2. The following data refer to a moving coil voltmeter : resistance 10 KΩ, dimension of coil 3 cm × 3 cm N = 100, flux density in gap 80 mWb/m^2, spring control constant K = 30 dyne-cm/degree. Find the deflection produced by 200 V.

Solution. Here a high resistance of 10 kΩ is connected in series with the galvanometer to work as a voltmeter. For measurement of 200 V, the current through the coil is $\dfrac{200}{10{,}000} = 20$ mA

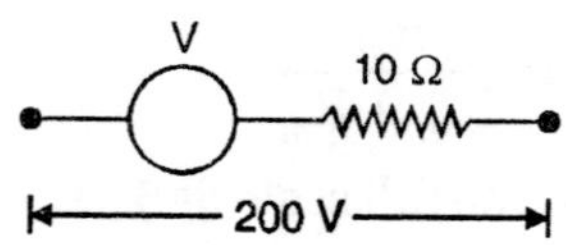

Fig. E4.2

Hence the deflecting torque is

$$T_D = NBIl.\,2r$$
$$= 100 \times 80 \times 10^{-3} \times 20 \times 10^{-3} \times 3 \times 3 \times 10^{-4}$$

At balance $\quad T_D = T_C$

or $\quad 30 \times 10^{-5} \times 10^{-2}\theta = 100 \times 80 \times 180 \times 10^{-10} = 8 \times 18 \times 10^{-6}$

or $$\theta = \frac{144 \times 10^{-6}}{30 \times 10^{-7}} = \frac{144 \times 10}{30} = 48° \textbf{ Ans.}$$

Example 4.3. Two ammeters are joined in series in a circuit carrying 10 A. Ammeter A has a resistance of 1000 Ω and is shunted by 0.02 Ω. The corresponding values for ammeter B are 1500 Ω and 0.01 Ω. What will the instruments read if the shunts are interchanged ?

Solution.

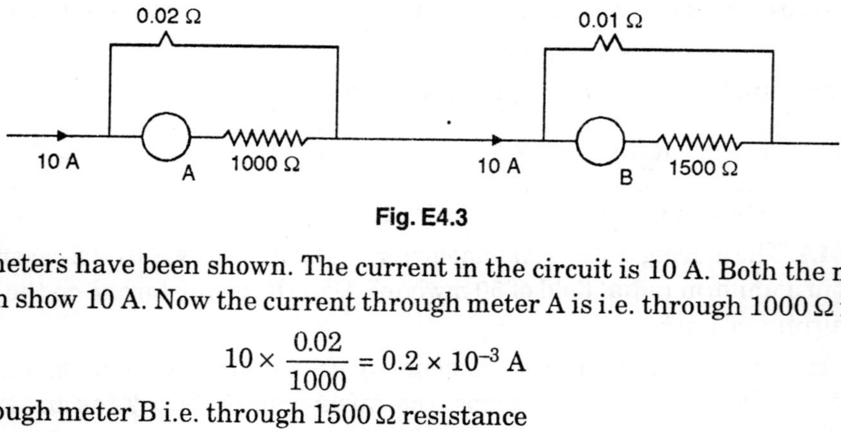

Fig. E4.3

Two meters have been shown. The current in the circuit is 10 A. Both the meters under this situation show 10 A. Now the current through meter A is i.e. through 1000 Ω resistance is

$$10 \times \frac{0.02}{1000} = 0.2 \times 10^{-3} \text{ A} \qquad ...(1)$$

and that through meter B i.e. through 1500 Ω resistance

$$10 \times \frac{0.01}{1500} \text{A} \qquad ...(2)$$

Now when these currents flow through the respective meters they indicate a current of 10 A. Please note that it is the current through the meter that decides the reading on that meter and not the current through its shunt.

Now if the shunts are interchanged i.e. the situation is as shown below.

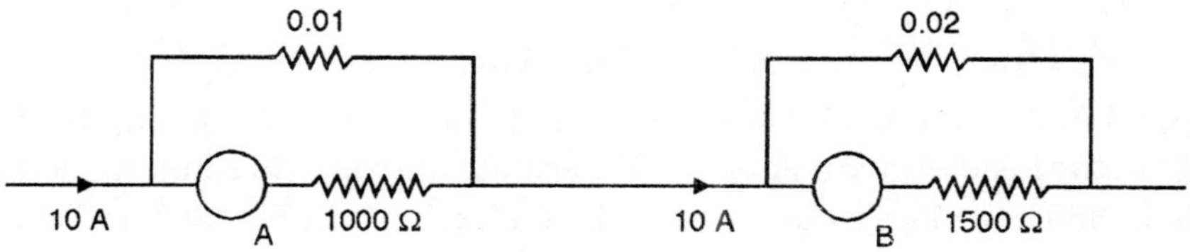

Fig. E 4.3.1

The current through meter A

$$10 \times \frac{0.01}{1000} \quad ...(3)$$

and that through B

$$10 \times \frac{0.02}{1500} \quad ...(4)$$

Comparing (1) and (3) and (2) and (4) we find that current in new situation (3) is half of that in (1) hence the meter will read half of its normal value of 10 A i.e. it will read 5 A and comparing (2) and (4), the meter B will read twice its previous value *i.e.* it will read 20A.

It is to be noted that since the instrument is PMMC, direct relations are possible.

Example 4.4. A moving coil instrument has a resistance of 10 Ω and gives full scale deflection when carrying 50 mA. Show how it can be adopted to measure voltages upto 750 V and current upto 100A.

Solution. The meter has 10 Ω internal resistance and gives full scale deflection when current through it is 50 mA.

For 750 V, A high resistance in series should be connected so that the current through the meter is 50 mA when 750 volt is applied across its terminals.

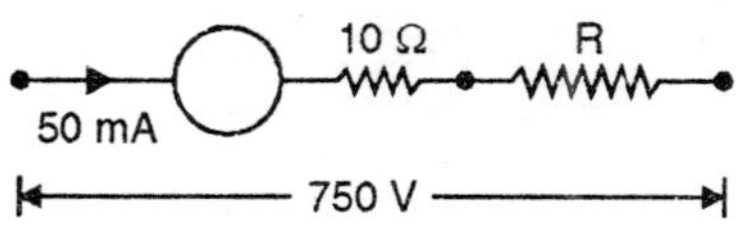

Fig. E 4.4

Hence $\quad 50 \times 10^{-3} = \dfrac{750}{R+10}$

or $\quad R + 10 = 15000\ \Omega \quad$ or $\quad R = 14990\ \Omega$

Hence a resistance of 14990 Ω should be connected to read 750 V on the meter.

To read 100 A

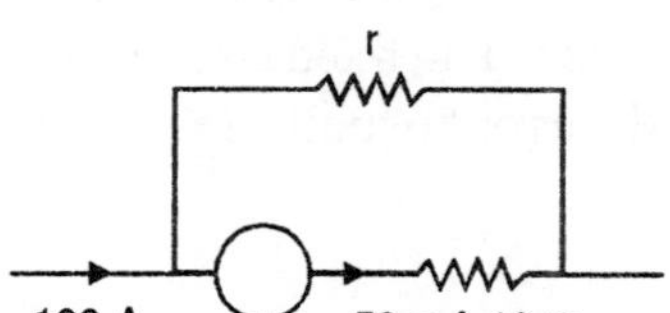

Fig. E 4.4.1

$$50 \times 10^{-3} = 100 \frac{r}{r + 10}$$

since $r << 10$

Hence $$50 \times 10^{-3} = 100 \frac{r}{10}$$

or $$r = 50 \times 10^{-4}\ \Omega \quad \textbf{Ans.}$$

Example 4.5. In a torsion head type electrodynamometer a current of 25 A requires a deflection of 90° in the head to give balance (*a*) Determine the range in Amperes of the instrument (maximum angle 360°) (*b*) For what current the deflection will be 180° (*c*) What will be the deflection for a current of 20 A ?

Solution. For this type of meter

$$\theta \propto I^2$$

(*a*) $$90 \propto 25^2$$

$$360 \propto I^2$$

or $$\left(\frac{I}{25}\right)^2 = \frac{360}{90} \quad \text{or} \quad \frac{I}{25} = 2$$

or I = 50 A in the full deflection range

(*b*) $$180 \propto I^2$$

$$90 \propto 25^2$$

or $$2 = \left(\frac{I}{25}\right)^2 \quad \text{or} \quad I = 25\sqrt{2} = 35.4 \text{ A} \quad \textbf{Ans.}$$

(*c*) $$90 \propto 25^2$$

$$\theta \propto 20^2$$

or $$\frac{90}{\theta} = \left(\frac{25}{20}\right)^2 = \frac{625}{400}$$

or $$\theta = 90 \times \frac{400}{625} = 57.6°. \quad \textbf{Ans.}$$

4.5 WATTMETER

There are basically three types of wattmeters

(*a*) Dynamometer

(*b*) Induction

(*c*) Electrostatic

Of these three, the first two are most commonly used and will be discussed here.

(*a*) **Dynamometer Type.** These instruments are similar in design and principle to the dynamometer ammeter already discussed. The fixed coils carry the currents as shown in Fig. 4.9 and are designed to handle the full load current usually 2.5, 5 or 10 A. The potential coil, the moving

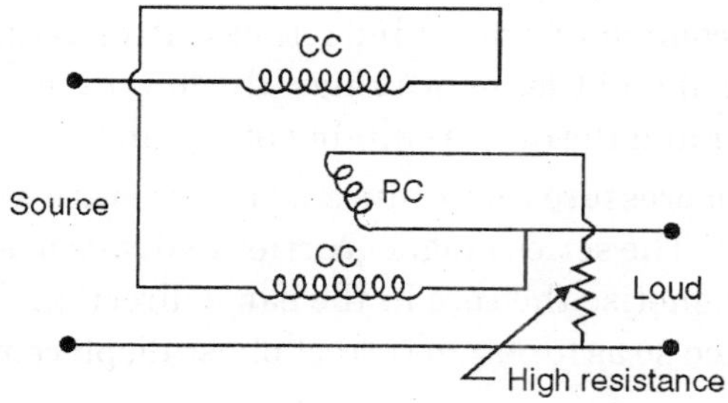

Fig. 4.9. Dynamometer wattmeter.

coil is in series with a high resistance usually designed for two voltage ranges 220/480 volt. Since one flux is proportional to load current and the other is proportional to load voltage, the torque on the pointer or the moving coil is proportional to the power. As a result, the power scale is not square but linear. The moving coil is carried on a pivoted spindle and the movement is spring controlled. The moving system carries a pointer and a damping vane, the latter moving in a sector-shaped box. The current coils are usually stranded or laminated especially when heavy currents are to be carried. For very heavy currents, a low range wattmeter is usually employed in conjunction with a current transformer. Similarly, if the circuit voltage is higher than the rated voltage of the wattmeter, potential transformer can be used in between the circuit and the wattmeter.

Fig. 4.10 shows the connection of current and potential transformers in conjunction with the wattmeter. Here CT is current transformer, PT the potential transformer, CC the current coil of the wattmeter and PC the pressure coil of the wattmeter.

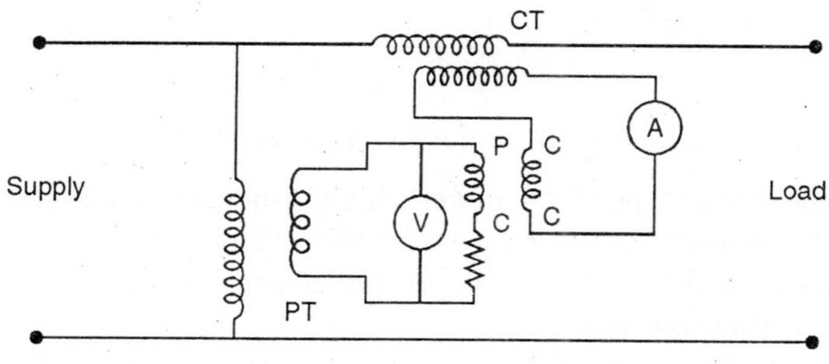

Fig. 4.10. Connection of current and potential transformers with wattmeter.

(*b*) **Induction Wattmeter.** The basic principle of operation of induction wattmeters is similar to the induction ammeters and voltmeters and can be used only on ac circuits.

Induction wattmeters have two laminated electromagnets, one is excited by the load current or a fraction of the load current and the other is excited by the current proportional to the load voltage. A thin aluminium disc is mounted so that it cuts the fluxes from both the electromagnets and the deflecting torque is produced by the interactions between these fluxes and the eddy currents which they induce in the disc. One or more copper bands or rings are fitted on one limb of the shunt magnet *i.e.* the magnet which is energized by the current proportional to load voltage (pressure coil), so that it causes the resultant flux in the magnet to lag in phase by exactly 90° behind the applied voltage. Fig. 4.11 shows the constructional details of the two types of induction wattmeters whereas Fig. 4.12 shows the phasor diagram for the wattmeter.

Fig. 4.11 shows two common forms of induction wattmeters. The upper one is a shunt magnet carrying pressure coil and the copper rings, the lower one is a series magnet carrying current coil of the wattmeter and the disc is suitably placed in between the two magnets.

In Fig. 4.11 (*a*) the two pressure coils connected in series are wound so that they both send flux through the middle limb. The series magnet carries two coils in series and are wound on the core such that they both magnetise the core in the same direction. The positions of the copper shading bands can be adjusted so as to obtain correct phase displacement between the shunt and series magnet fluxes.

However, in Fig. 4.11 (*b*) there is only one pressure coil and one current coil. A copper shading band, whose position is adjustable, surrounds the two projecting pole pieces of the shunt magnet for the purpose of correcting phase of the flux of this magnet. Both use spring to provide control torque and hence the advantage of a long and uniform scale upto 300°.

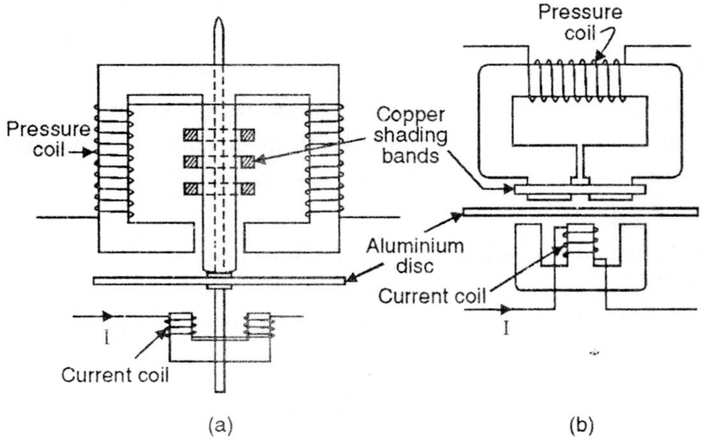

Fig. 4.11. Induction Wattmeters.

In case of dynamometer type of wattmeter a high resistance is connected in series with the pressure load so as to keep current in phase with the voltage. However, in induction type of wattmeters the pressure coil circuit is made as inductive as possible so that the flux due to pressure coil lags it's voltage by almost 90°.

Fig. 4.12 shows the phasor diagram for various quantities of the induction wattmeter.

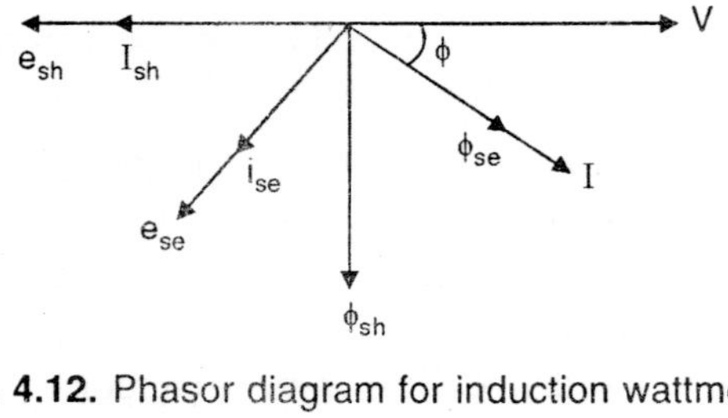

Fig. 4.12. Phasor diagram for induction wattmeter.

Let V be the applied voltage and I the current through the coil. ϕ_{sh} lags behind applied voltage V by 90° and this flux induces voltage e_{sh} in the disc which further lags the flux ϕ_{sh} by 90° as shown in Fig. 4.12. Assuming the aluminium disc to be resistive, current i_{sh} is in phase with

e_{sh}. Now ϕ_{se} is in phase with the current which produces it and e_{se} is the induced voltage in the disc which lags behind ϕ_{se} by 90° and again assuming disc to be resistive i_{se} is in phase with e_{se}.

Let $\quad V = V_m \sin \omega t \qquad I = I_m \sin (\omega t - \phi)$

$$\phi_{se} \propto I_m \sin (\omega t - \phi)$$

$$V \propto -\frac{d\phi_{sh}}{dt} \quad \text{or} \quad \phi_{sh} = -\int V_m \sin \omega t \, dt$$

or
$$\phi_{sh} = -\frac{V_m}{\omega} \cos \omega t$$

and
$$e_{sh} \propto \frac{d\phi_{sh}}{dt} \propto \frac{V_m}{\omega} \omega \sin \omega t$$
$$\propto V_m \sin \omega t$$

and
$$i_{sh} \propto V_m \sin \omega t$$

Similarly
$$e_{se} \propto \frac{d\phi_{se}}{dt} \propto I_m \omega \cos (\omega t - \phi)$$

and
$$i_{se} \propto I_m \omega \cos (\omega t - \phi)$$

The net torque is $\propto \phi_{se} i_{sh} - \phi_{sh} i_{se}$

$$T_D \propto I_m \sin (\omega t - \phi) V_m \sin \omega t + \frac{V_m}{w} \cos \omega t \, I_m \omega \cos (\omega t - \phi)$$
$$\propto V_m I_m [\sin (\omega t - \phi) \sin \omega t + \cos \omega t \cos (\omega t - \phi)]$$
$$\propto V_m I_m \cos (\omega t - \phi - \omega t)$$
$$T_D \propto V_m I_m \cos \phi$$
$$\propto VI \cos \phi$$
$$\propto \text{the power in the circuit.}$$

Since the deflecting torque is proportional to the power in the circuit and the instrument is spring controlled the scale is uniform. Similarly for dynamometer type wattmeter also the scale is uniform

Compared with dynamometer type wattmeters, the induction wattmeter have the advantage of a greater working torque and length of scale. However, they suffer from the inherent disadvantages of less accuracy greater weight of moving system, greater power consumption and also that these can be used only on ac circuits.

4.6 METHOD OF CONNECTION IN THE CIRCUIT

There are two ways in which a wattmeter could be connected in a circuit as shown in Fig. 4.13. There are four terminals in a wattmeter. Across terminals ML is the current coil and across V_1V_2 is the pressure coil. M stands for the mains (supply) side of the current coil and L stands for the load side of the current coil. If connections of any one coil is made opposite to what is shown in Fig. 4.13, the meter would give deflection in the opposite direction and hence the connection of any one of the coils should be reversed to obtain positive deflection. Even though the two connections shown above are normal, neither measures the power in the load accurately.

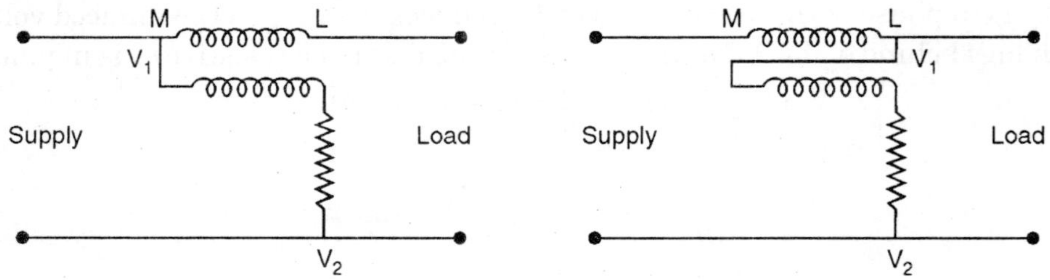

Fig. 4.13. Alternative wattmeter connections.

In Fig. 4.13 (*a*) where the pressure coil is connected on the supply (Main) side of the current coil, the voltage applied to the pressure coil is higher than that of the load by an amount equal to the drop in the current coil. However, in Fig. 4.13 (*b*) the current in the current coil is greater than that in the load by an amount equal to the current drawn by the pressure coil.

In the first case, the instrument measures I^2R_C, lost in the current coil and in the second case the watts lost in the pressure coil, as well as the power in the load.

If the load current is small, the volt drop in the current coil is small and hence first connection in Fig. 4.13 is suitable and would introduce small error. However, if the load current is large as compared to the current drawn by the pressure coil, the watts lost in the pressure coil will be small as compared to the watts consumed by the load and the second method of connection is preferred.

4.7 ENERGY METERS

The measurement of energy is the same process as measurement of power except that the instrument not merely indicates the power or rate of supply of energy but must take into account also the length of time for which the rate of energy is continued. There are basically three types of energy meters.

(*a*) Electrolytic meters

(*b*) Motor meters

(*c*) Clock meters.

Out of the above, motor meters are very widely used and among motor meters also induction type watthour meters are more commonly used and will be dealt with here.

4.7.1 Single Phase Induction Type Energy Meter

The construction of this meter is more or less similar to induction type wattmeter. The main alterations are the provision of only one pressure coil upon the central limb of the shunt magnet and only one copper shading band upon this limb. In addition there are two copper bands placed obliquely on the other two limbs of this magnet, their objective is to provide compensation against friction error in the meter (see Fig. 4.14).

In this meter the moving system is allowed to revolve continuously instead of being allowed merely to rotate through a fraction of one revolution as in an indicating instrument. The speed of revolution is proportional to the power in the circuit. It follows, therefore, that the no. of revolutions made by the revolving system in any given time is proportional to the energy supplied. The number of revolutions made by the meter is recorded by a counting mechanism consisting of a train of wheels to which the spindle of the rotating system is geared.

The control of speed is brought about by a permanent magnet called the brake magnet. The magnet is placed opposite to the electromagnet used for providing deflecting torque and it induces currents in the disc which produces retarding torque proportional to their magnitude which latter is proportional to the speed of the rotating system. This system attains a steady speed when the retarding torque exactly balances the driving torque produced by the power in the circuit. The braking torque produced by the brake magnet depends upon the strength of the magnet. If ϕ is the flux of the brake magnet, i the current induced by the rotation of the moving system in the field of the brake magnet and T_B is the braking torque, then

$$T_B \propto \phi\, i$$

If n is the speed of the rotating system (disc) and e the voltage induced in the disc of the meter.

$$e \propto n\phi$$

Let r be the resistance to the path of eddy current i in the disc, hence $i = \frac{e}{r}$

and
$$T_B \propto \phi \frac{e}{r}$$

$$\propto \frac{\phi^2 n}{r}$$

This braking torque equals the constant driving torque T_D when a steady speed of the disc is attained. Thus if N be the steady speed of the meter

$$T_B' \propto \frac{\phi^2 N}{r}$$

and at balance
$$T_B' = T_D \propto \frac{\phi^2 N}{r} \quad \text{or} \quad N \propto \frac{r}{\phi^2} T_D$$

Hence the steady speed attained by the meter for a constant driving torque T_D is directly proportional to the resistance of the path of the induced eddy currents and inversely proportional to the square of the flux of the brake magnet. Therefore, it can be seen that it is very important for brake magnet to have constant magnetic strength throughout the use of the energy meter. It is to be noted that the spring control and pointer of the wattmeter have been replaced in the energy meter by a brake magnet which provides braking torque as explained above.

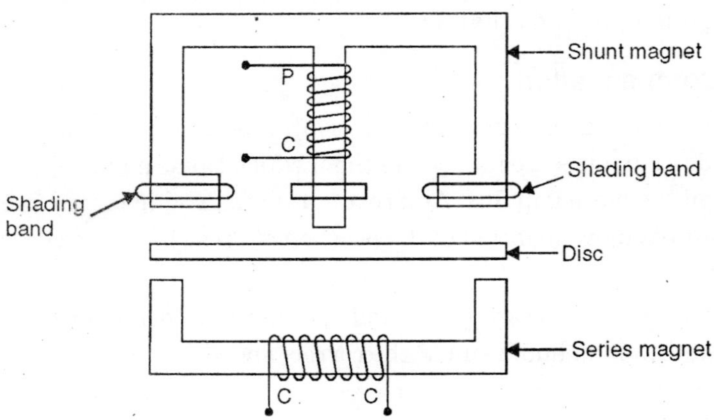

Fig. 4.14. Single phase induction type energy meter.

Using the theory of induction wattmeter.

Operating or driving torque $\propto$ VI cos ϕ

where V is the supply voltage across the shunt coil and I the load current through the series coil of the meter and ϕ the p.f. angle between V and I.

Now the braking torque has been shown to be proportional to the speed N of the disc *i.e.*

$$T_B \propto N$$

Since for a steady speed N, the driving torque T_D is equal to T_B, we have

$$N \propto VI \cos \phi$$

or power is proportional to speed N

Thus the total no. of revolutions which equals $\int N\, dt$ is proportional to

$$\int VI \cos \phi\, dt$$ *i.e.* proportional to energy supplied.

The speed of the disc can be adjusted by suitably positioning the brake magnet with respect to the spindle of the disc. If the brake magnet is moved towards the spindle the braking torque decreases and if taken away from the spindle the braking torque increases.

An energy meter is rated in terms of the supply voltage, the full load current and number of revolutions of the disc per kWhr (also known as constant of the energy meter). These are marked on the dial of the energy meter.

4.7.2 Creep

In some meters a slow but continuous rotation of the disc is obtained when the pressure coils are energized and there is no load current passing through the current coil *i.e.* current coil is not energized. This may be due to incorrect friction compensation to vibration, to stray magnetic field or to the fact that the supply voltage is in excess of the normal voltage.

To prevent such creeping of the meter two holes or slots are cut in the disc on opposite sides of the spindle. The disc tends to remain stationary when one of the holes comes under one of the poles of the shunt magnet. In some cases a small piece of iron wire is attached to the edge of the disc. The force of attraction of the brake magnet upon this iron wire is sufficient to prevent the creeping of the disc under no load condition.

4.7.3 Friction Compensation

The two shading bands embrace the flux contained in the two outer limbs of the shunt magnet and thus eddy currents are induced in them which cause a phase displacement between the enclosed flux and the main gap flux. As a result a small driving torque is exerted on the disc, this torque being adjusted by variation of the position of these bands to compensate for friction torque in the instrument.

The friction compensation can be checked whether it is correct or not by connecting light load under which condition it should run at the correct speed.

4.7.4 Meter Phase Angle Error

It is desired that the phase angle between the applied voltage to the pressure coil and the shunt magnet flux should be at 90° which is adjusted with the help of shading band on the shunt magnet. An error due to incorrect adjustment of the position of this shading band will be evident when the meter is tested on a load whose p.f. is less than unity. An error on the "fast" side under these conditions can be eliminated by bringing the shading band further drawn the limb of the shunt magnet *i.e.* nearer to the disc.

Example 4.6. For 20 A, 230 V meter the constant is 480 revolutions/kWhr. If upon test for a load of 4.6 kW the disc makes 40 revolutions in 66 sec. Calculate the error.

Solution. In 66 sec. the energy consumed by the load is $4.6 \times \frac{66}{3600}$ kWhr $= 0.0843$ kWhr

The energy recorded by the meter $\frac{40}{480} = 0.0833$ kwhr

Hence the meter is moving slow and it is slow by

$$\frac{0.0843 - 0.0833}{0.0843} \times 100$$

$$= \frac{0.001}{0.0843} \times 100 \simeq 1.2 \text{ percent slow.}$$

4.8 POLYPHASE INDUCTION WATT HOUR METERS

As in the case of power measurement with a two-element wattmeter, the energy supplied to a three phase circuit can be measured by a two element of the induction type, each element being similar to the construction to the single element meter.

There are two discs mounted on the same spindle and two separate brake magnets. The spindle drives a single counting train. In addition to the phase adjustment and friction error compensation in each element, one of the elements has an adjustable magnetic shunt across its shunt magnet, in order that the driving torque for the same watts may be the same in the two instruments. In carrying out this adjustment, the two current coils of the meter are connected in series and the two pressure coils in parallel, the polarities being such that the two driving torques are in opposition. Adjustment of the magnetic shunt is carried out until the meter ceases to rotate with full rated watts supplied to both elements.

The adjustments for phase displacement, friction compensation and for the correct position of the brake magnet are made upon the two elements separately, a single phase circuit being used. The adjustment of the brake magnet positions must be carried out on the two elements alternately, since both magnets are operative in producing the braking torque. Both the shunt elements should be energized meanwhile, to allow for interaction.

The connections to these polyphase meters are the same as those employed in the two wattmeter method of measuring three phase power.

Example 4.7. A dynamometer wattmeter with its voltage coil connected across the load side of the instrument reads 250 W. If the load voltage be 200 V. What power is being taken by the load ? The voltage coil has a resistance of 2000 ohm.

Solution. The connections are as shown

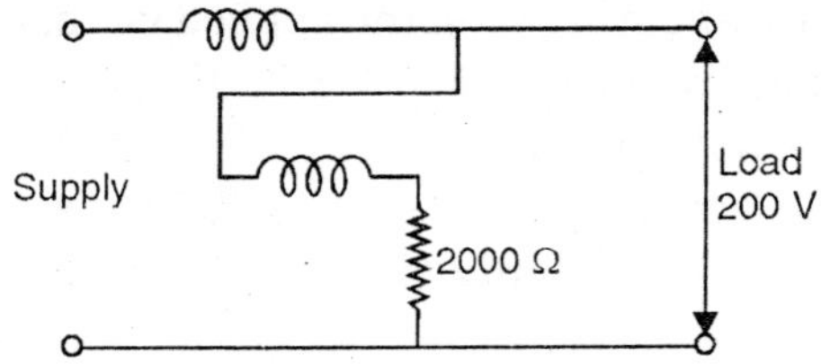

Fig. E4.7

Power taken by the load = Power recorded by the wattmeter–power consumed in the pressure coil.

The current through the pressure coil is

$$\frac{200}{2000} = 0.1 \text{ A}$$

and since the pressure coil is highly non-inductive, power absorbed by the pressure coil is VI = 200 × 0.1 = 20 watts. Hence power taken by the load is 250 – 20 = 230 watts.

Example 4.8. A 50 A, 230 V meter on full load test makes 61 revolution in 37 sec. If the normal disc speed is 520 revolutions per kWhr. What is the percentage error ?

Solution. The full load is 230 × 50 = 11.500 kW

In 37 secs the energy supplied $= 11.5 \times \frac{37}{3600} = 0.11819$ kwhr

The energy recorded by the meter $= \frac{61}{520} = 0.1173$

This means the disc or the meter is slow

The percentage error $= \frac{0.11819 - 0.1173}{0.11819} \times 100$

$= 0.75\%$ slow **Ans.**

4.9 PHANTOM OR FICTITIOUS LOAD

When the capacity of a meter under test is high, a test with the ordinary or actual loading would involve a considerable waste of power. To avoid this, phantom or fictitious loading arrangement is employed. In this, voltage across the pressure coil correspond to the rated voltage whereas the current coil is supplied from a separate low voltage source Fig. 4.15. This means that the total power supplied for the test is that due to the small pressure coil current at normal voltage plus that due to the load current at a low voltage and thus the total power supplied is only a comparatively small amount.

If a high capacity meter is to be listed whilst in service, the pressure coil is supplied from the line in the normal way but the current coil is removed from the consumer's load circuit and replaced by a short circuiting connection. The current coil is then supplied from a low voltage source for purposes of testing.

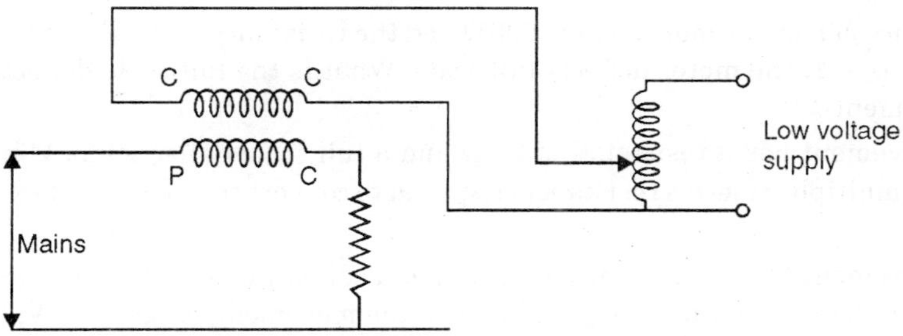

Fig. 4.15. Phantom loading of energy meter.

PROBLEMS

4.1. Describe the construction and principle of operation of a moving coil galvanometer.

4.2. Explain clearly how a galvanometer can be used as an ammeter or a voltmeter.

4.3. Explain the basic requirements of an instrument for measurement.

4.4. Explain clearly why a moving coil instrument can be used on dc only and not on a.c. circuits.

4.5. Explain why the resistance of an ammeter should be as low as possible where as that of voltmeter it should be as high as possible.

4.6. Explain how an ammeter and a voltmeter are connected in an electric circuit. What will happen if their positions are interchanged ?

4.7. Explain with neat diagrams the construction principle of operation and application of a moving iron ammeter.

4.8. Develop expressions for deflecting torque and control torque for a PMMC instrument and show that the scale of the instrument is linear.

4.9. Differentiate between a PMMC and dynamometer type moving coil instrument.

4.10. What do you mean by polarised and non-polarised instrument ? Classify the instrument : PMMC dynamometer moving coil. Attraction type MI, repulsion type MI.

4.11. Name which instrument can be used on ac only, on d.c. only, and on ac and dc both. Give reasons.

4.12. What are induction type of instruments ? Show that the deflecting torque in an induction disc type instrument is maximum when the two fluxes are 90° phase apart.

4.13. Describe the construction, principle of operation and application of an induction disc type of an ammeter. Discuss about its scale.

4.14. Explain with diagram the construction, principle of operation and application of (*a*) Induction type (*b*) dynamometer type wattmeters.

4.15. Explain why the potential coil of a dynamometer wattmeter should be highly non-inductive and that of induction wattmeter highly inductive.

4.16. What is meant by creeping of an energy meter ? How is this taken care of ?

4.17. What is phantom loading ? With a neat diagram explain how it is carried out.

4.18. Explain the construction and principle of operation of a 3-phase energy meter.

4.19. The resistance of a meter movement that has a full scale deflection of 1 mA is 50 Ω. Calculate the value of the shunt required to convert the movement to an ammeter having a full scale deflection of 2 A. [**Ans.** 25 mΩ.]

4.20. The resistance of a meter movement is 500 Ω and the resistance of it's shunt is 0.005 Ω. When the line current is 1 A, the meter deflects full scale. What is the full scale deflection current of the meter movement ? **[Ans.** 10 μA.]

4.21. A meter movement has a resistance of 25 Ω and a full scale deflection of 2 mA. Determine the value of the multiplier (series resistance) required to convert the movement to indicate 300 V full scale. **[Ans.** 149975 Ω.]

4.22. A meter movement has a resistance of 50 Ω and a full scale deflection of 1 mA. Determine the values of the series resistances required to give the movement ranges of 1 V, 3 V and 30 V. **[Ans.** 950 Ω, 2950 Ω, 9950 Ω.]

4.23. Explain two possible connections of a 1-phase wattmeter in an electronic circuit. Discuss their application.

4.24. A moving coil instrument gives a full scale reading of 24 mA when the p.d. across its terminals is 72 mV. Calculate (*a*) the shunt resistance for a full scale deflection corresponding to 120 A (*b*) the series resistance for full scale reading with 600 V. Find the power consumption in each case. **[Ans.** (*a*) 0.60012 mΩ, 8.64 w (*b*) 24997 Ω, 14.4 W]

4.25. The constant for a 3-phase, 2-element integrating wattmeter is 0.12 rev of disc per kWhr. If the meter is normally used with a potential transformer of ratio 22000/110 V and a current transformer of ratio 500/5A, determine the error, expressed as a percentage of the correct reading, from the following test figures for the instrument only : line voltage 110 V, current 5.25 A p.f. unity, time to complete 40 revolutions 61 secs. **[Ans.** – 1.68%]

CHAPTER

5 Transformer

5.1 INTRODUCTION

A transformer even though is not an energy conversion device, it forms a very important component of the energy conversion system. It is almost an indispensable component in both the electric and electronic systems. A transformer transforms electric energy from a certain voltage and current levels to another voltage and current levels keeping the frequency of the supply the same. It is because of the availability of the transformer that it is possible to generate energy from where it is available in abundance far away from the load centre and the energy so generated at economic voltage levels can be transmitted at economic voltage levels using transformers and the same energy can be utilised at economic and safe voltage levels as required by the devices using the electric energy. The importance of transformer in electric energy system can be gauged by the fact that a generation of 1 MW requires transformers of 3 to 3.5 MVA capacity from power plant to the consumer points.

The transformer is also very widely used in electronic circuits i.e. low power low current electronic and control circuits for performing such functions as matching the impedance of a source and the load for maximum power transfer. A transformer has normally two windings, the primary and the secondary. The primary is one where the source is connected and across the secondary load is connected. If the secondary voltage is higher than the primary voltage it is known as step-up transformer and it is known as step down if the secondary voltage is smaller than the primary voltage. However, if the two voltages are equal, it is known as an isolation transform. The one-to-one transformers are used when it is necessary or desirable to insulate the secondary side of the circuit from the primary circuit ; for though both the circuits will then have the same difference of potential between their terminals, they will not necessarily have the same difference of potential to ground. It is to be noted that either of the two windings can be considered as primary while the other then serves as the secondary. The primary is therefore defined as one which receives energy from the supply and the secondary as the one which delivers energy to the load. In general, however, the windings are designated as low tension (LT) or high tension (HT) windings.

In its simplest form the transformer consists of the insulated windings wound on a ferromagnetic core and so disposed with respect to each other as shown in Fig. 5.1 that a current through one of the windings will set up a magnetic flux linking more or less completely with the turns of the other. According to Faraday's laws of electromagnetic induction an emf is induced in the secondary which is proportional to the number of turns of the secondary and the flux linking the secondary. However, the frequency of the secondary voltage is same as that of the source.

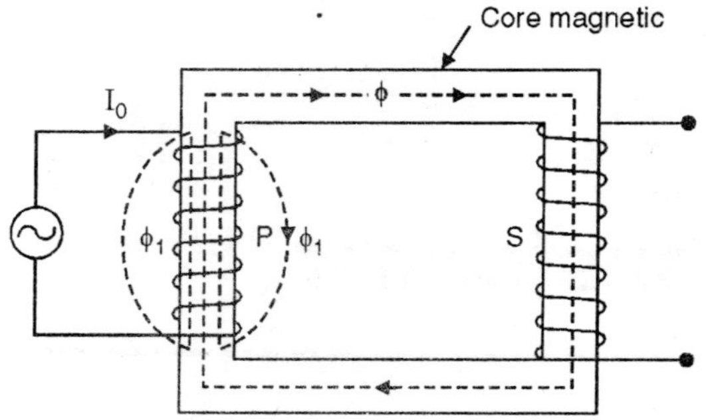

Fig. 5.1. A transformer (Basic).

In order to ensure the most effective magnetic linkage of the two windings, the core which serves to support them mechanically as well as to carry their mutual magnetic flux is usually made of a highly permeable iron or steel alloy designed to have a low reluctance. In some specially designed transformers the core may be of non-magnetic materials. Such transformers are known as air cored transformers. These transformers are normally used in radio devices and in certain types of measuring and testing instruments.

5.2 CONSTRUCTION

The main considerations in the construction of the transformer are simplicity and ease of construction, economy of material, provision of insulation to withstand electrical stresses, mechanical strength to withstand the forces caused by heavy short circuit currents and ventilation to get rid of the heat due to the losses in the coils and the core. Mainly there are two designs to meet these requirements (*i*) core type and (*ii*) shell type as shown in Fig. 5.2 (*a*) and (*b*) respectively.

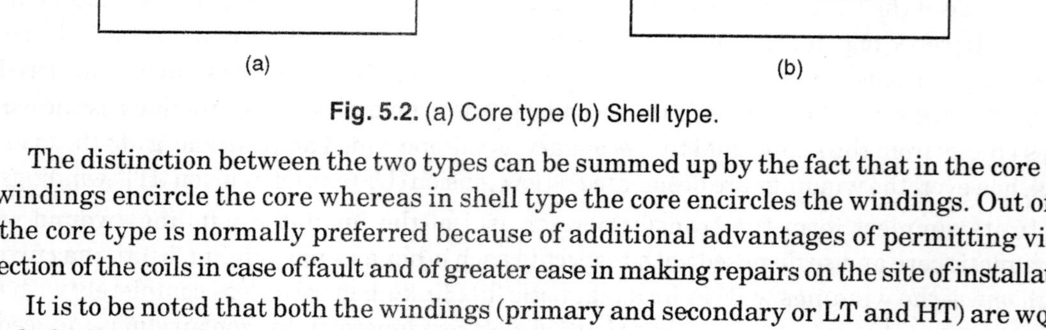

Fig. 5.2. (a) Core type (b) Shell type.

The distinction between the two types can be summed up by the fact that in the core type the windings encircle the core whereas in shell type the core encircles the windings. Out of the two the core type is normally preferred because of additional advantages of permitting visual inspection of the coils in case of fault and of greater ease in making repairs on the site of installation.

It is to be noted that both the windings (primary and secondary or LT and HT) are wound on both the limbs of the core type as well as shell type.

A portion of the LT winding in the form of a coil is placed close to the core from electrical stresses point of view and then over this we have HT coil and so on. Both the windings will be on both the limbs of the core type construction.

5.3 EMF EQUATION OF A TRANSFORMER

Refer to Fig. 5.1. Let N_1 be the no. of turns in the primary and N_2 turns in the secondary and assume that the secondary is open as shown in Fig. 5.1. Hence no current flows through the secondary of the transformer and under this condition the transformer acts as a reactance coil and the current I_0 flowing through the primary will depend upon the supply voltage V_1 and the impedance of the primary winding. Assume that the voltage V_1 varies harmonically with frequency f Hz and the core of the transformer is unsaturated *i.e.* flux ϕ increases linearly with the current. Note that ϕ is a function of mmf which is the product of current and no. of turns of the coil through which current flows and the resistance of the coil is negligible. The current is harmonically varying since the impressed voltage is harmonically varying and the transformer acts as the inductive reactance. As the current increases from its initial value of zero, the flux ϕ increases in direct proportion and since the flux is also varying harmonically, it will induce an emf in the primary winding according to Faraday's law of induction and the polarity of the emf will be as per Lenz's law in opposition of the impressed voltage V_1. Since by hypothesis the primary has no resistance, the resistance ohmic drop is zero and hence the primary winding is the seat of only two opposing emfs namely the impressed voltage V_1 and the counter emf E_1 induced by the alternating flux and in accordance with Kirchhoff's voltage law, these two emfs must at every instant be equal and opposite.

Since the impressed voltage is varying harmonically the counter emf also must vary harmonically and hence the flux. Suppose flux at any instant is given by

$$\phi = \phi_m \sin \omega t \qquad ...(5.1)$$

where ϕ_m is the maximum value of flux in webers (Wb), it follows that the instantaneous value of the induced emf. in N_1 turns of the primary is

$$e_1 = -N_1 \frac{d\phi}{dt} = -N_1 \omega N_1 \phi_m \cos \omega t$$

$$= 2\pi f N_1 \phi_m \sin (\omega t - \pi/2) \qquad ...(5.2)$$

This means the flux induces a sinusoidal voltage which lags behind the flux itself by 90° in time phase.

The maximum value of the induced voltage from equation 5.2 is

$$E_{1max} = 2\pi f \phi_m N_1 \qquad ...(5.3)$$

and hence its effective or rms value is

$$E_1 = \sqrt{2}\, \pi f N_1 \phi_m = 4.44\, f N_1 \phi_m \qquad ...(5.4)$$

and this also equals the impressed voltage in magnitude. Fig. 5.3 shows the phasor diagram of the various quantities studied so far (ideal transformer or lossless transformer).

Since I_0 lags the supply voltage V_1 by 90°, the power supplied from the source is $V_1 I_0 \cos 90°$ which is zero. This result is a natural consequence of our assumption that the transformer is lossless. Therefore, under these conditions $I_0 = V/X_0$ where X_0 is the reactance of the transformer.

Similarly emf induced in the secondary winding is

$$e_2 = -N_2 \frac{d\phi}{dt} = 2\pi f N_2 \phi_m \sin (\omega t - \pi/2) \qquad ...(5.5)$$

or

$$E_2 = 4.44\, f N_2 \phi_m \qquad ...(5.6)$$

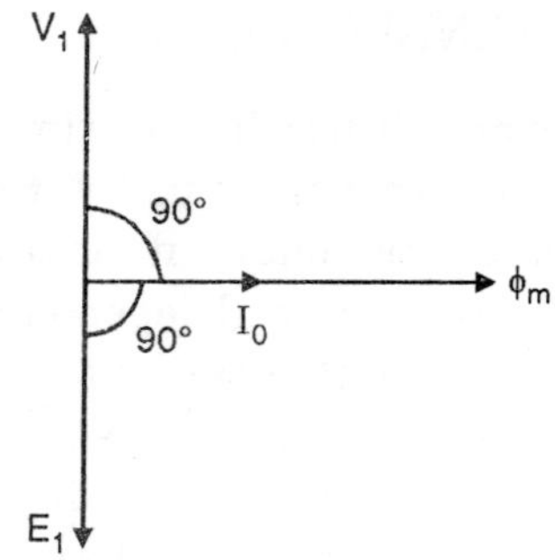

Fig. 5.3. Phasor diagram of ideal transformer, secondary open circuited.

Hence $$\frac{E_2}{E_1} \simeq \frac{V_2}{V_1} = \frac{N_2}{N_1} = n \qquad \text{...(5.6 A)}$$

where n is known as ratio of transformation.

5.4 NO LOAD OPERATION

Refer again to Fig. 5.1. A real life transformer has resistance R_1 of the primary winding and R_2 of the secondary winding and a copper loss of $I_0^2R_1$ will certainly take place in the primary winding of the transformer (No load condition). Also, since the flux is harmonically varying, it will cause hysteresis and eddy current loss in the ferromagnetic material of the core. Even though eddy current losses in the core can be reduced by laminating the core rather than using solid core and the hysteresis loss can be reduced by using a high permeability core material. However, these losses are always present and hence even though the secondary is open, the primary must supply power to meet the $I_0^2R_1$, eddy current and hysteresis losses (iron losses). Therefore, this loss must be supplied from the source and hence I_0 should not lag V_1 by 90° but an angle less than 90°. Rather it should lead ϕ by a small angle α known as no load angle or angle of hysteric advance as shown in Fig. 5.4

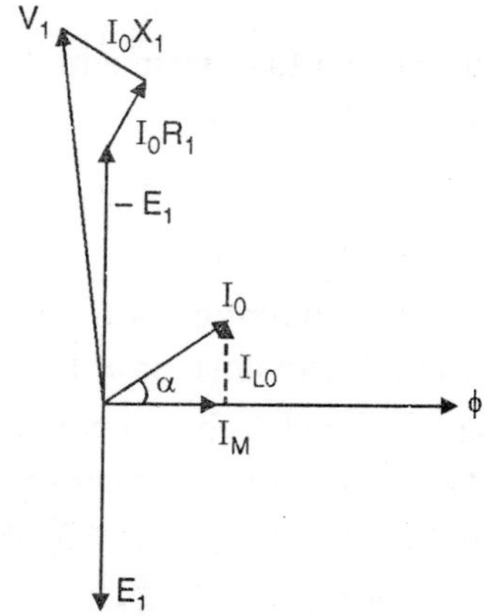

Fig. 5.4. Phasor diagram of transformer under no load condition.

The no load primary current may be decomposed into two components, one in phase with the flux ϕ and hence known as magnetising component of current while the other component I_w as the working component of the no load current in phase with $-E_1$ and is to meet the losses of the transformer.

Now the flux ϕ produced by the primary does not completely enclose the secondary winding through the core of the transformer. A part of the flux ϕ, completes its path through the air as shown in Fig. 5.1 and hence this component of flux does not contribute to the emf. in the secondary winding and hence this could be considered as a series voltage drop in the primary winding and is usually represented by I_0X_1 where X_1 is known as the primary leakage reactance. The quantity I_0X_1 leading I_0 by 90°, then represents the component of the impressed voltage required to balance it.

Finally, if we include the resistance component R_1 of the primary, it is evident that the impressed voltage V_1 must contain as one component the potential drop I_0R_1 in phase with I_0 as shown in Fig. 5.4 consequently the total supply voltage V_1 must be the phasor sum of (*i*) the component $-E_1$ equal and opposite to the emf induced by the mutual flux (*ii*) I_0X_1 the leakage reactance drop equal and opposite to the emf induced by the leakage flux ϕ_1 and (*iii*) the ohmic drop I_0R_1 in phase with I_0. The phasor sum of I_0R_1 and I_0X_1 is I_0Z_1 where Z_1 is called the leakage impedance of the primary. It is to be noted that in a real life situation I_0Z_1 forms a very very small fraction of the supply voltage and also $I_0^2R_1$ also is a small fraction of the core losses or iron losses and hence I_0Z_1 is neglected as compared to V_1 and $I_0^2R_1$ as compared to core losses and hence an equivalent circuit of the transformer under no load condition is shown in Fig. 5.5.

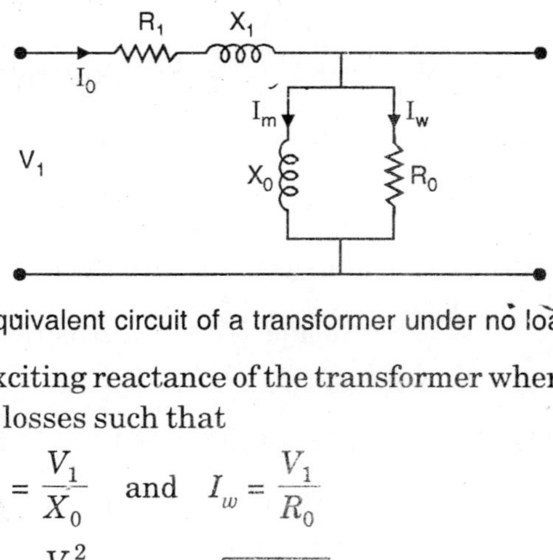

Fig. 5.5. Equivalent circuit of a transformer under no load condition.

Here X_0 represents exciting reactance of the transformer whereas R_0 represents the resistance to account for the core losses such that

$$I_m = \frac{V_1}{X_0} \quad \text{and} \quad I_w = \frac{V_1}{R_0} \qquad ...(5.6\text{ B})$$

$$\text{Core loss} = \frac{V_1^2}{R_0}, \quad I_0 = \sqrt{I_m^2 + I_w^2}$$

and
$$\cos\alpha = \frac{I_m}{I_0} \quad \text{or} \quad \alpha = \tan^{-1}\frac{I_w}{I_m} = \tan^{-1}\frac{X_0}{R_0}.$$

Example 5.1. A 1-phase transformer has 400 primary and 1000 secondary turns. The net cross sectional area of the core is 60 cm². If the primary is connected to a 50 Hz supply at 500 V, determine (*a*) the peak value of the flux density in the core and (*b*) the voltage induced in the secondary winding :

Solution. The emf equation of the transformer is

$$E_1 = 4.44\, fN_1\, B_m A$$

where B_m is the maximum or peak value of the flux density in the core of the transformer

Neglecting the leakage impedance drop

$$E_1 = V = 500 = 4.44 \times 50 \times 400 \times B_m \times 60 \times 10^{-4}$$

Hence $B_m = 0.938\ \text{Wb/m}^2$ **Ans.**

(*b*) To find out secondary voltage we make use of the relation

$$\frac{E_2}{E_1} = \frac{V_2}{V_1} = \frac{N_2}{N_1}$$

Substituting the values of V_1, N_2 and N_1 we have

$$V_2 = V_1 \frac{N_2}{N_1} = 500 \times \frac{1000}{400} = 1250 \text{ volt} \quad \textbf{Ans.}$$

Example 5.2. A single phase 50 Hz core type transformer has core of cross section area 400 sq. cms. The permissible maximum flux density is 1 Wb/m². Calculate the number of turns per limb on the high and low voltage sides for a 3000/220*v* ratio.

Solution. The area of cross-section of the core is

$$20 \times 20 \times 10^{-4} = 400 \times 10^{-4} \text{ sq m}$$

Neglecting the series voltage drop

$$V_1 \simeq E_1 = 4.44\, fN_1 B_m A$$

or $$3000 = 4.44 \times 50 \times N_1 \times 1 \times 400 \times 10^{-4}$$

or $$N_1 = 338 \text{ turns}$$

and $$N_2 = 338 \times \frac{220}{3000} \text{ using the relation}$$

$$\frac{V_1}{V_2} = \frac{N_1}{N_2}$$

Hence $N_2 = 24.78$ *i.e* 26 turns

Since there are two limbs the no. of turns per limb will be

$$\frac{338}{2} = 169 \text{ and } \frac{26}{2} = 13 \text{ turns respectively.}$$

5.5 OPERATION OF TRANSFORMER UNDER LOAD CONDITION

Refer to Fig. 5.6, when the secondary of a transformer is open an emf E_2 is induced in the secondary given by equation 5.6 which is a result of mutual flux ϕ_m between the primary and secondary windings.

Now, if the external circuit of the secondary is connected through the impedance represented by $Z = R + jX$ as shown in Fig. 5.6, the induced emf E_2 will set up in the secondary a current

$$I_2 = \frac{E_2}{(R_2 + R) + j(X_2 + X)} \qquad ...(5.7)$$

Where R_2 is the resistance of the secondary winding, X_2 is the equivalent reactance of the secondary winding corresponding to its leakage flux *i.e.* the part of the flux which links the secondary and not the primary. The current I_2 lags the voltage E_2 by an angle θ_2' given by

$$\theta_2' = \tan^{-1} \frac{X_2 + X}{R_2 + R} \qquad ...(5.8)$$

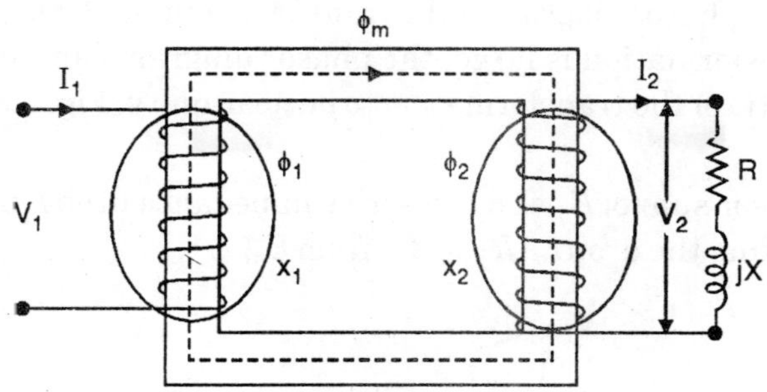

Fig. 5.6. Transformer under load condition.

The current I_2 induces current I_1' in the primary in such a way that the ampere turns I_2N_2 must equal $I_1'N_1$ and the flux produced due to I_2N_2 ampere turns must equal and oppose the flux produced by $I_1'N_1$ ampere turns leaving behind the flux ϕ_m in the core corresponding to the no load flux. Therefore, the current drawn by the secondary winding is reflected in the primary winding such that the net ampere turns remain I_0N_1 *i.e.* if the current in the secondary increases, there is increase in the primary current also. For all values of secondary currents which provides demagnetising effect as far as flux in the core is concerned, there is a primary current such that

$$I_1'N_1 = I_2N_2 \qquad ...(5.9)$$

Normally I_0 is a negligible fraction of I_1' and hence

$$I_1' \simeq I_1$$

and hence equation (5.9) becomes

$$I_1N_1 = I_2N_2 \cdot \qquad \text{or} \qquad \frac{I_1}{I_2} = \frac{N_2}{N_1} \qquad ...(5.10)$$

i.e. the current in the windings are in the inverse ratio of the turns of the windings whereas the voltages induced are directly proportional to the turns ratio *i.e.* larger the number of turns of a winding, larger is the induced voltage and smaller the current and vice versa.

Current I_1 is the phasor sum of I_1' and I_0

Current I_2 is also given by the equation

$$I_2 = \frac{V_2}{R + JX} \qquad ...(5.11)$$

where I_2 lags V_2 by an angle $\theta = \tan^{-1} \dfrac{X}{R}$ ...(5.12)

Fig. 5.7 shows ampere turn phasor diagram

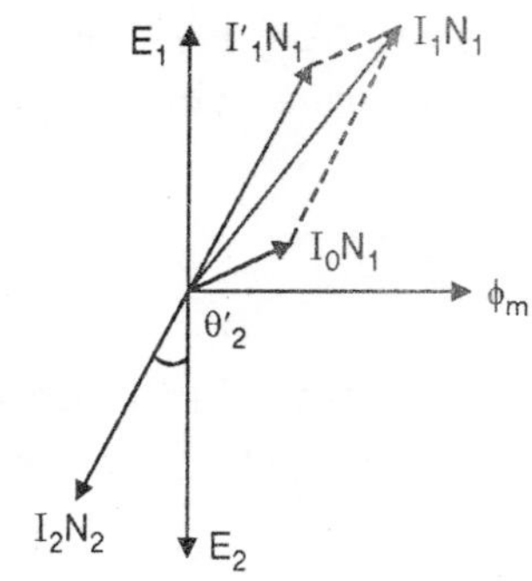

Fig. 5.7. Ampere turns phasor diagram for the transformer.

Fig. 5.8 shows the phasor diagram of the transformer under load condition. It is to be noted that if the ratio of transformation is large, the phasor diagram can't be drawn on the sheet to scale. Assuming the ratio of the transformation to be near unity, Fig. 5.8 gives relative locations of various phasors.

As V_1 is the phasor sum of E_1 and the series impedance drop $I_1(R_1 + jX_1)$, similarly V_2 is obtained after subtracting the drop $I_2 (R_2 + jX_2)$ from E_2.

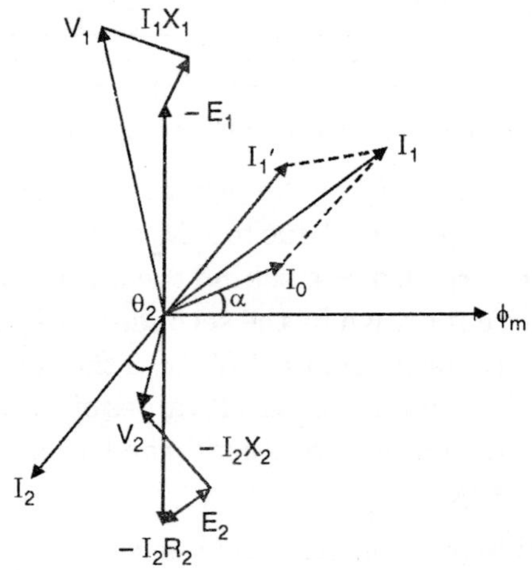

Fig. 5.8. Phasor diagram of transformer under load condition.

From equation (5.6 A) and (5.10) we have

$$\frac{N_2}{N_1} = \frac{I_1}{I_2} = \frac{E_2}{E_1} \qquad \text{...(5.13)}$$

We have already concluded that since the series impedance drop on the primary side is negligible as compared to the supply voltage $V_1 \simeq E_1$. Similarly on the secondary side also $V_2 \simeq E_2$ and hence equation (5.13) can be rewritten as

$$\frac{N_2}{N_1} = \frac{I_1}{I_2} = \frac{V_2}{V_1}$$

or

$$V_1 I_1 = V_2 I_2 \qquad \text{...(5.14)}$$

This means the volt-amperes supplied to the primary are equal to the volt-amperes developed in the secondary. This shows that the transformer transforms voltage and current at certain level to voltage and current at another level related through the transformation ratio n. There is no change in frequency.

The transformer is rated in terms of VA or KVA or MVA depending upon the size as the criterion for design is the heating of transformer which depends upon such rating rather than KW rating.

5.6 EQUIVALENT CIRCUIT

Fig. 5.9 shows equivalent circuit for a transformer for analysis purposes. However, there is a word of caution for this equivalent circuit especially in regard to various quantities indicated therein.

From the equivalent circuit, it appears

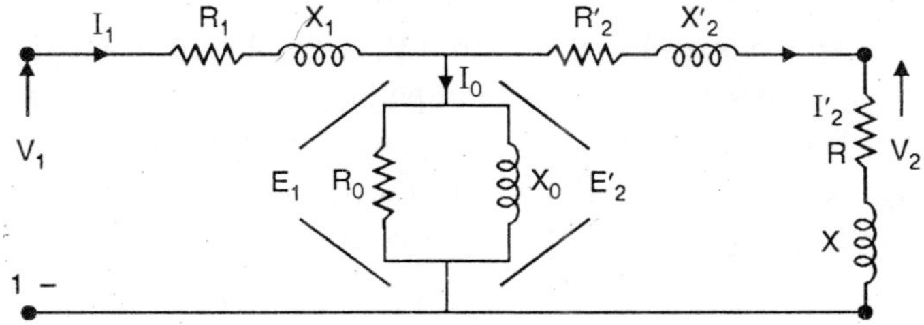

Fig. 5.9. Equivalent circuit of transformer.

$$E_2' = E_1 \quad \text{and} \quad I_1 = I_2' + I_0 \qquad ...(5.15)$$

whereas from equation 5.6 A

$E_2 = nE_1$ where n is the ratio of transformation. Similarly from equation (5.10)

$$\frac{I_1}{I_2} = n \quad \text{or} \quad I_2 = \frac{I_1}{n} \quad \text{or} \quad I_1 = nI_2 \qquad ...(5.16)$$

In equation 5.6 A and equation (5.10) the induced voltage E_2 is the actual voltage induced in the secondary whereas in the equivalent circuit of Fig. 5.9. E_2' is the secondary voltage when referred to the primary winding and is equal to $\frac{E_2}{n}$ whereas the current in actual secondary winding is I_2 but when referred to primary winding it equals nI_2. So voltage quantities on the secondary side are referred to the primary side by dividing actual secondary voltage by the ratio of transformation (n) whereas the current quantities on the secondary side are referred to primary side by multiplying the actual secondary current by ratio of transformation n.

However, if we want the equivalent circuit referred to secondary side, the ratio of transformation will be used in the reverse fashion *i.e.* $E_1' = nE_1$ and $I_1' = \frac{I_1}{n}$ where n is again as defined in equation (5.6 A).

In Fig. 5.9 R_2', X_2', R' and X' are the resistances and reactances of the secondary side when referred to the primary side whereas R_2, X_2, R and X are the actual secondary quantities. To find out the referred quantities in terms of the actual quantities we proceed as follows :

The actual drop in the resistance of the secondary is

$$V_r = I_2R_2 \quad V_r' = \frac{V_r}{n} \text{ and } I_2' = nI_2$$

Hence
$$\frac{V_r'}{I_2'} = \frac{V_r}{n}\frac{1}{nI_2} = \frac{V_r}{I_2}\cdot\frac{1}{n^2} = \frac{R_2}{n^2} = R_2' \qquad ...(5.17)$$

Similarly it can be shown that $X_2' = \frac{X_2}{n^2}$, $R' = \frac{R}{n^2}$ and $X' = \frac{X}{n^2}$.

It is to be noted that electrical connection of primary and secondary windings as shown in Fig. 5.9 is possible only when the electrical quantities are referred to one particular winding otherwise not, as the two windings are coupled only through magnetic circuit.

Let us see the validity of quantities from secondary to primary side. Assume secondary is an HT winding and primary the LT winding *i.e.* it is a step up transformer and hence as defined in equation 5.6 A, $n > 1$ and $E_2 > E_1$. Also $N_2 > N_1$, therefore $R_2 > R_1$ and $X_2 > X_1$. If now E_2 is to be referred to primary side, it must be divided by n and R_2 and X_2 must be divided by n^2 so as to have these quantities more or less of the same order as the primary quantities and hence after referring the secondary quantities to primary side, electrical connection of the two windings is possible.

In the preceding analysis we have seen that $V_1 \simeq E_1$, the difference is very small due to the drop of potential through the primary leakage impedance. Also the exciting current I_0 is a small fraction of the primary full load current, so that I_2' (or I_2 referred to primary) is practically equal to I_1. Therefore, no appreciable error will be introduced by transferring the shunt branch in Fig. 5.9 to the one shown in Fig. 5.10 which is the approximate equivalent circuit of the transformer.

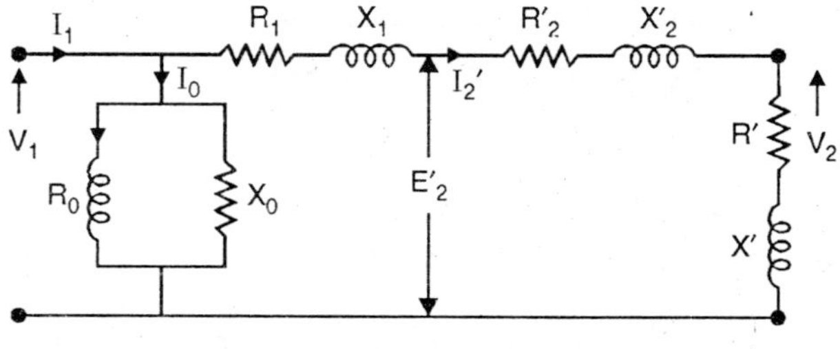

Fig. 5.10. Approximate equivalent circuit of the transformer.

Since I_0 is negligibly small as compared to I_1 or I_2'

$$I_1 \simeq I_2'$$

and the phasor diagram for the approximate equivalent circuit in Fig. 5.10 is shown in Fig. 5.11.

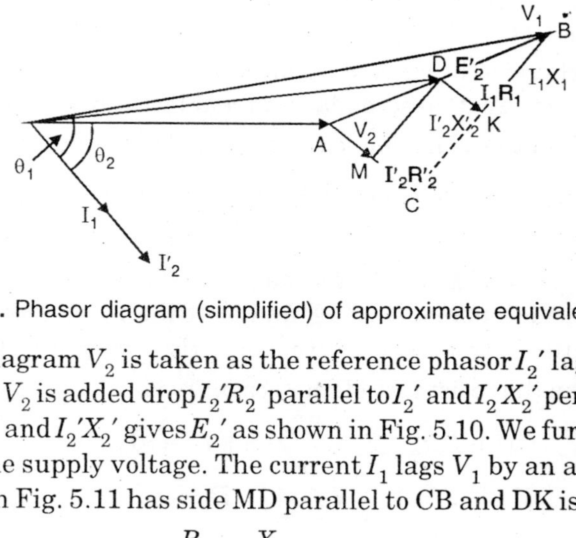

Fig. 5.11. Phasor diagram (simplified) of approximate equivalent circuit.

In this phasor diagram V_2 is taken as the reference phasor I_2' lags V_2 by an angle θ_2, the p.f. angle of the load. To V_2 is added drop $I_2'R_2'$ parallel to I_2' and $I_2'X_2'$ perpendicular to $I_2'R_2'$. The phasor sum of V_2, $I_2'R_2'$ and $I_2'X_2'$ gives E_2' as shown in Fig. 5.10. We further add drops $I_1 R_1$ and $I_1 X_1$ to E_2' to get V_1, the supply voltage. The current I_1 lags V_1 by an angle θ_1 as shown in Fig. 5.11. In triangle ABC in Fig. 5.11 has side MD parallel to CB and DK is parallel to AC, hence

$$\frac{R_1}{R_2'} = \frac{X_1}{X_2'}. \qquad ...(5.18)$$

i.e. the ratio of primary resistance to secondary resistance referred to primary equals the ratio of their corresponding reactances. This is very important and proves quite useful when the transformers are connected in parallel and share a load. The secondary current of the two transformers will be in phase with each others.

Example 5.3. A single phase transformer has 180 and 90 turns respectively in its secondary and primary windings. The respective resistances are 0.233 Ω and 0.067 ohm. Calculate the equivalent resistance of (*a*) the primary in terms of the secondary winding (*b*) the secondary in terms of the primary winding and (*c*) the total resistance of the transformer in terms of the primary.

Solution. Here $\frac{N_2}{N_1} = n = \frac{180}{90} = 2$

$$R_1 = 0.067 \text{ and } R_2 = 0.233$$

(*a*) Resistance of primary in terms of the secondary.

Since resistance of LT winding is to be referred to the HT winding. The ratio of transformation $n = 2$, hence $n^2R_1 = R_1' = 4 \times 0.067 = 0.268\ \Omega$.

(*b*) Resistance of secondary in terms of primary

$$R_2' = \frac{R_2}{n^2} = \frac{0.233}{4} = 0.058\ \Omega$$

(*c*) Total resistance of the transformer in terms of the primary winding

$$R_1 + R_2' = 0.067 + 0.058 = 0.125\ \Omega \quad \textbf{Ans.}$$

Example 5.4. A 40 KVA transformer with a ratio of 2000/250 V has a primary resistance of 1.15 Ω and a secondary resistance of 0.0155 Ω. Calculate

(*a*) the total resistance in terms of the secondary winding

(*b*) total copper loss on full load.

Solution. The ratio of transformation $\frac{250}{2000} = \frac{1}{8} = n$

The total resistance in terms of secondary winding is

$$R_2 + R_1'$$

Now $R_1' = R_1n^2 = 1.15 \times \frac{1}{8^2} = 0.0179$

Hence total resistance on secondary side is

$$0.0155 + 0.0179 = 0.0334\ \Omega.$$

(*b*) Now total copper loss on full load

$$= I_1^2R_1 + I_2^2R_2$$

Where I_1 and I_2 are full load current on the primary and secondary side respectively.

The rating of the transformer is 40 KVA or 40,000 VA and primary voltage is 2000 V.

Hence full load primary current $= \frac{40{,}000}{2000} = 20$ A

Similarly full load secondary current $= \frac{40{,}000}{250} = 160$ A

Hence full load copper loss $= 20^2 \times 1.15 + 160^2 \times 0.0155$

$$= 460 + 396.8 = 856.8$$

Alternatively, the same result can be obtained if we refer the total resistance to one side and corresponding to the full load current on that side we can find out the total copper loss.

In part (a) we have found out the total equivalent resistance of the transformer on the secondary side to be 0.0334 Ω and full load current on secondary side is 160 A. Hence total copper loss.

$$160^2 \times 0.0334 \simeq 855 \text{ watts}$$

It is same as obtained above. The small difference is because of round off error.

From the above analysis it is clear that we can study the performance of a transformer if we know the various parameters of the transformer *viz* $R_0, X_0, R_1, R_2 X_1$ and X_2. To find of R_0, X_0 we perform on the transformer what is known as No load or open circuit test and to find out the leakage impedances we perform short circuit test. These tests and the procedure to obtain these parameters are discussed in the following articles.

5.7 NO LOAD OR OPEN CIRCUIT TEST

As the name suggests the test is to be performed when the transformer is not loaded *i.e.* the secondary winding is open circuited. The following connections are made as shown in Fig. 5.12.

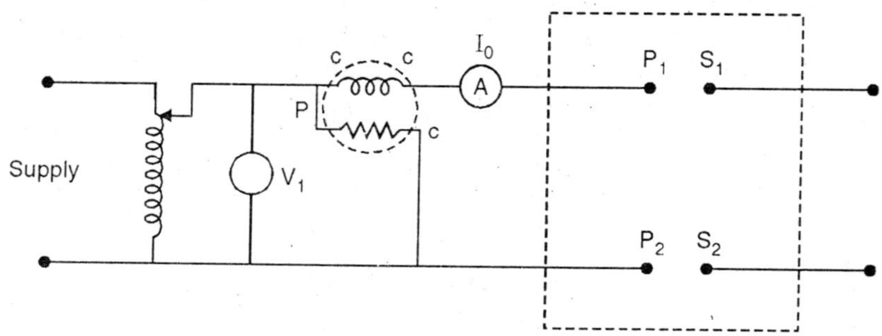

Fig. 5.12. Circuit diagram for O.C. test.

For O.C. test rated voltage should be applied to the transformer and a wattmeter to measure loss (power) and an ammeter to measure no load current should be connected as shown in Fig. 5.12 and the secondary should be left open. Now for a transformer, as mentioned earlier there is no fixed primary or secondary winding. The winding where the supply is connected becomes the primary and the other the secondary. Since the rated voltage is to be applied, it is safer from operation point of view to use LT winding as the primary winding. Also the meters required for conducting the O.C. test will have low ratings and are easily available in testing laboratories. So we use LT winding as the primary winding. Since I_0 is about 5% of full load current, a suitable range ammeter can be used. Under no load condition the pf is low, therefore, a low p.f. wattmeter should be used. Wattmeter also has both potential coil (voltage rating) and current coil (current rating), these must be considered while selecting wattmeter. In order to have good results it is very important to select proper range meters.

Suppose the observation of the meters after rated voltage is applied to the LT winding are as follows.

Voltage V_1, current I_0 and wattmeter reading is P_i watts. Here suffix i is used to indicate iron losses which includes eddy current and hysteresis losses. These losses are a function of voltage for a given frequency and core dimensions and hence are sometimes known as constant losses since the primary under normal conditions is connected across rated supply voltage.

$$\text{Eddy current loss} \propto V^2$$

and $$\text{Hysteresis loss} \propto V^x$$

Where x varies between 1.6 to 2. Under the no load condition as shown in Fig. 5.12, the wattmeter also measures copper loss corresponding to $I_0^2R_1$ in addition to constant iron loss, but as I_0 is about 5% of full load current the copper loss under this condition is negligibly small as compared to iron loss and hence it is neglected.

Referring to Fig. 5.10, the approximate equivalent circuit of the transformer, and using equation (5. 6 B)

$$P_i = \frac{V_1^2}{R_0} \quad \text{or} \quad R_0 = \frac{V_1^2}{P_i}$$

and $$I_w = \frac{V_1}{R_0} \quad \text{Hence} \quad I_m = \sqrt{I_0^2 - I_w^2} \qquad \text{...(5.19)}$$

and $$X_0 = \frac{V_1}{I_m}$$

Therefore, R_0 and X_0 can be obtained from the observations of the no load test.

To find out R_1, X_1, R_2 and X_2 parameters of the transformer we perform what is known as short circuit test which is described in the next article.

5.8 SHORT-CIRCUIT TEST

As the name suggests one of the windings would be short circuited and across the other winding supply would be connected.

The connection diagram is as shown in Fig. 5.13.

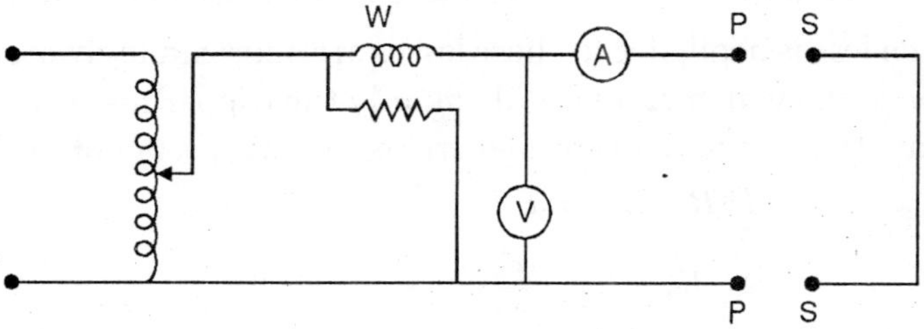

Fig. 5.13. Connection diagram for S.C. Test.

The equivalent circuit of the transformer under short circuit condition is shown in Fig. 5.14.

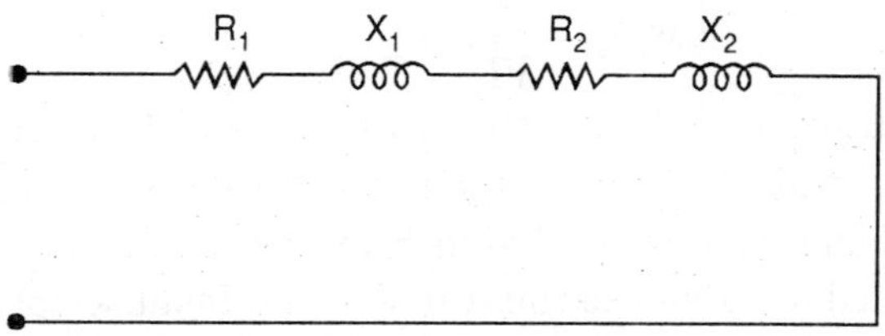

Fig. 5.14. Equivalent circuit of transformer under S.C. condition.

Here the shunt branch is omitted as the current drawn by the shunt I_0 is negligibly small as compared to full load current and when short circuit is carried out the requirement is that the supply voltage should be such that rated current flows in the two windings. Now if rated voltage is supplied, the current is limited by the series impedances and the short circuit current is found to be as high as ten times the full load current which will overheat both the windings, and insulation of the windings will be damaged and hence it is required to make measurements with such a reduced voltage that rated currents circulate in the two windings and with this voltage which is approximately $\frac{1}{10}$th of the rated voltage I_0 is going to be $\frac{1}{10}$th of what it would be when rated voltage is applied and hence the shunt branch can be eliminated from the equivalent circuit.

The short circuit test is most conveniently made by short-circuiting the LT winding, for this means that the current to be measured will be the moderate amount corresponding to full load on the HT side and at the same time the voltage to be supplied will be only a small fraction of the normal voltage of the HT winding, therefore, within the range of meters normally available in the testing laboratories.

While selecting ranges of the various meters, following points must be kept in mind. The range of the ammeter should be slightly higher than the full load current of the transformer on HT side, the voltmeter should be such that its range is about 10% of the rated voltage of the transformer on HT side and the voltage and current ratings of the wattmeter should correspond to the voltmeter and ammeter ratings. Since the p.f. of the transformer is low under short circuit as p.f. angle θ_{sc} is given by $\tan^{-1}\frac{X_1 + X_2}{R_1 + R_2}$ and since this ratio is normally more than 4, the p.f. angle is large and hence p.f. is low, a low p.f. wattmeter should be used.

Suppose the readings of the meters are P_C, V and I, the parameters can be obtained as follows. It is to be noted that under the testing conditions since the voltage applied is about $\frac{1}{10}$th of rated voltage, the iron loss is negligible as these losses are approximately proportional to square of the voltage and hence the wattmeter records mainly the copper losses and hence suffix c is attached to P to indicate that the losses recorded are copper loss predominantly.

Now $$I^2(R_1 + R_2) = P_c$$

or $$R_1 + R_2 = \frac{P_c}{I^2} \quad \text{...(5.20)}$$

Also $$Z = \frac{V}{I}$$

and hence $$X_1 + X_2 = \sqrt{Z^2 - (R_1 + R_2)^2}$$

$$\cos\theta_{sc} = \frac{P_c}{VI}$$

It is to be noted with caution that since the measurements have been made on the HT side of the transformer, $(R_1 + R_2)$ and $(X_1 + X_2)$ represent the total resistance and reactance respectively of the transformer when referred to HT side. Similarly R_0 and X_0 as measured using O.C. test are the quantities referred to LT side of the transformer. If subscript 1 stands for the LT and 2 for the HT side, here R_1 means the resistance of the LT side when referred to HT side.

In order to study the performance of the transformer *i.e.* voltage regulation and efficiency, we need to know the equivalent resistance and reactance. However, if we want to find out resistance and reactance of individual winding following procedure is usually adopted.

The resistance of individual winding can be obtained by applying low dc voltage such that rated current flows in the corresponding winding and with the help of ammeter and voltmeter readings, the resistance of the winding can be obtained and using equation (5.20)

$$X_1 + X_2 = \sqrt{Z^2 - (R_1 + R_2)^2} \quad ...(5.21)$$

where R_1 is the resistance of LT winding referred to HT side. Again using equation (5.18) and the above equation X_1 and X_2 can be obtained. Sometimes approximate value of X_1 and X_2 are each taken half of the value obtained from equation (5.21).

It is to be noted that the values of R_1 and R_2 as obtained through SC test are slightly higher than the actual values whereas X_1 and X_2 are slightly lesser than the actual values. However, the values obtained provide sufficiently accurate results from engineering considerations and are used to study the performance of the transformer.

Fig. 5.15 shows phasor diagram of the transformer under short circuit condition.

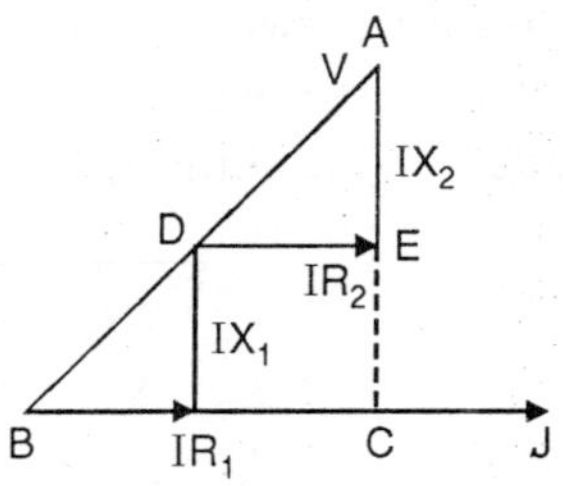

Fig. 5.15. Phasor diagram under SC condition.

Here V is the impressed voltage and I is the current in the winding. Since DE is parallel to BC

$$\frac{R_2}{R_1 + R_2} = \frac{X_2}{X_1 + X_2}$$

or
$$\frac{R_1 + R_2}{R_2} = \frac{X_1 + X_2}{X_2} \quad \text{or} \quad \frac{R_1}{R_2} = \frac{X_1}{X_2}$$

as given in equation 5.18.

Next we study the performance of the transformer in terms of voltage regulation and efficiency.

5.9 VOLTAGE REGULATION

When a transformer is supplied at a constant voltage and frequency, the secondary voltage keeps on changing depending upon the loading condition. If there is no load, the secondary terminal voltage corresponds to the induced voltage and as we load the transformer, the secondary voltage changes depending upon the load and its p.f. Fig. 5.16 shows the equivalent circuit of the transformer under inductive and capacitive loading conditions and Fig. 5.17 shows the corresponding phasor diagrams for the two loading conditions.

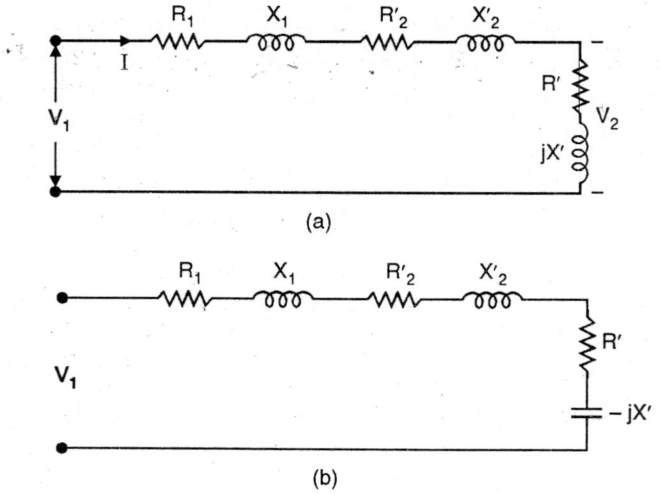

Fig. 5.16. (*a*) Equivalent circuit for inductive loading (*b*) Equivalent circuit for capacitive loading.

To draw the phasor diagrams for the two loading conditions we take V_1 fixed and since it is a series circuit we take I as the reference phasor. In the inductive circuit I lags V_1 by say an angle α and in capacitive circuit I leads V_1 by an angle β.

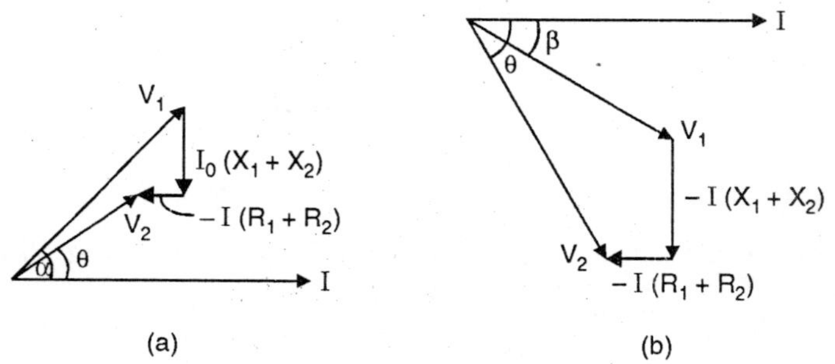

Fig. 5.17. Phasor diagram for (*a*) inductive load (*b*) capacitive load.

In both the phasor diagrams, the secondary voltage V_2 is obtained after subtracting the drops due to $I(R_1 + R_2)$ and $I(X_1 + X_2)$ from the supply voltage V_1 and we find that in case of inductive load the magnitude of secondary voltage is less than the supply voltage and for capacitive loads it is otherwise.

Voltage regulation of a transformer for a particular load is defined as the change in secondary voltage from the no load to the load under consideration expressed as a ratio of the rated voltage of the load. If E_2 in the rated secondary terminal voltage and V_2 is the secondary terminal voltage for a particular load, then percentage regulation is expressed as $\frac{E_2 - V_2}{V_2} \times 100$ keeping the primary voltage magnitude and frequency fixed.

Please note that $V_1 \simeq E_1$ and E_2 when referred to primary side $E_2 = E_1$

Let us develop an expression for voltage regulation making use of the phasor diagram for the inductive load

$$V_1 \cos \alpha = V_2 \cos \theta + I(R_1 + R_2) \text{ and } V_1 \sin \alpha = V_2 \sin \theta + I(X_1 + X_2)$$

Let $R_1 + R_2 = R$ and $X_1 + X_2 = X$

Hence $V_1^2 \cos^2 \alpha = (V_2 \cos \theta + IR)^2$

and $V_1^2 \sin^2 \alpha = (V_2 \sin \theta + IX)^2$

or $$V_1^2 = V_2^2 + 2IRV_2 \cos \theta + 2\, V_2 IX \sin \theta + I^2(R^2 + X^2) \quad ...(5.22)$$

or $$V_1 = V_2 \sqrt{1 + \frac{2IR}{V_2} \cos \theta + \frac{2IX \sin \theta}{V_2} + \frac{I^2}{V_2^2}(R^2 + X^2)} \quad ...(5.22\text{ A})$$

Normally rated load voltage V_2 is known, the parameter of transformer R and X are known and p.f. $\cos \theta$ of the load is also known, hence V_1 can be calculated and then

$$\% \text{ regulation} = \frac{V_1 - V_2}{V_2} \times 100 \quad ...(5.23)$$

From equation (5.22 A) exact value of regulation can thus be obtained. However, in practice the last term under the square root sign in equation 5.22 is generally negligible, therefore

$$V_1 = V_2 \left\{1 + \left(\frac{2IR}{V_2} \cos \theta + \frac{2IX}{V_2} \sin \theta\right)\right\}^{1/2} \quad ...(5.24)$$

Further simplification can be made as the terms within the simple brackets is small as compared to unity and hence using binomial expansion and limiting only to second term

$$V_1 \simeq V_2 + IR \cos \theta + IX \sin \theta$$

or $$\text{regulation} = \frac{V_1 - V_2}{V_2} = \frac{IR}{V_2} \cos \theta + \frac{IX}{V_2} \sin \theta \quad ...(5.25)$$

The quantity $\frac{IR}{V_2} = V_r$ is dimensionless and is known as per unit resistance of the transformer. Similarly $\frac{IX}{V_2} = V_x$ is the per unit reactance of the transformer. Equation (5.25) has been derived for an inductive load where θ is positive and hence $\sin \theta$ is positive. However, for a capacitive load θ is negative and hence $\sin \theta$ is negative and hence, the general expression for regulation is given by equation as follows

$$\text{regulation} = \frac{IR}{V_2} \cos \theta \pm \frac{IX}{V_2} \sin \theta \quad ...(5.26)$$

where plus sign is to be used for inductive load and minus sign for a capacitive load.

From the equation 5.26 it is seen that the voltage regulation of a transformer can be zero if the load is capacitive and the p.f. angle θ is such that

$$\frac{IR}{V_2} \cos \theta - \frac{IX}{V_2} \sin \theta = 0$$

$$R \cos \theta = X \sin \theta$$

or $$\frac{\sin \theta}{\cos \theta} = \frac{R}{X}$$

or $$\theta = \tan^{-1} \frac{R}{X} \quad ...(5.26\text{ A})$$

and the p.f. is leading.

In fact the voltage regulation would be negative if the p.f. is leading and it is such that

$$X \sin \theta > R \cos \theta$$

or
$$\theta > \tan^{-1} \frac{R}{X} \quad ...(5.26\text{ B})$$

If p.u. resistance and reactance of a transformer are given instead of ohmic values, the expression for voltage regulation is given as

$$\text{p.u. voltage regulation} = V_r \cos \theta \pm V_x \sin \theta \quad ...(5.27)$$

Example 5.5. A 5 kVA transformer has a nominal voltage rating of 1100/110 volts. With the low voltage winding short circuited, it is found that 33 volts is required to circulate rated full load current and the corresponding power input is 85 watts. Determine the per cent voltage regulation when the load takes rated current at 0.8 p.t. lagging.

Solution. The full load current $= \dfrac{5 \times 1000}{1100} = 4.55$

Since the supply is connected to HT side we will refer all quantities to HT side

$$I^2(R_1 + R_2) = P_c = 85$$

$$4.55^2 R = 85 \text{ or } R = 4.1\ \Omega$$

and
$$Z = \frac{V}{I} = \frac{33}{4.55} = 7.25\ \Omega$$

or
$$X = \sqrt{7.25^2 - 4.1^2} = \sqrt{35.8} = 5.96\ \Omega$$

Using equation 5.22

$$V_1^2 = V_2^2 + 2V_2 IR \cos \theta + 2\, V_2 IX \sin \theta + I^2(R^2 + X^2)$$

$$1100^2 = V_2^2 + 2V_2 \times 4.55 \times 0.8 \times 4.1 + 2V_2 \times 4.55 \times 5.96 \times 0.6 + 4.55^2(4.1^2 + 5.96^2)$$

$$= V_2^2 + V_2(29.85 + 32.54) + 52.56$$

$$1210000 = V_2^2 + 62.4V_2 + 52.56$$

or
$$V_2^2 + 62.4V_2 - 1210000 = 0 \text{ neglecting } 52.56$$

$$V_2 = \frac{-62.4 \pm \sqrt{3893 + 4840000}}{2}$$

$$= \frac{-62.4 \pm 2200}{2} = 1068.8 \text{ volt}$$

Hence percent regulation $\dfrac{(1100 - 1068.8) \times 100}{1068.8} = 2.9\%$

Alternatively using approximate expression (5.25) for the regulation

$$\left(\frac{IR}{V_2} \cos \theta + \frac{IX}{V_2} \sin \theta\right) \times 100 = \left(\frac{4.55 \times 4.1 \times 0.8}{1068.8} + \frac{4.55 \times 5.96 \times 0.6}{1068.8}\right)_{100}$$

$$= (0.01396 + 0.01522)\ 100 = 2.9\% \quad \textbf{Ans.}$$

The two formulae give almost the same voltage regulation. Voltage regulation sometimes is also defined as

$\frac{E_2 - V_2}{E_2}$ where E_2 is the rated secondary voltage. In that case for the above problem regulation comes out to be

$$(1100 - 1068.8)/1100 = 2.84\%$$

Example 5.6. Determine the regulation of a transformer in which ohmic loss is 1% of the output and the reactance drop 5% of the voltage when p.f. is

(*a*) 0.8 lag (*b*) unity (*c*) 0.8 lead.

Solution. Since ohmic loss is 1% of the output

i.e. $$\frac{I^2R}{VI} = 0.01 \quad \text{or} \quad \frac{IR}{V} = 0.01$$

and reactance drop is 5% of the voltage

i.e. $$IX = 0.05\ V \quad \text{or} \quad \frac{IX}{V} = 0.05$$

(*a*) Using equation 5.25 for percent regulation we have

$$\%\ \text{regulation} = \left(\frac{IR}{V}\cos\theta \pm \frac{IX}{V}\sin\theta\right)100$$

For 0.8 lag p.f. taking plus sign we have

$$(0.01 \times 0.8 + 0.05 \times 0.6)\ 100 = 3.8\%$$

(*b*) For unity p.f. $\sin\theta = 0$ Hence

$$\%\ \text{regulation}\ \frac{IR}{V}\ 100 = 0.01 \times 1 \times 100$$
$$= 1\%$$

(*c*) For 0.8 p.f. lead $\sin\theta = -0.6$ Hence

$$(0.01 \times 0.8 - 0.05 \times 0.6)\ 100 = -2.2\%$$ **Ans.**

5.10 LOSSES IN A TRANSFORMER

Transformer being a static device, the friction and windage losses are absent. There are mainly two types of losses in a transformer as mentioned earlier (*i*) Copper loss and (*ii*) Core loss or iron loss. Iron loss is further divided into two (*a*) hysteresis loss and (*b*) eddy current loss.

(*i*) **Copper loss.** This is due to the flow of current through the conductors of the winding *i.e.* it is an ohmic loss I^2R where R is the resistance of the conductor. In addition to this, there is loss caused by nonuniform distribution of the current density in the conductors. This nonuniformity may be considered due to the flow of eddy currents superimposed upon the theoretically uniform current density (assuming I to be flowing uniformly across the section of the conductor of the winding) which of itself would cause the pure ohmic loss.

(*ii*) (*a*) **Hysteresis loss.** This is caused by similar to molecular friction as the ultimate particles of the core tend to align themselves first in one direction, then in the other, as the magnetic flux alternates periodically as shown in Fig. 5.18. The area of the loop ABCDEA

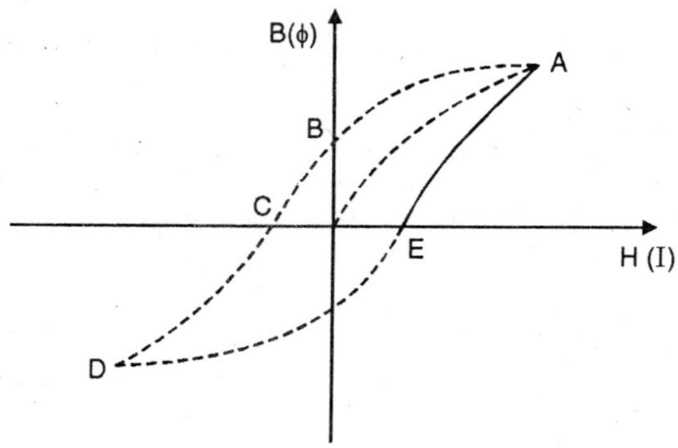

Fig. 5.18. B-H curve.

represents loss in one cycle of variation of ac current. The curve is known as B-H curve. The larger the value of permeability (μ) the smaller is the loop and hence smaller the hysteresis loss in the core. Also hysteresis loss is proportional to frequency of supply and proportional to B^x where x varies between 1.5 and 2. So smaller the B the lower will be the hysteresis loss. However from transformer emf equation, for certain operating voltage if B is low the area of cross section of the core should be large. It is really a design problem and is out of the scope of this book. Normally silicon steel is used for the core material.

(*b*) **Eddy current loss.** The alternating flux due to alternating current flowing through the winding induces voltage in the stampings (laminations) of the core just as it does in the coil of the windings. This voltage circulates current in the stampings in the form of eddies. These currents, therefore, heat the stampings and results in power loss. This is known as eddy current loss. The eddy current loss is proportional to $B^2f^2t^2$ where t is the thickness of the stampings. The smaller the thickness, the lower the eddy current losses. This is why the core is laminated so as to reduce the eddy current loss. The thickness is usually 0.5 mm.

5.11 EFFICIENCY OF TRANSFORMER

The efficiency of a transformer is defined as the ratio of output power to input power and is denoted by η *i.e.*

$$\% \, \eta = \frac{\text{output power} \times 100}{\text{Input power}} = \frac{\text{output} \times 100}{\text{output} + \text{loss}} = \frac{(\text{Input} - \text{loss})\, 100}{\text{Input}} \qquad \text{...(5.28)}$$

The losses in the transformer have already been discussed in the previous article. Since transformer is a static device it does not have rotational and windage loss and hence the efficiency of a transformer, in general, is very high, being higher in units of larger units than in smaller units.

The expression for efficiency of a transformer is given as

$$\eta = \frac{V_2 I_2 \cos\theta_2}{V_2 I_2 \cos\theta_2 + P_i + P_c} \qquad \text{...(5.29)}$$

where $V_2I_2 \cos\theta$ is the output power, P_i the constant iron loss and P_c the variable copper loss. P_i is constant as the supply voltage is constant rated voltage and P_c the copper loss is variable depending upon the loading condition. If the load is zero, P_c is practically zero however iron losses are taking place and are constant under no load condition, the output is zero and input corresponds to iron loss and hence efficiency is zero. Assuming V_2 to remain constant over certain loading condition, with increase in load, I_2 increases and hence for certain p.f. of load, the output increases and the copper loss also increases, the iron loss remains constant corresponding to no load loss, the efficiency increases. It will be shown shortly that when the loading is such that $P_i = P_c$ the efficiency is maximum and if we increase the load further $P_c > P_i$ and the increase in output is not as much as the increase in copper loss, as a result the efficiency decreases. The variation of efficiency as a function of loading is shown in Fig. 5.17.

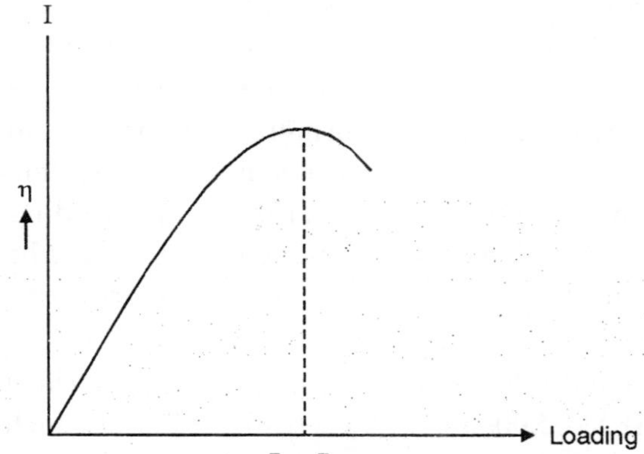

Fig. 5.19. Efficiency V_s loading.

For a certain p.f. of the load, the increase in load means effective decrease in the load impedance. In fact when the effective impedance is zero which means the secondary is short circuited, again the output is zero, even though there is substantial increase in current ; the secondary voltage is zero and hence output is zero and the complete input goes as iron and copper losses and the efficiency is zero. Please note that the efficiency of the transformer under no load and short circuit conditions is zero. Let us find out the loading condition for which the efficiency is maximum. Let P be the full load power and x be the fraction of full load for which the efficiency is maximum. Assuming secondary voltage to remain constant over the range of consideration of loading, the iron losses remain P_i whereas copper loss will become x^2P_c where P_c is the copper loss corresponding to full load. Hence

$$\eta = \frac{xP}{xP + P_i + x^2 P_c} = \frac{P}{P + \dfrac{P_i}{x} + xP_c} \quad ...(5.30)$$

For efficiency to be maximum $\dfrac{P_i}{x} + xP_c$ should be minimum

Let $$f(x) = \frac{P_i}{x} + xP_c$$

For $f(x)$ to be minimum $f'(x) = 0 = -\dfrac{P_i}{x^2} + P_c$

or $$P_c = \frac{P_i}{x^2} \quad \text{or} \quad x^2 P_c = P_i \qquad ...(5.31)$$

or $$x = \sqrt{\frac{P_i}{P_c}} \qquad ...(5.32)$$

From equation (5.31) it is clear that maximum efficiency occurs at a loading when copper loss equals the iron loss and the loading is given by equation (5.32). Normally the power transformers are so designed that the maximum efficiency occurs some where between 75% to 85%. Of the full load.

The all day efficiency of a transformer is the ratio of the total energy output (kwhrs) in a 24-hr day to the total energy input during the same time. Since transformers on distribution circuits are connected permanently to the supply line, the core loss goes on continuously, so that it is important to design distribution transformers with low iron loss consistence with other requirements. In general, low core loss means keeping low flux density which can be brought about either by using a relatively large number of turns in the winding or by increasing the cross-section of the core or both ways. The apportionment of the losses as between core loss on the one hand and copper loss on the other is, therefore, of much importance in determining all day efficiency.

Example 5.7. Calculate the values of R_0, X_0, R_t and X_t in the diagram for the equivalent circuit of a single phase 4KVA, 200/400 V 50 Hz transformer of which the following are the test results

O.C. test 200 V, 0.7 A, 70 W on low voltage primary side

S.C. test 15 V, 10 A, 80 W on HT side.

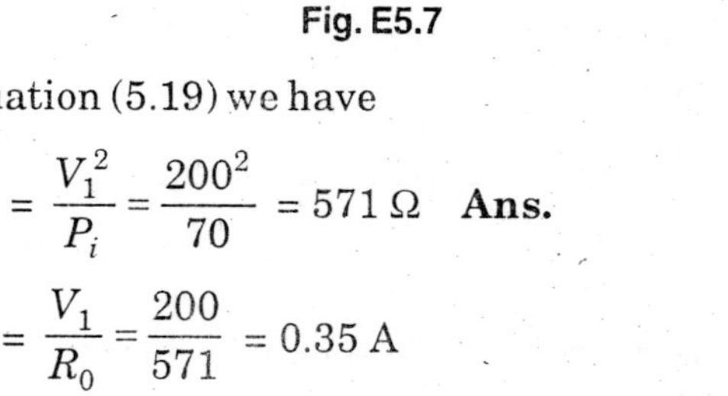

Fig. E5.7

Solution. Using equation (5.19) we have

$$R_0 = \frac{V_1^2}{P_i} = \frac{200^2}{70} = 571\ \Omega \quad \textbf{Ans.}$$

$$I_w = \frac{V_1}{R_0} = \frac{200}{571} = 0.35\ \text{A}$$

Hence $$I_\mu = \sqrt{0.7^2 - 0.35^2} = 0.6\ \text{A}$$

Hence $$X_0 = \frac{V_1}{I_\mu} = \frac{200}{0.6} = 333\ \Omega \quad \textbf{Ans.}$$

Similarly using equation 5.20

$$R_t = \frac{P_c}{I^2} = \frac{80}{10^2} = 0.8\ \Omega$$

and $$Z = \frac{V}{I} = \frac{15}{10} = 1.5\ \Omega$$

Hence $$X_t = \sqrt{2.25 - 0.64} = 1.269$$

R_t and X_t are referred to HT since whereas the equivalent circuit required is on LT side. Hence

$$R_t(LT \text{ side}) = 0.8 \times \frac{1}{n^2} = 0.8 \times \frac{1}{2^2} = 0.2 \quad \textbf{Ans.}$$

Similarly $X_t\ (LT \text{ side}) = \dfrac{1.269}{4} = 0.317\ \Omega$ **Ans.**

Example 5.8. Fig E 5.8 shows the equivalent circuit for a 1-phase transformer. Fig. gives resistance and reactance in ohms in terms of the primary side. The ratio of secondary to primary turns is 10. Determine (*a*) the secondary terminal voltage (*b*) the primary current (*c*) the efficiency.

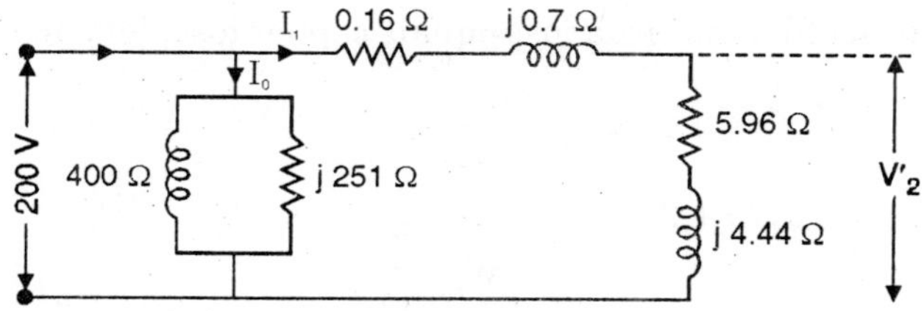

Fig. E 5.8.

Solution. We solve the circuit as referred to primary side.

$$I_1 = \frac{200}{(5.96 + 0.16) + j\,5.14} = \frac{200}{6.12 + j\,5.12} = \frac{200}{7.979\,\underline{/39.9}}$$

$$= 25.06\,\underline{/-39.9^\circ}$$

$$I_w = \frac{200}{400} = 0.5 \text{ A and } I_\mu = \frac{200}{j\,251} = -j\,0.797$$

Hence primary current $= 0.5 - j\,0.797 + 19.22 - j\,16.07$

$$= 19.72 - j\,16.867 = 25.95\,\underline{/-40.3} \quad \textbf{Ans.}$$

Now current $\quad I_1 = 25.06\,\underline{/-39.9}$

and load impedance is $5.96 + j\,4.44$

$$= 7.43\,\underline{/36.68^\circ}$$

Hence secondary terminal voltage referred to primary side is $25.06 \times 7.43\,\underline{/-3.22}$ = $186.2\,\underline{/-3.22}$ volts.

Hence actual secondary terminal voltage 186.2×10

$$= 1862\angle -3.22^\circ \quad \textbf{Ans.}$$

Now output power $= 25.06^2 \times 5.96 = 3743$ watts

Iron loss $= \dfrac{V_1^2}{R_0} = \dfrac{200^2}{400} = 100$ watts

Copper loss $25.06^2 \times 0.16 = 100.48$ watts

Hence efficiency $= \dfrac{3743 \times 100}{3743 + 100 + 100.48} = 94.9\%$ **Ans.**

Example 5.9. The maximum efficiency of a 500 KVA, 3300/500 V 50 Hz single phase transformer is 97% and occurs at $\frac{3}{4}$ full load unity p.f. If the impedance is 10%. Calculate the regulation at full load, p.f. 0.8 lag.

Solution. The output at $\frac{3}{4}$th full load unity p.f. is

$$500 \times 1 \times \frac{3}{4} = 375 \text{ kW}$$

Since at maximum efficiency iron loss equals copper loss, let each be x kW, hence

$$0.97 = \frac{375}{375 + 2x}$$

or $375 + 2x = 386.6$

or $x = 5.8$ kW

The load current at maximum efficiency

$$\frac{375 \times 1000}{500} = 750 \text{ A}$$

Hence $R = \dfrac{5.8 \times 1000}{750^2} = 0.0103$

Now full load current is $\dfrac{500 \times 1000}{500} = 1000$ A

Hence percent resistance $= \dfrac{1000 \times 0.0103}{500} \times 100 = 2.06$

Hence percent reactance $= \sqrt{10^2 - 2.06^2} = \sqrt{95.76} = 9.78$

Hence percent regulation

$$2.06 \times 0.8 + 9.78 \times 0.6$$

$$= 1.648 + 5.87 = 7.51\% \quad \textbf{Ans.}$$

Example 5.10. In a 25 KVA 2000/200 V transformer the iron and copper losses are 350 and 400 W respectively. Calculate the efficiency on unity p.f. at (*a*) full load and (*b*) half load (*c*) Determine the load for maximum efficiency and the iron and copper loss in this case.

Solution. The copper loss at full load given is 400 W

Hence (*a*) $$\% \eta = \frac{25000 \times 1 \times 100}{25000 + 350 + 400} = 97.08\%$$

(*b*) At half load $$P_i = 350 \text{ watts } P_c = \frac{400}{2^2} = 100 \text{ watt}$$

Hence $$\% \eta = \frac{12500 \times 1 \times 100}{12500 + 100 + 350} = 96.52\%$$

(*c*) The load at which maximum efficiency will occur = $\sqrt{\frac{350}{400}} \times 25$ KVA = 23.38 KVA **Ans.**

and the losses are each 350 watt. **Ans.**

Example 5.11. Calculate the efficiencies at half, full and $1\frac{1}{4}$ load of a 100 KVA transformer for p.f. of (*a*) unity (*b*) 0.8. The copper loss is 1000 watt at full load and the iron loss is 1000 watts.

Solution. (*a*) Unity p.f.

(*i*) Half full load $= \frac{100}{2} \times 1 = 50$ kW

Copper loss at $\frac{1}{2}$ load $= \frac{1000}{4} = 250$

Iron loss = 1000 watts

Hence $$\% \eta = \frac{50000 \times 100}{50000 + 1000 + 250} = 97.56\%$$

(*ii*) Full load $$\% \eta = \frac{100000 \times 100}{1000 + 1000 + 100000} = 98.04\%$$

(*iii*) At $\frac{5}{4}$ full load copper loss $= 1000 \times \frac{25}{16} = 1562$ watt

$$\frac{100000 \times \frac{5}{4} \times 100}{125000 + 1000 + 1562} = 97.99\%$$

(*b*) 0.8 p.f.

(*i*) Half full load output $= \frac{100}{2} \times 0.8 = 40$ kW

$$\% \eta = \frac{40000 \times 100}{40000 + 1000 + 250} = 96.96\%.$$

Example 5.12. A transformer has it's maximum efficiency of 0.98 at 15 KVA at unity p.f. During the day it is loaded as follows :

12 hrs – 2 kW at p.f. 0.5

6 hrs – 12 kW at p.f. 0.8

6 hrs – 18 kW at p.f. 0.9

Determine the all day efficiency.

Solution. To find copper and iron loss

$$0.98 = \frac{15 \times 1}{15 \times 1 + 2x} \quad \text{where } x \text{ is iron loss and also copper loss at 15 KVA loading}$$

$$15 + 2x = 15.306$$

or $$x = 153 \text{ watts}$$

Iron loss for 24 hrs are 153 watts

Copper loss, however, depend upon loading

During first 12 hrs when load is 2 kW at 0.5 p.f. *i.e.*

the KVA is 4 and hence copper loss $= 153 \times \left(\frac{4}{15}\right)^2$

$= 10.88$ Watts

During next 6 hrs $153 \times \left(\frac{15}{15}\right)^2 = 153$ watts

and During next 6 hrs $153 \times \left(\frac{20}{15}\right)^2 = 272$ watts

Hence

$$\% \eta = \frac{(12 \times 2 + 6 \times 12 + 6 \times 18) \times 100 \times 1000}{12 \times 10.88 + 6 \times 153 + 6 \times 272 + 24 \times 153 + 204 \times 1000}$$

$$= \frac{(24 + 72 + 108)\, 1000 \times 100}{130.56 + 918 + 1632 + 204 \times 1000}$$

$$= \frac{204 \times 1000}{206681} \times 100 = 97.5\%.$$

PROBLEMS

5.1. Calculate in terms of the primary the equivalent resistance and the leakage reactance of a transformer which gave the following data on test with the secondary terminals short circuited ; Applied voltage 60 V, current 100 A, power input 1.2 kW. **Ans.** 0.12 Ω, 0.59 Ω.

5.2. The required no-load ratio in a 1-phase 50 Hz core type transformer is 6000/250 V. Find the no. of turns in each winding if the flux is to be about 0.06 wbs. **Ans.** 480, 20.

5.3. A 50 Hz 1-phase transformer has a turn-ratio of 6. The resistances are 0.90 ohm and 0.03 ohm and the reactances 5 ohm and 0.13 Ω for high voltage and low voltage respectively. Determine (*a*) the voltage to be applied to the high voltage side to obtain full load current of 200 A in the low voltage winding on short circuit (*b*) the p.f. on short circuit. **Ans.** (*a*) 330 V (*b*) 0.2.

5.4. The high and low voltage windings of a 6600/250 V 50 Hz 1-phase transformer have resistances of 0.21 ohm and 2.72×10^{-4} ohm and reactances of 1.0 ohm and 1.3×10^{-3} ohm respectively. Determine the current and power input when the high voltage winding is connected to a 400 V 50 Hz supply, the low voltage winding being short-circuited. **Ans.** 206 A ; 17 kW.

5.5. The equivalent circuit shown refers to a 200/400 V 1-phasor 50 Hz 4 KVA transformer, the values given being reduced to the low voltage side. For a high voltage current of 10 A at a lagging p.f. of 0.8, calculate

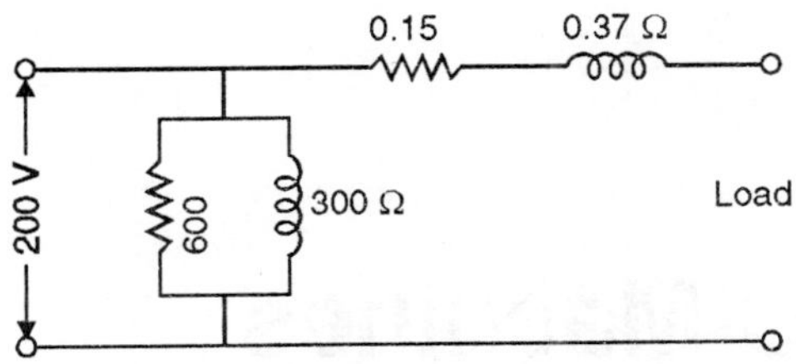

Fig. P 5.5

(*a*) The low voltage input current

(*b*) The efficiency

(*c*) The voltage at the terminals of the high voltage side. **Ans.** (*a*) 20.65 A (*b*) 96.08% (*c*) 386 V.

5.6. The efficiency, at unity p.f. of a 6600/384 V 200 KVA, 1-phase transformer is 98% both at full load and half load. The p.f. on no load is 0.2 and the full load regulation at a p.f. 0.8 lag is 4%. Determine the equivalent parameter values referred to the LT side.

Ans. R_0 = 108 Ω X_0 = 22.1 Ω R = 0.01 Ω X = 0.036 Ω.

5.7. A 100 KVA 6600/330 V, 50 Hz single phase transformer took 10 A and 436 watt at 100 V in a short circuit test ; the figures referring to the HT side. Determine the voltage to be applied to the HT side on full load at p.f. 0.8 lag when the secondary terminal voltage is 330 V. **Ans.** 6734 V.

5.8. Determine the efficiency of a 150 KVA transformer at 25%, 33% and 100% full load (*a*) at unity p.f. (*b*) at 0.8 p.f. lag if the copper loss is 1600 W at full load and the iron loss is 1400 W.

Ans. (*a*) 96.15, 96.94, 98.04% (*b*) 95.23, 96.21, 97.56%.

5.9. The efficiency of a 400 KVA 1-phase transformer is 98.77% when delivering full load at 0.8 pf and 99.13% at half full load and unity p.f. Calculate (*a*) the iron loss (*b*) the full load copper loss.

Ans. (*a*) 1 kW (*b*) 3 kW.

5.10. A 50 KVA 2000/200 V transformer has impedance drop of 10% and resistance drop of 4% (*a*) Find regulation at full load 0.8 lag (*b*) Find p.f. for which the regulation is zero. **Ans.** 8.7% 0.917 lead.

5.11. A 230/2300 volt single phase transformer is excited by 230 V ac supply. The equivalent resistance and reactance when referred to primary winding are 0.2 and 0.8 ohm respectively. R_0 and X_0 are 500 Ω and 200 ohm respectively. The load impedance is 400 + *j* 600 ohms. Determine (*a*) primary current (*b*) secondary terminal voltage (*c*) power factor at primary terminals.

Ans. $30\angle -58.7$, 2075.4 V, 0.52 lag.

5.12. A 100 KVA distribution transformer is supplying the following loads

80 kW at 0.8 p.f. for 8 hrs.

40 kW at unity p.f. for 6 hrs.

No load for the remaining 10 hrs.

Determine all day efficiency if full load copper losses are 2 kW and iron loss 1 kW. **Ans.** 95.5%.

CHAPTER

6 DC Machines

6.1 INTRODUCTION

The first electromagnetic machines to be developed were dc machines. The first central power station developed by Thomas A. Edison to serve a part of New York City had dc generators in 1882 which operated at 110 V d.c. The use of transformers for transmitting power over longer distances and at higher voltages justified the use of a.c. especially where the electric energy was to be harnessed from water power which usually is available far from the load centres. When alternating current supplemented dc as the mode for the generation and transmission of electrical energy, induction and synchronous machines supplemented dc machines as motors and generators to a large extent. Still today dc machines account for about 40 per cent of the dollar volume in electrical-machine manufacturing and sales. This is because

(*i*) most highway vehicles use lead-acid batteries for electrical energy storage. The starter motors, windshield wiper motors, fan motors and motors to drive other accessories in vehicles must be dc motors. Literally millions of dc motors are manufactured every year for such purposes.

(*ii*) dc motors have very accurate speed/torque characteristics as compared to other drives. DC motors are, hence, nearly the universal choice for driving power shovels, steel and aluminium rolling mills, electric elevators, railroad locomotives and large earth moving equipment.

6.2 CONSTRUCTION

A d.c. machine consists of two parts : the stator *i.e.* the stationary part and the rotor *i.e.* the rotating part. Fig. 6.1 shows a sectional view of a 4-pole dc machine. The field poles are located on the stator and project inward from the inside surface of the iron cylinder that forms the stator yoke. The yoke serves as a return path for the pole flux. Each iron pole consists of a narrower part, called the pole core around which is placed the exciting winding called field winding or coil. This coil may consist of two or more separate windings to provide for controlling the strength of the field flux. A pole shoe, usually laminated, distributes the pole flux over the rotor surface. The surface of the pole shoe opposite the rotor is called the pole face. The space between the pole face and the rotor surface is called the air gap. As the teeth of the rotor core sweep past the pole faces, localised variations in the flux density within the pole shoe are produced. These flux density variations produce hysteresis and eddy current losses. Laminating the pole shoes greatly reduces the eddy current loss in the iron of the pole shoes.

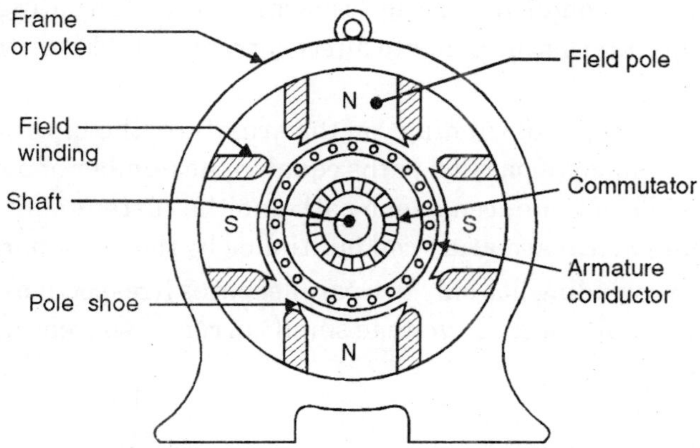

Fig. 6.1. Sectional view of a 4-pole dc machine.

Fig. 6.2 shows a longitudinal view of armature *i.e.* rotor along with two coils of the winding. The armature winding is composed of coils embedded in slots in the rotor. The armature has a cylindrical steel core consisting of a stack of slotted laminations. The slots in the laminations are aligned to form axial slots in the outside surface of the core, in which the coil sides of the armature winding are placed. In small machines, the slots are sometimes skewed. Skewing makes winding a little more difficult, but it results in quieter operation and a slight reduction in losses. The laminated iron between the slots forms the teeth of the core. The armature coils are often held in place by wood or fibre 'wedges' driven into the slots from one end of the core stack.

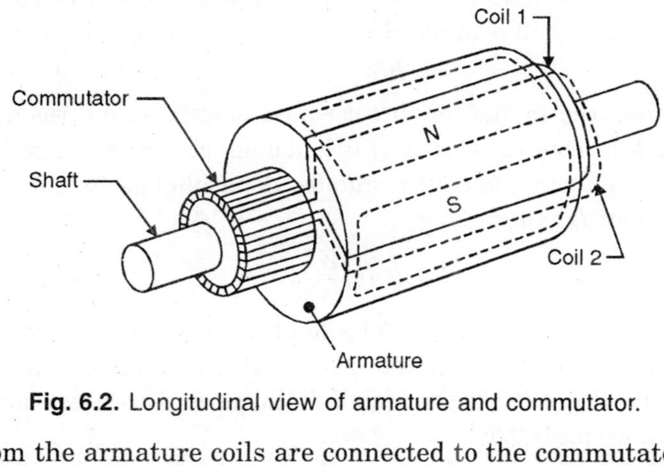

Fig. 6.2. Longitudinal view of armature and commutator.

The leads from the armature coils are connected to the commutator. The commutator consists of radial copper segments separated from each other by insulating material, usually mica. Current is conducted to the armature coils by carbon brushes that are held against the cylindrical surface of the commutator by the force of springs. The brushes should have proper conductivity and sliding-friction characteristics for a given motor. Brushes should be fitted to the surface of the commutator. They are held in brush holders and must be free to slide radially in these holders to maintain contact with the commutator as they wear away. Current is carried into the brush by a 'pigtail' or "brush shunt". This is a very flexible wire embedded in the brush material and connected to the brush holder. Brushes are normally placed along the magnetic neutral axis. Brushes must be inspected regularly and replaced as necessary, as they are worn away. The emf induced in the conductors of the dc machine is an alternating emf. The commutator helps in converting this alternating emf into direct current emf. This is, therefore, also known as mechanical converter.

Mainly there are two types of windings on the armature, the lap and the wave winding. The lap winding has the number of parallel paths equal to the number of poles or no. of brushes. The total induced emf is the emf induced in one path and the current delivered to the external circuit equals the current in each armature coil multiplied by the no. of parallel paths.

In contrast the wave winding has only two parallel paths irrespective of the no. of poles and hence wave winding is suitable for large voltage small currents whereas lap winding for small voltage large currents.

6.3 EMF EQUATIONS

A dc machine may be a dc generator or a dc motor. The dc generator converts mechanical energy into the electrical energy whereas a dc motor converts electrical energy into mechanical energy. A dc generator works on the principle that the motion of a conductor (mechanical energy) through a magnetic field causes an emf to be induced in the conductor.

However when a current carrying conductor is brought in a magnetic field and the conductor cuts the flux lines (not paralled to flux lines), a force acts on the conductor and thus torque is developed and the electric energy is converted into mechanical energy.

The pole faces are so designed that the magnetic flux lines cut the conductors in the armature slots radially and hence when the armature is rotated with the help of a prime mover the flux lines cut the conductor and an emf is induced in the conductor which is given by

$$e = Blv \tag{6.1}$$

where e is the voltage induced in one conductor whose length is l metres and is moving with velocity v *metres*/sec and the conductor is moving in a magnetic field of flux density B Wb/m^2. If Z is the total number of conductors and A the number of parallel paths, the no. of conductors in series is Z/A. Therefore emf E generated is

$$E = B.lv.\frac{Z}{A} \qquad \text{...(6.2)}$$

Let ϕ be the flux per pole, P the number of poles, the total flux in the air gap is $P\phi$. This gap is in the form of a cylinder with length equal to the armature conductor length and r the radius of the air gap. Since the flux in the air gap is radial, the normal cross-section of the flux is the surface of the cylinder and equals $2\pi rl$

Hence flux density B

$$B = \frac{\phi P}{2\pi rl} \qquad \text{...(6.3)}$$

If N is the speed of rotation of the armature in rpm, the linear velocity is

$$v = \frac{N 2\pi r}{60} \qquad \text{...(6.4)}$$

Substituting the values of B and v in equation (6.2) we have

$$E = \frac{P\phi}{2\pi rl} . l . \frac{N 2\pi r}{60} . \frac{Z}{A}$$

$$= \frac{P\phi NZ}{60A} \qquad \text{...(6.5)}$$

$$= K\phi N$$

where $K = \frac{ZP}{60A}$ and depends upon the design of the machine and is constant for a particular machine.

When current is flowing out of the + brushes, the torque opposes the rotations. The machine is acting as a generator and the voltage generated is given by equation (6.4) and is a source voltage for the current. However, when current is flowing into the + brushes, the torque is in the direction of rotation and the machine is acting as a motor. The generated voltage is again given by equation (6.4) but this is known as counter voltage *i.e.* voltage acting in opposition to supply voltage.

The polarity of voltage induced in the generation mode of operation of dc machine is given by Fleming's right hand rule and is stated as follows :

Hold the thumb, forefinger and the central finger of the right hand at right-angles to each other. If the thumb points to the direction of motion and forefinger to the direction of field, the central finger will point in the direction of induced emf *i.e.* the central finger will point to the positive terminal of emf or will indicate the direction of current flow if the ends of the conductor are connected to the external circuit (load).

Similarly the direction of force on a current carrying conductor in the motoring mode of operation of a dc machine, when conductor is situated in a magnetic field is given by Fleming's left-hand rule and is stated as follows :

Hold the thumb, the forefinger and the central finger at right angle to each other of the left hand. If the forefinger points in the direction of magnetic field and the central finger to the direction of current, the thumb will point to the direction of force or motion.

In motoring action when a conductor of length l carrying a current I amperes is brought in a magnetic field of strength B Wb/m^2, it experiences a force BIl and thus a torque is developed which rotates the armature. As a result of rotation (motion) of armature conductor in a magnetic field due to stator poles voltage in the rotor conductor is induced which is known as a counter emf or back emf which opposes the applied voltage (supply voltage). It is through the mechanism of back emf that a motor adjusts its electrical input to meet an increase or decrease in mechanical load on shaft. It is to be noted that the commutator acts as a mechanical inverter for motoring action and hence the current in armature conductors is alternating and not direct current. Interaction of the two fluxes the stator magnetic field and the armature conductor current magnetic field which are displaced in time and space develop the torque which is directly proportional to the product of the magnitude of the two fluxes.

6.4 CLASSIFICATION OF DC MACHINES

It is to be noted that there is no difference in the construction of dc motor or generator. A dc machine which is being used as a dc generator can also be used as a dc motor and *vice versa*.

Hence dc machines in general are classified based on the connection of the field coil and the armature winding terminals. Refer to Fig. 6.3 for different connections.

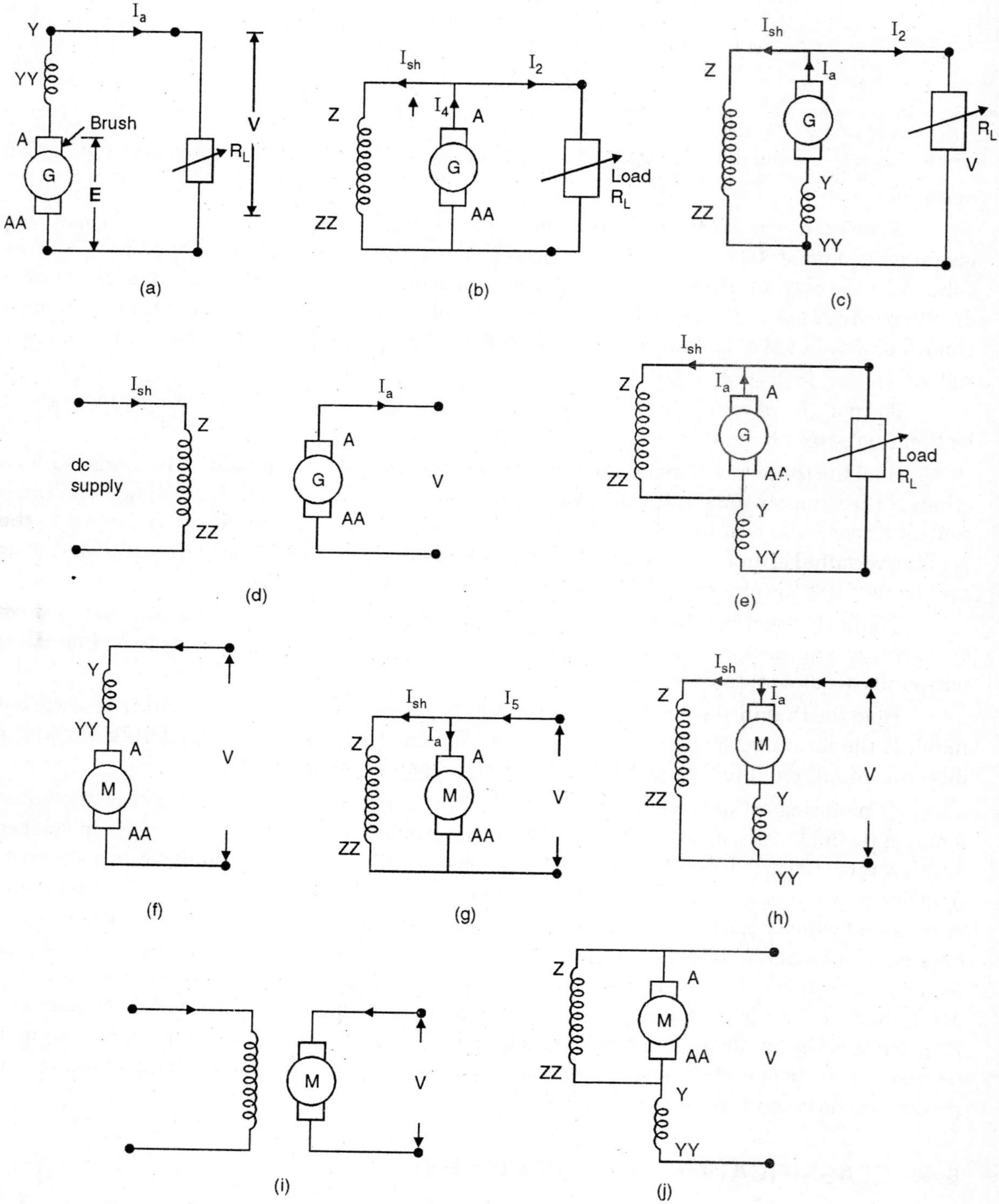

Fig. 6.3. (*a*) to (*e*) dc generators, (*f*) to (*j*) dc motors.

In Fig. 6.3 *A*, *AA* stand for armature terminals

Z,*Z Z* for shunt field terminals

Y, *Y Y* for series field terminals

In Fig. 6.3 (*a*) since the field winding is connected in series with the armature, it is known as dc series generator. Similarly in Fig. 6.3 (*f*) the machine is known as dc series motor. In Fig. 6.3 (*b*) and (*g*) the field winding is connected in shunt with the armature terminals, these machines are called d.c. shunt generator and motor respectively.

In Fig. 6.3 (*c*) and (*f*) we have both the field windings series as well as shunt winding. If the two windings are so connected that their fluxes add, the connection is known as cumulatively compounded dc generator and motors respectively. However, if the two fluxes oppose each other, the connection is known as differentially compounded dc generator and motors respectively. Also if *ZZ* is connected to YY of the series field it is known as long shunt connection whereas if *ZZ* is connected to *Y* as shown in Fig. 6.3 (*e*) and *j* it is known as short shunt connection. In Fig. 6.3 (*d*) and (*i*) the field winding is not connected to the armature terminals and hence these are known as separately excited dc generator and motor respectively.

The series field winding has few turns and since it is connected in series with the armature and the load in case of generator, the current through the field winding is large and hence the field winding has thicker conductor. The resistance R_{se} is in terms of fraction of an ohm. The armature resistance R_a is also in terms of fraction of an ohm. However, the shunt field winding has large number of turns and the conductor cross-section is very small. It's resistance is in terms of hundreds of ohms. Let E be the induced emf in case of generators and E_b the back emf for a dc motor, I_a armature current, I_L load current, I_{sh} the shunt field current. Following relations can be obtained for the generators and motors after referring to the circuit diagram shown in Fig. 6.3. Fig. 6.3 (*a*) is for series generator

$$E = V + I_a(R_{se} + R_a) \quad ...(6.6)$$

where V is the terminal voltage after load R_L is connected. In fact without connecting load in case of dc series generator no voltage can be induced as series field circuit is open. Fig. 6.3 (*b*) is for shunt generator

$$I_a = I_{sh} + I_L \quad ...(6.7)$$

and

$$E = V + I_aR_a \quad ...(6.8)$$

Similar relations can be written for other generator circuits. For series motor

$$V = E_b + I_a(R_a + R_{se}) \quad ...(6.9)$$

For shunt motor

$$I_s = I_a + I_{sh} \quad ...(6.10)$$

and

$$V = E_b + I_aR_a \quad ...(6.11)$$

Replacing the generator by its emf and internal resistance (armature resistance) and motor by its back emf and internal resistance we have the equivalent circuits for the series, shunt generators and motor as shown in Fig. 6.4.

Series, shunt and compound generators are self-excited dc generators *i.e.* they do not require any external source for excitation. However, as the name suggests separately excited dc generator requires an external dc source to excite its field winding.

Before we study conditions for self-excitation we first obtain magnetisation characteristics of these machines.

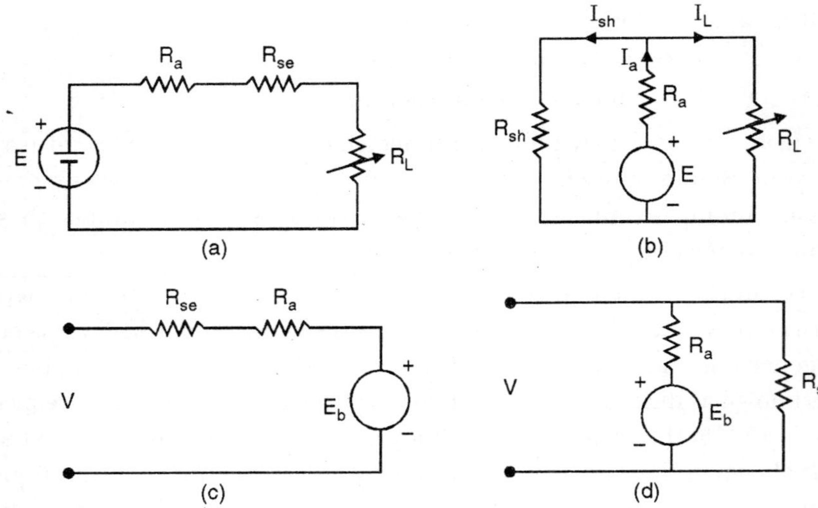

Fig. 6.4. (*a*) and (*b*) generator (*c*) and (*d*) motor.

6.5 MAGNETISATION CHARACTERISTICS

This characteristic is also known as open circuit characteristic *i.e.* when the machine is not loaded. It gives relation between the variation of generated voltage or no load voltage when the field current is varied and the machine is run at constant speed.

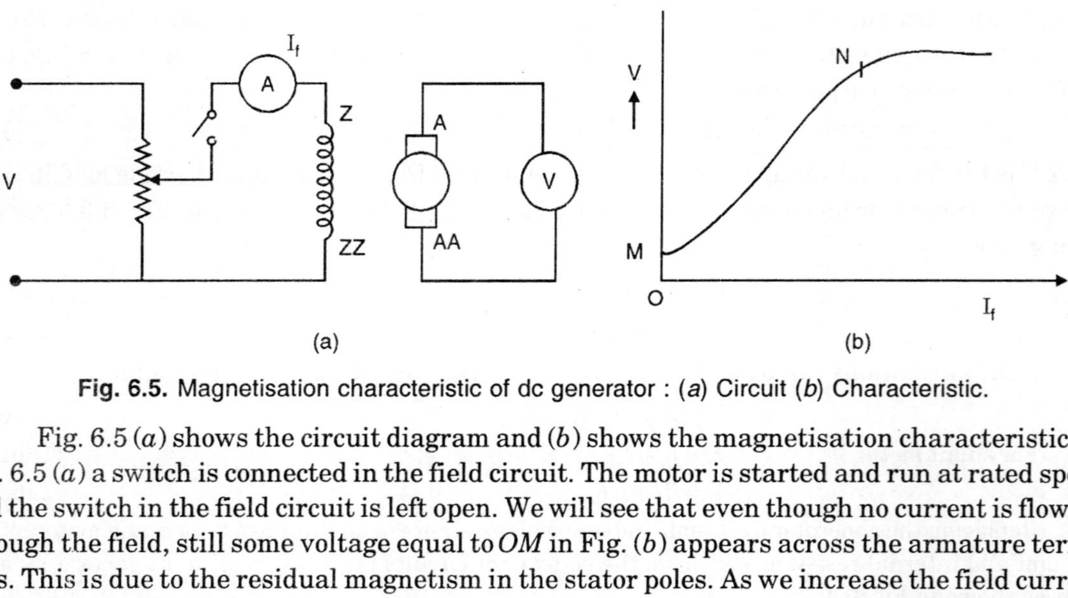

Fig. 6.5. Magnetisation characteristic of dc generator : (*a*) Circuit (*b*) Characteristic.

Fig. 6.5 (*a*) shows the circuit diagram and (*b*) shows the magnetisation characteristic. In Fig. 6.5 (*a*) a switch is connected in the field circuit. The motor is started and run at rated speed and the switch in the field circuit is left open. We will see that even though no current is flowing through the field, still some voltage equal to *OM* in Fig. (*b*) appears across the armature terminals. This is due to the residual magnetism in the stator poles. As we increase the field current the voltage recorded increases almost linearly. However, at some value of I_f (corresponding to point *N* in Fig. 6.5 (*b*)) the magnetic circuit starts getting saturated and hence further increase in I_f does not increase the flux proportionately and the voltage almost gets flattened.

Since the generated voltage is proportional to the speed, magnetisation characteristic for any other speed can be obtained from the test curve at one speed.

We now mention the necessary conditions for voltage build-up due to self-excitation :

1. Residual magnetism must be present.
2. The connection of the field winding with respect to armature terminals should be such that the field current strengthens the residual magnetism.
3. The resistance of the field winding should be less than the critical resistance.
4. The speed of the machine should be higher than critical speed.
5. In a series generator, the load should be connected and load current should flow. The resistance of the load should be less than critical resistance and speed should be higher than critical speed.

Because of paucity of space we will be very brief here we draw the circuit diagram and the characteristics which can be obtained using the circuit diagram Fig. 6.6 (*a*) circuit for self-excitation of dc shunt generator and (*b*) suggests the process of build up of voltage starting from residual magnetism OM.

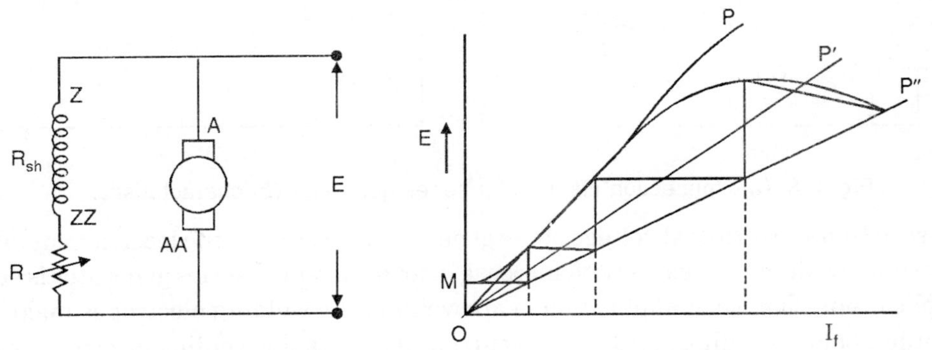

Fig. 6.6. Self-excitation (*a*) circuit (*b*) process of build-up of voltage.

Here OP′ or OP″ represent the total resistance of the field circuit including R. The intersection of the magnetisation characteristic and the resistance line OP′ or OP″ gives the final voltage induced.

It is seen that OP is the critical resistance line and the build-up of voltage is possible only if the field resistance is less than the critical resistance.

6.6 TERMINAL CHARACTERISTIC OF SHUNT GENERATOR

Fig. 6.7 (*a*) shows the circuit and Fig. 6.7 (*b*) the terminal characteristic of the dc shunt generator.

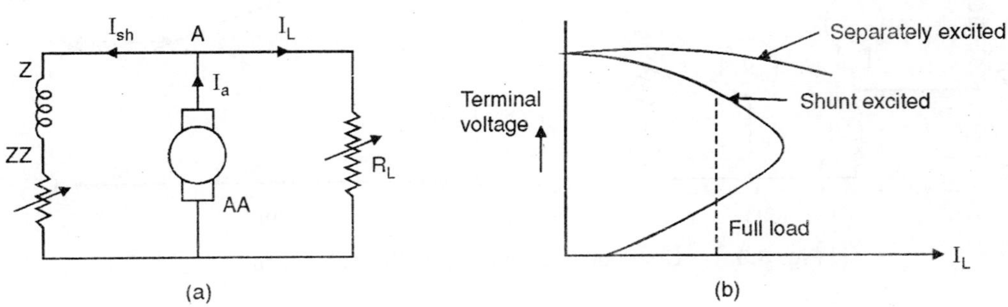

Fig. 6.7. (*a*) Connection diagram (*b*) Characteristic.

As the load on the shunt generator increases the armature drop $I_a R_a$ increases. This has a double effect. The terminal voltage V is given as

$$V = E - I_a R_a$$

Here E decreases due to increase in armature reaction and also because the field current decreases as V decreases. The voltage regulation is poor for most applications. However field rheostat may be adjusted to obtain the desired voltage at a given load. This is normally used as an exciter with synchronous machines.

6.6.1 Series generator characteristic

Fig. 6.8 (*a*) gives the connection diagram and 6.8 (*b*) the load characteristic.

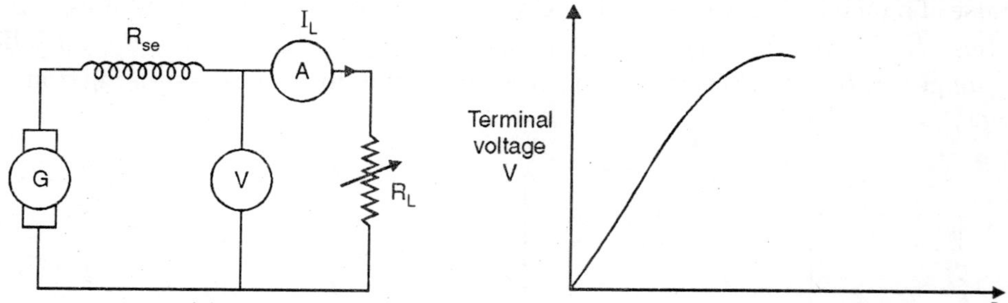

(a) (b)

Fig. 6.8. (*a*) Connection diagram for series generator (*b*) characteristic.

It is seen that the terminal voltage of the generator increases as the load current increases. Since the terminal voltage increases with increase in load current, a series generator is suited for special applications. One such application is arc welding. As welding electrodes make contact with each other, before welding starts, a high current flows. As the welding electrodes separate, there is a steep rise in terminal voltage because armature reaction decreases due to slight decrease in current. The high terminal voltage ensures that arc will be maintained in air across the contacts so that welding will be accomplished.

Another application is as a series booster. It's connected in series with the line to compensate for the series voltage drop as the current in the line increases.

6.6.2 Load characteristics of compound generator

Fig. 6.9 (*a*) shows the connection diagram and 6.9 (*b*) the characteristics.

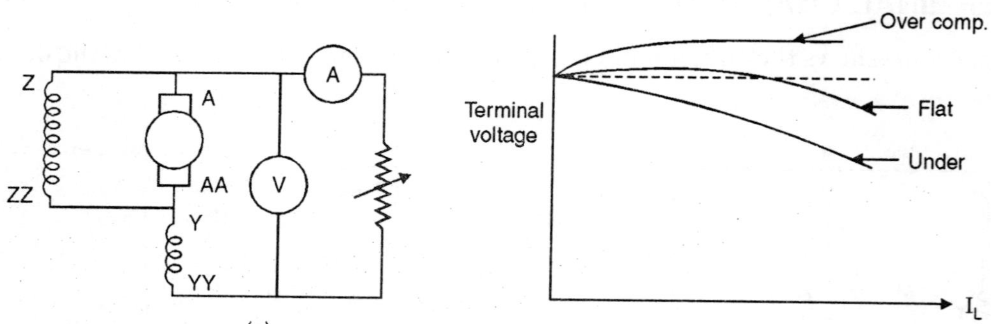

(a) (b)

Fig. 6.9. (*a*) Connection diagram (*b*) Characteristics.

A cumulative compound generator is capable of much better voltage regulation. Since the load current flows through the series field, the pole mmf actually increases with load and so does the generated voltage E.

If the increase in E is not sufficient to overcome $I(R_a + R_s)$ drops in the armature and series field resistances at full load, there will be positive voltage regulation and the machine is said to be undercompounded.

If the increase in E exactly compensates for the armature and series field $I(R_a + R_{se})$ drops at full load, the voltage regulation will be zero, and the generator is said to be flat compounded.

If the terminal voltage at full load is greater than that at no load (negative regulation) the generator is overcompounded.

An overcompounded generator is useful in mining applications, where the load may be at some distance from the generator. The rising voltage characteristic of the generator may be made to compensate for the $I(R_a + R_{se})$ drop in the line, so that the voltage at the load may be virtually constant.

A differential compound generator is often used in power shovel drives.

Example 6.1. A 6-pole 2 circuit wave connected armature has 250 conductors and runs at 1200 rpm. The electromotive force generated on open circuit is 600 V. Determine the useful flux per pole.

Solution. Using equation (6.5) we have

$$E = \frac{\phi ZNP}{60A}$$

Here $Z = 250$, $N = 1200$ rpm, $E = 600$ V, $P = 6$, $A = 2$, $\phi = ?$

Substituting these values

$$600 = \frac{\phi \,.\, 250 \,.\, 1200 \,.\, 6}{60 \times 2}$$

or

$$\phi = \frac{600 \times 60 \times 2}{250 \times 1200 \times 6} = 0.04 \text{ Wb} \quad \textbf{Ans.}$$

Example 6.2. An 8-pole lap connected armature has 960 conductors, a flux of 40 mWb per pole and a speed of 400 rpm. Determine the emf generated.

Solution. Using again equation (6.5) we have

$$E = \frac{\phi ZNP}{60\,A}$$

Here ϕ =40 mWb, $Z = 960$, N = 400 rpm, $P = 8$ and since it is lap connected $A = P = 8$.

Substituting these values we have

$$E = \frac{40 \times 10^{-3} \times 960 \times 400 \times 8}{60 \times 8}$$

$$= 256 \text{ volts} \quad \textbf{Ans.}$$

Example 6.3. If in example 6.2 the armature were wave connected, at what speed it must be driven to generate 400 V.

Solution. Since the armature is wave connected the no. of parallel paths is 2 independent of no. of poles, hence

$$400 = \frac{40 \times 10^{-3} \times 960 \times N \times 8}{60 \times 2}$$

or $$N = \frac{400 \times 60 \times 2}{40 \times 10^{-3} \times 960 \times 8} = \frac{10^4 \times 60 \times 2}{960 \times 8}$$

$$= 156 \text{ rpm}$$

Example 6.4. A shunt machine connected to 250 V mains has an armature resistance (including brushes) of 0.12 Ω and the resistances of the field circuit is 100 Ω. Find the ratio of speed as a generator to the speed as a motor, the line current in each case being 80 A.

Solution. Shunt machine as a generator and as a motor is shown here.

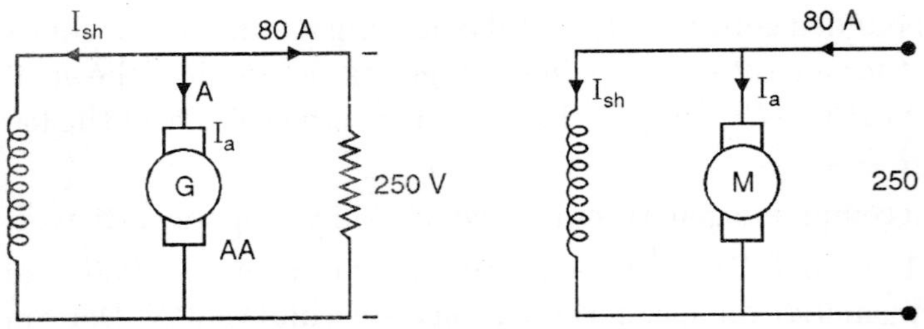

Fig. E 6.4

Since the terminal voltage is same and line current is same, the induced emf and back emf will be different.

For generator using equations (6.7) and (6.8)

$$I_a = I_L + I_{sh} = 80 + \frac{250}{100} = 82.5 \text{ A}$$

$$E = V + I_a R_a = 250 + 82.5 \times 0.12 = 259.9 \text{ volts}$$

For motor using equation (6.10) and (6.11)

$$I_a = I_L - I_{sh} = 80 - 2.5 = 77.5$$

$$E_b = V - I_a R_a = 250 - 77.5 \times 0.12 = 240.7 \text{ volts}$$

Since the emfs are directly proportional to speed if all other parameters are same, hence

$$\frac{\text{Generator speed}}{\text{Motor speed}} = \frac{259.9}{240.7} = 1.08 \quad \textbf{Ans.}$$

Example 6.5. A shunt generator has an induced voltage on open circuit of 127 V. When the machine is on load the terminal voltage is 120 V. Find the load current if the field circuit resistance be 15 Ω and the armature resistance 0.02 Ω. Ignore armature reaction.

Solution. Referring to Fig. E 6.5 and equations (6.7) and (6.8) we have

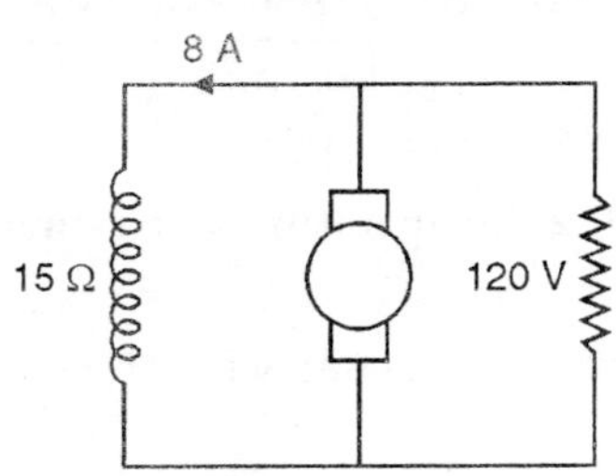

Fig. E 6.5

$$I_{sh} = \frac{120}{15} = 8 \text{ A}$$

$$I_a = I_L + I_{sh} = (I_L + 8)$$

$$E = V + I_a R_a$$

$$127 = 120 + (I_L + 8) \times 0.02$$

or $$350 = I_L + 8 \quad \text{or} \quad I_L = 342 \text{ Amp} \quad \textbf{Ans.}$$

6.7 DC MOTORS

Assuming the losses to be negligibly small the power developed by the dc motor is given by

$$P = E_b I_a$$

Now $$E_b = \frac{\phi ZNP}{60 A}$$

and $$P = \frac{2\pi NT}{60} \text{ where } T \text{ is the torque in Nm}$$

Hence $$\frac{2\pi NT}{60} = \frac{\phi ZNP}{60 A} . I_a$$

or $$T = \frac{\phi ZP}{2\pi A} . I_a = K\phi I_a \qquad ...(6.12)$$

where $$K = \frac{ZP}{2\pi A} \quad \text{constant for a dc machine}$$

Hence torque in a dc machine is proportional to the product of flux per pole in the air gap and the armature current

Also we know that $$E_b = \frac{\phi ZNP}{60 A} = K\phi N \qquad ...(6.13)$$

Assuming the armature drop to be negligibly small *i.e.* $V \simeq E_b$, the speed of a dc motor is inversely proportional to the flux per pole. We now study the torque-speed characteristics of d.c. series and shunt motors.

6.7.1 DC shunt motor

From equation (6.11)

$$V = E_b + I_a R_a$$

and from equation (6.13) we have

$$V = K\phi N + I_a R_a$$

or $$N = \frac{V - I_a R_a}{K\phi} \qquad ...(6.14)$$

Now from equation (6.12) $T = K\phi I_a$,

substituting for I_a in equation (6.14) we have

$$N = \frac{V}{K\phi} - \frac{R_a}{K\phi} . \frac{T}{K\phi}$$

$$= \frac{V}{K\phi} - \frac{R_a}{K^2\phi^2} T \qquad ...(6.15)$$

Under no load condition *i.e.* when load torque is zero

$$T = 0 \quad \text{hence} \quad N = \frac{V}{K\phi}$$

and as the load torque increases the speed decreases. Equation (6.15) can be rewritten as

$$y = c - mx$$

where $m = \frac{R_a}{K^2\phi^2}$ and indicates the slope of the speed torque characteristic

If a resistance R is inserted in series with the armature circuit of the shunt motor for controlling the speed torque characteristic, the equations are

$$V = E_b + I_a(R_a + R)$$

and $$N = \frac{V}{K\phi} - \frac{T\,(R_a + R)}{K^2\phi^2} \quad ...(6.16)$$

The nature of the characteristic is identical to the one given in equation (6.15) except the slope is more negative and of course the no load speed is the same. These characteristics have been shown in Fig. 6.10.

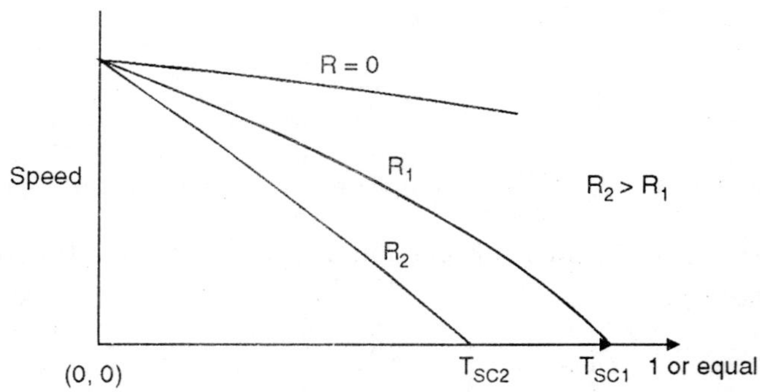

Fig. 6.10. T-speed characteristic of dc shunt motor.

In fact for a dc shunt motor without any control (neither armature nor field) the drop in speed from no load to full load is so small that the shunt motor can be considered as a constant speed motor. In Fig. 6.10 T_{sc1} and T_{sc2} indicate the stalling torques of the motor for resistances R_1 and R_2 respectively and the corresponding armature currents will be

$$I_{sc1} = \frac{V}{R_a + R_1} \quad I_{sc2} = \frac{V}{R_a + R_2} \quad ...(6.17)$$

Since dc shunt machine is a constant flux machine the relation between speed and I_a and T and I_a are shown in Fig. 6.11, using relation (6.14) and (6.12) respectively

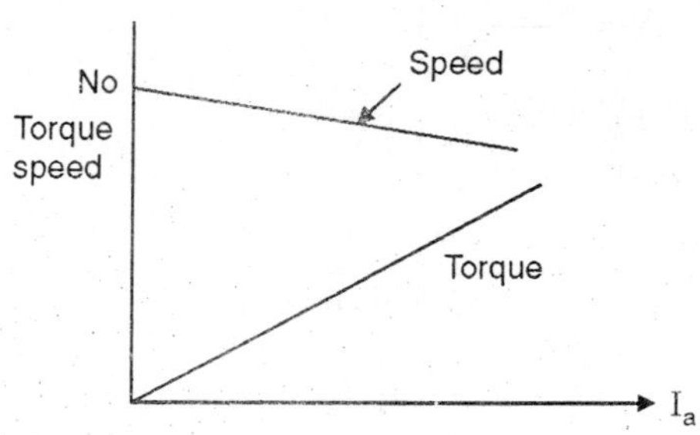

Fig. 6.11. Speed, I_a and Torque/I_a characteristic of dc shunt motor.

The speed vs armature current characteristic can be obtained by connecting a rheostat in series with armature circuit and keeping the load constant. This is known as armature control of speed. The other method is field control, by inserting a resistance in the field circuit. In the armature control, speeds less than the no load speed are obtained whereas in field control (Fig. 6.12) the speeds more than the normal speed are obtained.

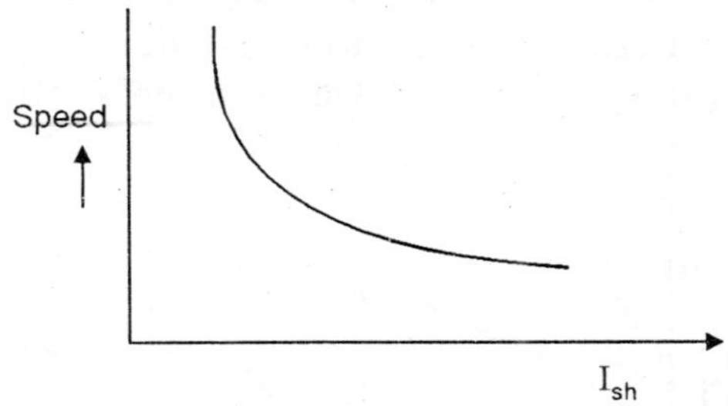

Fig. 6.12. Field control of speed for shunt motor.

6.7.2 DC series motor

In case of dc series motors, under unsaturated conditions, the field flux is proportional to the armature current. The equation describing the performance under such conditions is given as

$$V = E_b + I_aR_a = K\phi N + I_aR_a$$

and $$T = K_1\phi I_a = K_1I_a^2 \quad \text{or} \quad I_a = \sqrt{\frac{T}{K_1}}$$

Therefore, $$V = K\phi N + I_aR_a = KI_aN + I_aR_a$$

$$= I_a(KN + R_a) = \sqrt{\frac{T}{K_1}}(KN + R_a)$$

or $$N = \frac{V}{K}\sqrt{\frac{K_1}{T}} - \frac{R_a}{K}$$

$$= \frac{A}{\sqrt{T}} - B \qquad \text{...(6.18)}$$

where A and B are constants. The characteristic shown in Fig. 6.13 is a hyperbola. From the characteristic it is clear that when a d.c. series motor is started on no load ($T = 0$) the speed is infinite *i.e.* very high speed and the motor may fly into pieces because of mechanical stresses. Therefore a dc series motor is never started on no load.

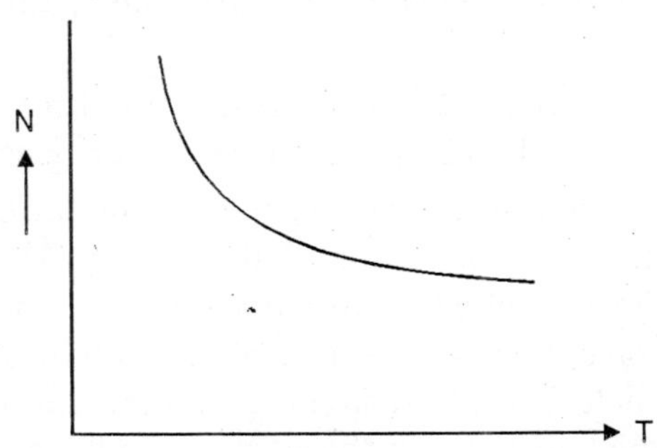

Fig. 6.13. Torque-speed characteristic of dc series motor.

However, if the load current is large *i.e.* the core is saturated

$$V = K\phi N + I_a R_a$$

and
$$T = K\phi I_a$$

Therefore
$$N = A_1 - B_1 T$$

The machine will behave as a shunt machine. The series machine characteristics for speed vs I_a and torque vs I_a are shown in Fig. 6.14. Initially, under unsaturated condition when I_a is small, $T \alpha I_a^2$ and after large value of I_a when the machine gets saturated $T \alpha I_a$

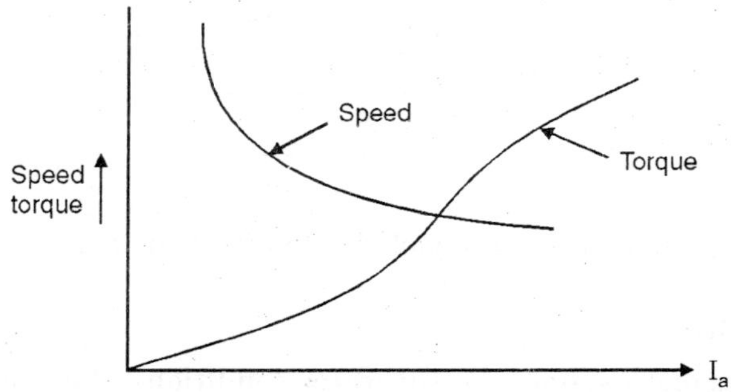

Fig. 6.14. Speed $V_s I_a$ and torque $V_s I_a$ characteristics for series motor.

The speed torque curve of a dc series motor is very suitable for electric traction and cranes. The starting torque is very high ($T \alpha I^2$). Therefore, heavy masses can be accelerated very fast. The decrease in speed with increase in torque is a stable operating condition for this motor. When the motor has to go up hill, the torque required increases, hence the speed automatically reduces.

6.8 LOSSES IN DC MACHINES

Mainly following losses take place in a dc machine :

1. **Copper loss.** These are the I^2R loss and take place in armature conductors (I_a^2Ra), series field winding (I^2R_{se}) and shunt field winding $\frac{V^2}{R_{sh}}$ or $I_{sh}^2R_{sh}$.
2. **Rotational losses.** In a dc machine, the core loss is a true rotational loss. Except in the pole faces, the flux in the stator remains constant in steady state operation. The flux in the armature core is constantly alternating as the machine rotates and at zero speed the core loss is zero. The rotational loss then consist of the sum of the friction windage and core losses. In a generator or constant speed motor these are considered part of the constant losses.
3. **Brush drop loss.** Brush contact loss occurs due to voltage drop across the brush and is equal to $2V_dI_a$ where V_d is the voltage drop across one brush.
4. **Stray load loss.** We calculate armature copper loss on the basis of dc resistance whereas the current in the armature winding is alternating. An incremental loss should be included to account for skin effect, eddy currents in the conductors and core loss due to leakage flux in the iron surrounding the armature conductors. The hysteresis and eddy current losses in the pole faces are also somewhat increased at high armature currents. There is no way by which to measure these "stray load" losses. They are often considered negligible in small dc machines. For large machines stray load loss may be taken as 1 per cent of the output power.

6.9 EFFICIENCY OF DC MACHINES

Since it is difficult to measure input for the generator and slightly difficult to measure output at the shaft of a motor, the efficiency of generator and motor is defined as follows :

$$\eta \text{ of a generator} = \frac{\text{Output power}}{\text{Output} + \text{Losses}}$$

and

$$\eta \text{ of a motor} = \frac{\text{Input} - \text{Losses}}{\text{Input power}}$$

The variation of efficiency as a function of load is shown in Fig. 6.15. It is seen that initially the efficiency increases as the load increases and it reaches maximum value around full load and again starts declining.

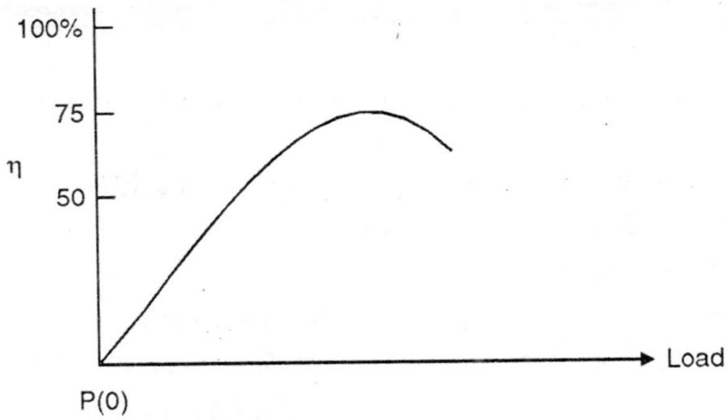

Fig. 6.15. Efficiency curve of a d.c. machine.

Example 6.6. Find the useful flux per pole on no load of a 250 V, 6-pole shunt motor having wave connected armature winding with 110 turns. The armature resistance including brush is 0.2 Ω. The armature current is 13.3 A at no load speed of 908 rpm.

Solution. Total no. of conductors 110 × 2 = 220 = Z. Since the armature current under no load is 13.3 A, the load current is zero and hence 13.3 A current flows in the shunt field winding.

The back emf is

$$250 - 13.3 \times 0.2 = 247.36$$

Here $P = 6, Z = 220, N = 908, A = 2$

Hence
$$247.36 = \frac{\phi ZNP}{60\,A} = \frac{\phi \times 220 \times 908 \times 6}{60 \times 2}$$

Hence
$$\phi = 24.8 \text{ mWb} \quad \textbf{Ans.}$$

Example 6.7. A direct current shunt motor develops 10 hp at 600 rpm when drawing a line current of 18A at 500 volts. Find the efficiency at this load and the useful torque.

Solution. $\eta = \frac{\text{Output}}{\text{Input}} = \frac{10 \times 735.5}{18 \times 500} = 0.817 \quad \text{or} \quad 81.7\%$

Now output $= 2\pi NT = 735.5$

where N is in rps and T in N-m

Hence
$$T = \frac{735.5}{2\pi \times 10} = 117 \text{ N-m.}$$

Example 6.8. A series motor of resistance 1 Ω between terminals runs at 800 rpm at 200 V with a current of 15 A. Determine the speed at which it will run when connected in series with a 5 Ω resistance and taking the same current at the same supply voltage.

Solution. The back emf 200 – 15 × 1 = 185 V

In the second case the back emf is = 200 – 15 × 6 = 110 volts

Since the speed is directly proportional to back emf if ϕ is same. Hence the speed in the second case is

$$\frac{110}{185} \times 800 = 476 \text{ rpm} \quad \textbf{Ans.}$$

PROBLEMS

6.1. Describe the construction of a dc machine and explain what materials are used for each component of the machine.

6.2. Explain the functions of brushes and commutator in a dc machine.

6.3. Why are some parts of a dc machine laminated ?

6.4. State the purpose of the following :

(*a*) Armature winding (*b*) Armature core

(*c*) Field winding (*d*) Pole shoes

(*e*) Pole core (*f*) Commutator.

6.5. What things distinguish motor action from generator action in a dc machine ?

6.6. Derive emf equation for a d.c. machine.

6.7. Derive an expression for the torque developed by a dc motor.

6.8. Classify the dc machine based on excitation systems.

6.9. Explain with a neat diagrams how you obtain magnetisation characteristic of a dc machine and explain the significance of the characteristic.

6.10. What is meant by self excitation of a dc generator ? Discuss requirements for self-excitation.

6.11. Draw and discuss external (load) characteristics of (*i*) shunt (*ii*) series and (*iii*) compound dc generators.

6.12. Derive mathematically torque-speed characteristic of various dc motors. Draw the curves and discuss them.

6.13. Discuss various losses in a dc machine.

6.14. A 4-pole generator has a flux of 40 mWb per pole and a lap connected armature with 740 conductors. Determine the emf generated on open circuit at 1000 rpm. [**Ans.** 494 V.]

6.15. A 4-pole shunt generator with lap connected armature having field and armature resistances of 50 Ω and 0.1 Ω respectively supplies sixty 100 V 40 watt lamps. Calculate the total armature current, the current per armature path and the generated emf. Allow a contact drop of 2 volts.

[**Ans.** 26 A, 6.5 A 104.6 V.]

6.16. A 4-pole, 500 V, shunt motor has 720 wave connected conductors on it's armature. The full load armature current is 60 A and the flux per pole 0.03 Wb. The armature resistance is 0.2 Ω and the contact drop is 1 V per brush. Calculate the full load speed of the motor. [**Ans.** 675 rpm.]

CHAPTER

7

Three-Phase Synchronous Motors

7.1 INTRODUCTION

Synchronous machines are called "synchronous" as their speed is directly related to the line frequency.

$$n_s = \frac{120 f}{p} \text{ rev/m}$$

where n_s is the synchronous speed in rpm, p the number of magnetic poles designed into each machine and f is the frequency in Hz of the supply voltage. Thus when two or more synchronous machines are connected to the same supply line, they will all run in synchronism because they are all operating at the same supply frequency. Of course, if one machine has 2 poles (Turbo-alternators) and another has 12 poles, the 12-pole machine will run at precisely $\frac{1}{6}$th the speed of the 2-pole machine.

7.2 CONSTRUCTION

Rotating machines have an outside (stationary) part called the stator. The inner (rotating) part is called the rotor. The rotor is centred within the stator so that the rotor axis is concentric with that of the stator. The space between the outside of the rotor and inside of the stator is called the air gap.

The rotor is mounted on a stiff rod (usually steel) called a shaft. The shaft is supported in bearings so that the rotor is free to rotate. The shaft extends through one or both of the bearings to provide a means to connect the machine to the mechanical system. It is to be noted that the rotor is solidly fastened to the shaft so that the rotor and the shaft rotate at the same speed. Therefore, the terms 'rotor speed', 'shaft speed' or machine speed all mean the same thing and are used interchangeably.

The rotor and stator each have three basic parts, the core, the windings and the insulation. Thus, it is proper to speak of "the rotor core", 'the stator windings' 'the rotor insulation' and so on. The purpose of the rotor and stator cores is to conduct the magnetic field through the coils of the windings. The cores are almost always made of iron or steel.

The stator winding is known as armature winding as it carries the load current whereas the rotor winding is known as field winding as this winding provides exciting current or magnetising current. Since the armature winding carries load current, it has to handle all of the power being converted or transformed by the device, however, the magnetising power requirement is relatively small. The steady state power input to a field winding is only about $\frac{1}{2}$ to 2 per cent of the rated power of the machine. The input power to dc field winding is all consumed as I^2R loss except during transient period which lasts for almost a second or two during which energy is being stored in magnetic field.

The winding conductors are either of copper or aluminium and that could be in the form of coils of wire or heavy bars depending upon the current carrying capacity required. Also each winding consists of several coils or bars in series or in series-parallel combination depending upon the voltage and current requirements of the machine. The ends of certain of the windings are brought out to terminals to allow easy connection to the electrical system.

In some machines, the windings of either the rotor or the stator may be placed around projecting magnetic pole pieces, called salient poles (salient means prominent). When a core has salient poles the coils of the winding are wound around the waists of the pole pieces. These narrower parts of the salient poles are called the pole cores. The shaped ends of the poles are called the pole shoes. Their purpose is to provide the correct flux density distribution in the air gap. Salient poles are used on the stator cores of dc machines and on the rotor cores of many synchronous machines and in both cases they carry dc field windings. The mechanical weakness and air resistance of salient poles prohibit their use on rotors of large high speed synchronous generator designed to be driven by steam or gas turbines. Turboalternators as these are usually called always have cylindrical rotors with the field windings embedded in slots cut into the rotor surface. A machine having both its rotor and stator windings in slots is known as round rotor machine or non-salient pole machine.

The insulation system consists of (*i*) the conductor or wire insulation *e.g.* superenamelled wires or conductors ; (*ii*) the coil insulation usually some kind of tape or several layers of tapes depending upon operating voltage ; and (*iii*) the slot lines when the windings consist of coils that are located in slots, the coils are held in place in the slots by slot wedges. The insulation system thus prevents short circuits between turns of a given winding coil and insulate the winding from the iron core which is always grounded for safety's sake. The insulation system must protect the machine against damage due to overvoltages and overcurrents that may occur during the operation of the machine.

7.2.1 Magnetic poles in machines

The torque developed by electrical machines is the result of attraction and repulsion between magnetic poles on the rotor and stator. We know that magnetic poles of like polarity repel each other whereas magnetic poles of unlike polarity attract each other. In machines, the windings are so designed that they develop magnetic poles on the inside surface of the stator and the outside surface of the rotor. In other words, these poles appear on the inner and outer surfaces of the air gap. Repulsion and attraction between the rotor poles and those of the stator produce a torque on the rotor and a reaction torque against the stator.

To produce a unidirectional and continuous torque, the no. of rotor poles must equal the number of stator poles. The no. of poles on either the rotor or the stator must be even. If a machine has four stator poles, there must be four rotor poles and the machine is said to have four poles. If the machine is in the motor mode, the torque on the rotor causes it to rotate.

Since the rotor poles have constant polarity, their windings must be supplied with direct current. This current may be provided by an external dc generator or by a rectifier. In this case the leads from the field winding are connected to insulated rings mounted concentrically on the shaft. Stationary contacts called brushes ride on these slip rings to carry current to the rotating field windings from the dc supply. The brushes are made of a carbon compound to provide a good contact with low mechanical friction. An external d.c. generator used to provide field current is called an exciter.

7.2.2 Steady state operation

As mentioned earlier the synchronous motor as such is not self-starting. This is explained as follows : Suppose we feed dc supply to the field winding on the rotor through an exciter. As a result north and south poles are set up on the rotor. These north and south poles are stationary in space. If we feed three-phase supply to the stator winding, this gives rise to a rotating magnetic field of constant magnitude and rotating at synchronous speed. The north and south poles of the stator magnetic field are rotating at synchronous speed. Suppose at the instant of switching, the north pole of stator coincides with the south pole axis of the rotor. The tendency of the rotor pole will be to get attracted and start rotating along the stator magnetic field. But since the rotor has certain inertia and by the time the rotor tries to go along the stator magnetic field the south pole of the stator magnetic field coincides with the same south pole of the rotor and the rotor is repulsed. The variation is so fast that the rotor is not able to respond to the instantaneous variation of the stator magnetic field and the rotor thus is not able to start rotation and it is a non self-starting machine.

7.2.3 Power and Torque

Using the equivalent circuit in Fig. 7.1 of the synchronous motor we derive expressions for power and torque. Since we are considering the motor circuit, electric power is being fed into the motor and the mechanical power is available at the shaft of the motor.

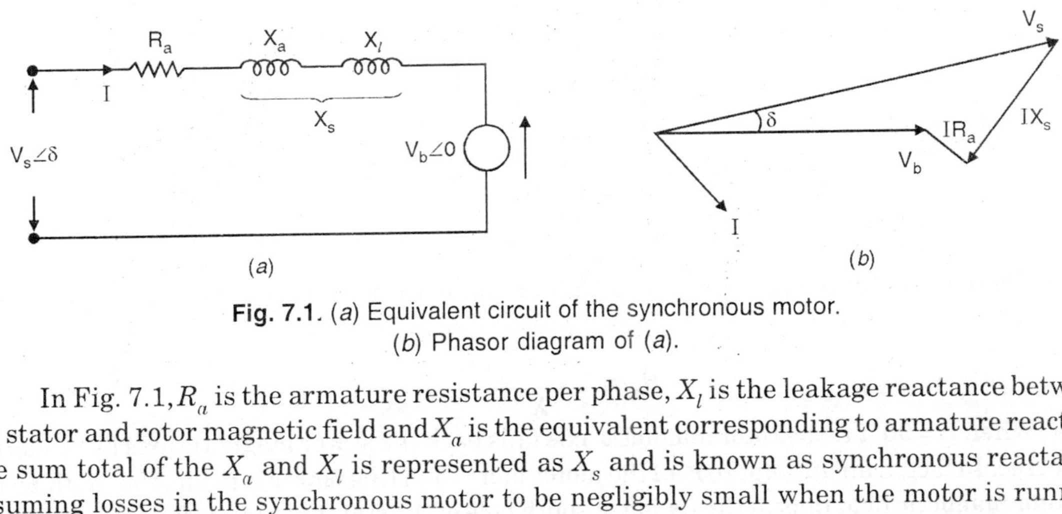

Fig. 7.1. (*a*) Equivalent circuit of the synchronous motor.
(*b*) Phasor diagram of (*a*).

In Fig. 7.1, R_a is the armature resistance per phase, X_l is the leakage reactance between the stator and rotor magnetic field and X_a is the equivalent corresponding to armature reaction. The sum total of the X_a and X_l is represented as X_s and is known as synchronous reactance. Assuming losses in the synchronous motor to be negligibly small when the motor is running under no load condition, the stator magnetic field axis coincides with the rotor magnetic field axis. However if some load is added to the shaft of the motor, under steady state operating condition, the rotor magnetic field axis falls back from the stator magnetic field axis by an angle δ

known as torque angle or load angle. Obviously, this angle will be decided by the load added to the shaft. In Fig. 7.1 (*b*), V_s is shown to be leading V_b by an angle δ which is the torque angle. Here V_b the counter emf is taken as reference and V_s is the supply per phase voltage leading V_b by an angle δ. In fact this angle is an indicator of power flow from supply terminals to motor terminals. Power will always flow from leading voltage towards lagging voltage. The current I

$$I = \frac{V_s \angle\delta - V_b \angle 0}{j X_s} = \frac{V_s \angle\delta - 90 - V_b \angle -90}{X_s} \qquad \text{...(7.1)}$$

neglecting resistance of the armature winding.

Now, Power $P = R_e\ [V_b I^*]$

$$= R_e \left[V_b \frac{V_s \angle 90 - \delta - V_b \angle 90}{X_s} \right]$$

$$= \left[\frac{V_b V_s \cos (90 - \delta)}{X_s} - V_b^2 \cos 90° \right]$$

$$= \frac{V_b V_s}{X_s} \sin \delta \qquad \text{...(7.2)}$$

$$Q = I_m \left[V_b \frac{V_s \angle 90 - \delta - V_b \angle 90}{X_s} \right]$$

$$= \frac{V_b V_s}{X_s} \cos \delta - \frac{V_b^2}{X_s} \qquad \text{...(7.2 A)}$$

A plot of equation 7.2 is shown in Fig. 7.2. Assuming $|V_s|$ and $|V_b|$ to remain constant, one important conclusion is drawn.

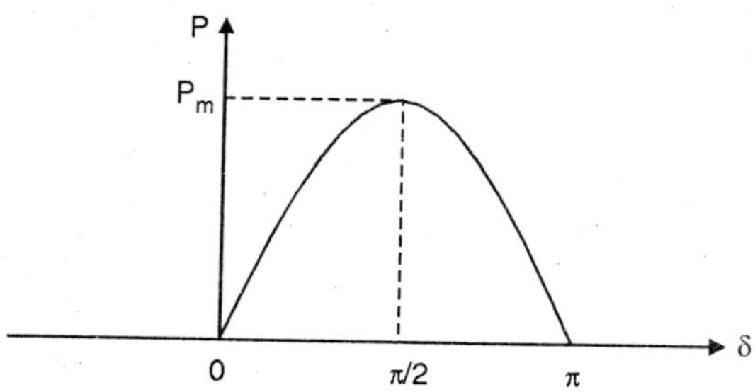

Fig. 7.2. Power angle curve for a motor connected to the supply system.

There is a definite upper limit to the power flow to the motor from the source and is equal to

$$P_m = \frac{|V_s||V_b|}{X} \qquad \text{...(7.3)}$$

When $\delta = 90°$ *i.e.* the rotor magnetic axis falls back the stator magnetic axis by 90 electrical degree. In fact, if an attempt to load the motor more than this power is made, the south pole of the rotor magnetic field falls in the region ot south pole of the stator magnetic field and there will be repulsion and the rotor will fall out of step from the stator magnetic field and the rotor comes to a stop. This is also known as pull out power of the motor

Now $$P = 2\pi NT$$

or $$T = \frac{P}{2\pi N}$$

Since synchronous motor, under steady state conditions either runs at synchronous speed or does not run at all. Hence $N = n_s$

and $$T = \frac{|V_b||V_s|}{X_s} \sin\delta \cdot \frac{1}{2\pi n_s} \quad ...(7.4)$$

where n_s is the synchronous speed in rps

From phasor diagram of Fig. 7.1 (*b*) we can write

$$V_s = V_b + I(R_a + jX_s)$$

i.e. the supply voltage equals phasor sum of the counter emf and the drop across the synchronous impedance. The drop is normally negligible as compared to $|V_s|$ or $|V_b|$ and hence $|V_s|$ and $|V_b|$ are practically equal to each other. We make use of the observation and study the effect of excitation on the armature current.

7.3 EFFECT OF EXCITATION ON ARMATURE CURRENT

We know that a synchronous motor is fed dc supply to the rotor and 3-phase ac to the stator. The rotor field winding gives rise to magnetic field in the air gap and induces voltage in the stator winding. However, the supply from the stator winding induces voltage in the stator conductor (counter emf) and the resultant of the supply voltage and the counter emf circulates current in the stator winding. These currents give rise to magnetic flux in the air gap which either demagnetises the flux due to dc rotor winding or cross-magnetises it depending upon whether the current in the stator is a leading current or lagging current respectively with respect to supply voltage. In fact, it is the net flux (flux due to dc field and due to armature current) which decides the counter emf in the stator winding.

Now the counter emf is proportional to the speed and to the field flux, and since the speed of the motor is constant at synchronous speed, the field flux likewise be substantially constant within the normal limits of operation. It follows then that if the field excitation is increased, the flux due to field current increases in the air gap. In order for the net flux in the air gap to remain constant, the current change in the stator conductor be such as to nullify the change (increase) in flux. This means the stator must draw leading component of current as the leading component of current in armature demagnetises the flux in the air gap *i.e.* with this increase in excitation if the armature current earlier was lagging with some p.f., now it will be less lagging and hence higher p.f. Similarly, if the field excitation is decreased which decreases flux in the air gap, the armature will draw additional component of lagging current to strengthen the field in the air gap. For any set of operating conditions, there will be some value of field excitation which will cause the p.f. to be unity *i.e.* the current to be in phase with the terminal voltage. When this condition exists while the motor is carrying its rated load, the motor is said to have normal excitation. Assuming losses to be negligible, for certain power input or output, it can be demonstrated that by varying excitation of the field of the motor, the current will vary from lagging p.f. to leading p.f. as shown in Fig. 7.3 and this is known as 'V' curve of the synchronous motor.

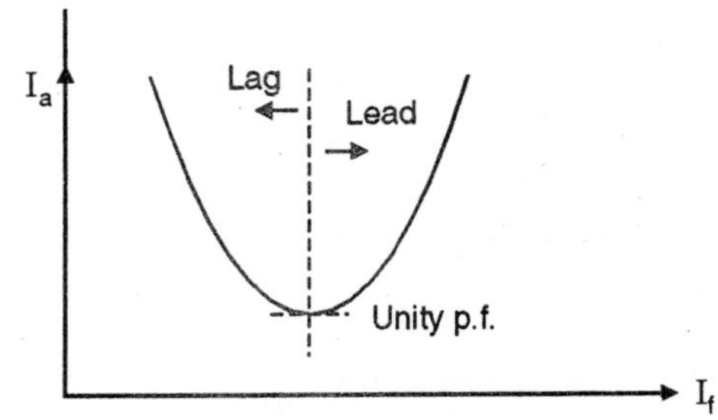

Fig. 7.3. '*V*' curve for the synchronous motor.

When synchronous motor is used exclusively for drawing leading current from the system (or supplying lagging current to the system) to compensate for the other inductive loads, the motor is known as synchronous condenser. It can, therefore, be seen that a synchronous motor is a very good neighbour. It can be used to correct a lagging p.f. and it helps keep line voltages constant and balanced. It helps keep line voltages constant because it can draw lagging and leading current by changing excitation suitably.

If the terminal voltages supplied to a synchronous motor are unbalanced, the air gap flux will vary as it passes the different phase groups. The damper winding (about this winding we will discuss a little later) will have according to Lenz's law currents flowing in them which will tend to make flux constant. This improves the voltage balance.

7.4 CHANGE OF LOAD

Suppose a synchronous motor is connected to the grid and is carrying certain load and the load angle or torque angle is δ_0. Now, let us add some more load to the shaft of the motor. Suppose corresponding to this load the load angle under steady state operating condition is δ_1. The motor does not come to this angle δ_1 instantly because of inertia of the rotor. In fact while falling further back from the stator magnetic axis from δ_0 to δ_1, it overshoots and keeps on oscillating around its new position δ_1 as shown in Fig. 7.4.

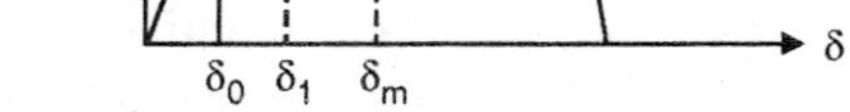

Fig. 7.4. Increase of load.

When the rotor moves in its forward journey from *a* to *c* the speed is less than the synchronous speed. At *a* and *c* the speed equals the synchronous speed. Between *a* to *b* the speed is decreasing and is less than synchronous speed and is minimum at *b*. Between *b* and *c* the speed increases but is still less than synchronous speed. During its return journey from *c* to *a*, the speed is more than synchronous speed. Between *c* and *b* the speed increases and is above the synchronous speed. At

b, during return journey the speed is maximum and more than synchronous speed. Between b and a, the speed decreases but is more than the synchronous speed. At a the speed is again the synchronous speed. Finally, of course, the motor will settle at b corresponding to δ_1 after the oscillations are damped out.

7.5 STARTING OF SYNCHRONOUS MOTOR

We know that a synchronous motor is not-self starting. Some additional devices should be used to start the motor. Basically there are two methods :

1. Induction motor starting (Damper winding)
2. Auxiliary motor starting.

7.5.1 Damper winding starting

Most of the synchronous motors have salient pole structure. As mentioned earlier a winding consisting of heavy copper bars is installed in slots in the pole faces. These bars are all shorted together at both ends of the rotor. When 3-phase supply is fed to the stator, a rotating magnetic field is produced in the air gap which induces currents in the bars which further produces flux. The interaction of the fluxes produces torque in the direction of field rotation. In other words, the motor is started as an induction motor, the bars in the pole face slots forming a sort of squirrel cage rotor. Induction motor action will bring the motor to nearly synchronous speed. At synchronous speed there is no relative motion between the poles of the air gap field and the pole face bars. No current is induced in the bars at synchronous speed, and no torque would be produced by them. However, the maximum speed developed on induction motor action is very close to synchronous speed and the rotor falls into step when the dc field current is switched on.

The field winding terminals are usually shorted through a resistor during starting until such time that the field is excited. This has two advantages. First it protects the slip ring insulation from the high ac voltage induced in the field during starting. Second the current circulating in the field winding would provide a small additional accelerating torque.

The heavy bar winding in the pole faces serve an additional purpose. As mentioned in the previous article that when there is sudden change of load the load angle δ changes to adjust to the new load requirement. The magnetic attraction between the air gap flux and the rotor mmf poles has a springlike quality and the rotor has considerable inertia. As a result there is an oscillatory motion of the rotor superimposed over the normal synchronous rotation of the shaft. This relative motion is called hunting. The relative speed between the air gap flux and the bars induces currents in these bars. The field set up by these currents produces a torque that opposes any change in δ and the angular oscillation is damped out fast. For this reason the pole face bars are called damper bars and the winding consisting of these heavy bars and their end connections is called the damper winding or by their French name, the amortisseur winding (Roughly translated this means "killer winding"). Because of high degree of hunting present in these machines, the damper windings are installed in nearly all synchronous machines, generators as well as motors having salient poles. Cylindrical rotor cores cannot be laminated and still maintain the required mechanical strength. Eddy currents, therefore, can circulate freely in the pole faces. Since the flux through the pole faces remains constant under steady state condition, these eddy currents are induced only during transients (*e.g.* during change of load). The interaction of the eddy currents with the air gap flux provides the necessary damping action in cylindrical rotor synchronous machines.

7.5.2 Auxiliary motor starting

The auxiliary motor may be a dc shunt motor or an induction motor having the same no. of poles as the synchronous motor or two poles less as compared to synchronous motor. The job of the auxiliary motor is to bring the synchronous motor to synchronous speed or near synchronous speed. The auxiliary motor is mechanically coupled to the synchronous motor. No load is put on the synchronous motor during starting. Therefore, the auxiliary motor has to overcome only the inertia of the synchronous motor and may have its rating much smaller than that of the synchronous motor being started.

When the speed is near synchronous speed, the 3-phase supply is switched on to the armature and dc supply to the field circuit of synchronous motor. The synchronous motor pulls into step, its speed rises to synchronous speed and it continues to run at this speed.

If the induction motor has two poles less than synchronous motor, the induction motor speed is higher than the synchronous speed of the synchronous motor. Once the higher speed is attained, the supply to induction motor is switched off and 3-phase supply to synchronous motor is switched on. When the speed of the synchronous motor almost equals the synchronous speed, it's field winding is energized. The synchronous motor pulls into step and starts running at synchronous speed. However, in modern days this method of starting is hardly used. It is mainly the damper winding which is utilised for starting the synchronous motor.

7.6 SPEED TORQUE CHARACTERISTIC

A synchronous motor runs at a constant speed, known as synchronous speed and depends upon the no. of poles and frequency of supply. It is usually connected to a large system also known as an infinite bus which has constant voltage and frequency. The torque equation (7.4) is rewritten here

$$T = \frac{|V_b||V_s|}{X_s} \sin \delta . \frac{1}{2\pi n_s}$$

The speed torque characteristic of synchronous motor is shown in Fig. 7.5.

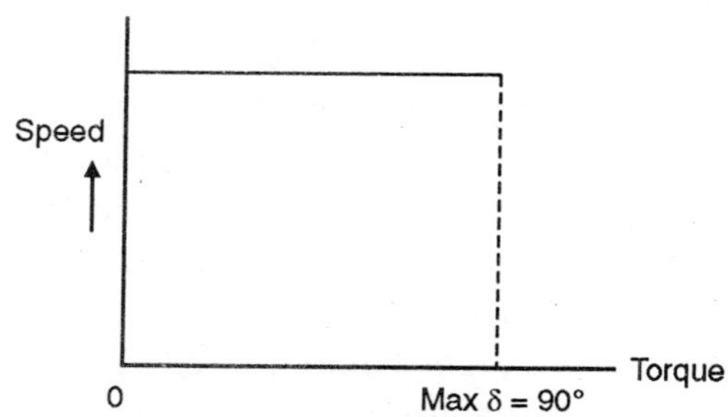

Fig. 7.5. Speed torque characteristic of a synchronous motor.

The speed remains constant from no load to the maximum load (pull out load) which occurs at $\delta = 90°$. The full load torque is normally 50% ($\delta = 30°$) of the pull out torque ($\delta = 90°$).

As explain earlier when the load on the motor is more than the pull out load ($\delta = 90°$) the north of the rotor comes against the north pole of the stator magnetic field and they repel each other which finally results into the motor losing synchronism with the supply system and the motor comes to a stop.

7.7 SYNCHRONOUS CONDENSER

As mentioned earlier when a synchronous motor is overexcited it acts as a synchronous condenser. Of course there is an upper limit for excitation due to heating of the field winding. The special feature of the motor when used as synchronous condenser is that it is run under no load condition, thereby $\delta = 0$ and the induced emf due to excitation is greater than the supply voltage *i.e.*, reactive power Q can be continuously and simply controlled by controlling induced emf by varying the excitation. However, if the synchronous motor is used over the complete range of excitation as shown in Fig. 7.3, the machine is then known as synchronous phase modifier.

A synchronous phase modifier has a smaller shaft (as no mechanical load is to be connected) and bearing and higher speeds as compared to a synchronous motor used for mechanical loads. A synchronous phase modifier has a higher overall efficiency as compared to a synchronous motor.

A great advantage of the synchronous capacitor is its flexibility for use for all load conditions because it supplies reactive power when overexcited *i.e.* during peak load conditions when the system voltage is likely to go down below acceptable limits if p.f. correcting measures are not adopted and it consumes reactive power when underexcited during light load conditions. During light load condition the system voltage may go up, more than the accepted value due to Ferranti effect. Here the phase modifier may be under excited thereby absorbing the reactive power and maintaining the system voltage within permissible limits. There is smooth variation of reactive power by synchronous capacitor as compared with step by step variation by the static capacitors.

Synchronous motors can be overloaded for short periods whereas static capacitors cannot. For large outputs the synchronous capacitors are much better than the static capacitors from economic point of view because otherwise a combination of shunt capacitors and reactors is required which becomes costlier and also the control is not smooth as is achieved with synchronous capacitors.

The main disadvantage of the synchronous capacitor is the possibility of its falling out of step which will thus produce a large sudden change in voltage. Also, these machines add to the short-circuit capacity of the system during fault condition.

An important statement is made about the synchronous machines.

A synchronous machine (irrespective of whether it is a generator or a motor) delivers lagging reactive power to the system when overexcited and it draws lagging reactive power from the system when underexcited.

To understand this statement we take up a very simple illustration shown in Fig. 7.6.

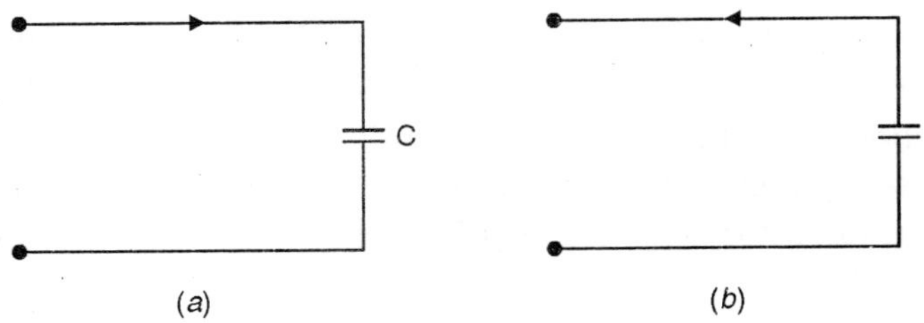

Fig. 7.6. Capacitor connected to supply.

In Fig. (*a*) the capacitor is taking leading current and that is what actually is, as it is acting a sort of load. However, if we consider capacitor a source of reactive power, Fig. (*b*) is appropriate and Fig. (*b*) means the capacitor is supplying lagging reactive power to the source. Therefore if a load takes leading current it is equivalent to saying that it is supplying lagging current.

Similarly synchronous motor is a load and when it is overexcited, it operates at leading p.f. *i.e.* it takes leading reactive power from the source which is equivalent to saying that it is supplying lagging reactive power to the system. For a generator which generates and feeds to the system, operates at lagging p.f. and hence it delivers lagging reactive power to the system when overexcited. Similarly, it can be justified that if the synchronous machine is underexcited it takes lagging reactive power from the system.

7.8 APPLICATIONS

Synchronous motors have many advantages when a constant speed is desirable. Typical applications are to drive pumps, fans, and dc generators. They can be operated at unity p.f., thus minimising the line current. In sizes above 2000 hp synchronous motor's cost less than induction motors having static capacitors for p.f. correction. A major advantage of synchronous motors is that it can be used as a phase modifier and hence it can take care of the overloading and no load voltage variations by overexciting and underexciting its field winding respectively. During overexcitation it draws leading current and thus corrects the overall p.f. of a plant having other equipment drawing lagging current.

Example 7.1. A 2300 V, 3-phase synchronous motor driving a pump is provided with a line ammeter and a field rheostat. When the rheostat is adjusted so that the ac line current is a minimum, the ammeter reads 8.8 A. What approximate power is being delivered to the pump? How should the rheostat be adjusted so that the motor is operating at 0.8 pf lead? How many KVArs is the motor supplying to the system at 0.8 p.f. lead?

Solution. From the 'V' curve we know that for a given power, the current is minimum at unity p.f. Hence power drawn from the line is

$$P = \sqrt{3}\, V_L I_L = \sqrt{3}\ .\ 2300\ .\ 8.8 = 35 \text{ kW}$$

In order to obtain 0.8 lead p.f., the excitation current should be increased *i.e.* the field rheostat resistance should be decreased such that the armature current I is given as

$$\sqrt{3}\ V_L\, I \cos\phi = 35000$$

or

$$I = \frac{35000}{\sqrt{3} \times 2300 \times 0.8} = 10.98 \text{ A} \simeq 11 \text{ Amp.}$$

The kVAr supplied by the motor

$$\sqrt{3}\,VI \sin\phi = \sqrt{3} \times 2300 \times 11 \times 0.8$$
$$= 26.29 \text{ kVAr.} \quad \textbf{Ans.}$$

Example 7.2. A manufacturing plant presents an electrical load to the power system, of 5000 kW at 0.8 p.f. lag. It has been dicided to replace a 500 hp induction motor that drives a pump. The motor operates at an efficiency of 96% and a p.f. of 0.9 lag. If a synchronous motor is purchased as a replacement, which is capable of operating at 0.8 p.f. lead. What will be the new plant p.f. ? Assume the synchronous motor to have the same efficiency. What per cent decrease in line current will result from the improved p.f.

Solution.

Original power (complex) $= 5000 + j\,5000 \tan \cos^{-1} 0.8$

$$= 5000 + j\,5000 \times \frac{6}{8} = 5000 + j\,3750 \text{ kW}$$

The power input to induction motor

$$\frac{500 \times 735}{0.96} = 382.8 \text{ kW}$$

Hence complex power of induction motor

$$382.8 + j\,382.8 \tan \cos^{-1} 0.9$$
$$= 382.8 + j\,382.8\,\frac{0.436}{0.9}$$
$$= 382.8 + j\,185.4 \text{ kVA}$$

Complex power of the synchronous motor

$$382.8 - j\,382.8 \times \frac{0.6}{0.8}$$
$$= 382.8 - j\,287.1 \text{ kVA}$$

Now plant requirement

$$5000 + j\,3750 - 382.8 - j\,185.4 + 382.8 - j\,287.1$$
$$= 5000 + j\,3277.5 \text{ kVA} = 5978 \text{ kVA}$$

Now p.f. $\quad \cos \tan^{-1} \dfrac{3277.5}{5000} = 0.836$

Original current $= \dfrac{5000 \times 1000}{\sqrt{3}\,V_L \times 0.8} = \dfrac{6250}{\sqrt{3}\,V_L}$

New current $= \dfrac{5978 \times 1000}{\sqrt{3}\,V_L}$

Hence percentage reduction $\dfrac{6250 - 5978}{6250} \times 100$

$= 4.3\%.$ **Ans.**

Example 7.3. In the plant of example 7.2, what should be the kVAr rating of a synchronous condenser to correct the original p.f. to 0.9.

Solution. The original complex power of the plant is $5000 + j\,3750$

At 0.9 p.f. the plant Q would be

$$5000 \tan \cos^{-1} 0.9 = 5000 \times \frac{0.436}{0.9}$$

$$= 2422 \text{ kVAr}$$

The difference must be supplied by the synchronous condenser

$$3750 - 2422 = 1328 \text{ kVAr.}$$ **Ans.**

Example 7.4. A 2300 V, 3-phase star connected synchronous motor has synchronous reactance of 10 ohm per phase. When motor delivers 255 hp the efficiency is 90%. The power angle is 20°. Calculate (*a*) E per phase (*b*) I (*c*) p.f. Neglect resistance.

Solution. Neglecting resistance the power expression is

$$P = \frac{3EV}{X_s} \sin \delta$$

Now power input to the motor is $\dfrac{255 \times 735.5}{0.9}$

Hence $$3E \times \frac{2300}{\sqrt{3}} \cdot \frac{1}{10} \sin 20^\circ = \frac{255 \times 735.5}{0.9}$$

$$136 \text{ E} = 208391$$

or $$E = 1532 \text{ volts per phase}$$

$$\text{Current } I = \frac{1328 \angle 0 - 1532 \angle -20}{j\,10} = \frac{1328 - 1439 + j\,524}{j\,10}$$

$$= \frac{-111 + j\,524}{j\,10} = j\,11.1 + 52.4$$

$$\phi = 11.9^\circ$$

Hence p.f. 0.98 lead. **Ans.**

7.9 STEPPER MOTOR

A stepper motor is a polyphase synchronous motor having salient stator poles. The name *stepper* derives from the most common application for these machines *i.e.* rotating a fixed angular step in response to each input pulse received by their controller when this type of motor is fed from an electronic drive, accurate position control and precise rotational speeds are obviously obtained.

Based on principle of operation, stepper motors may be divided into three categories. These are, permanent magnet, variable reluctance and hybrid type which utilize both principle for torque development. Fig. 7.7 shows the three rotor types and their respective torque/displacement characteristics.

7.9.1 Permanent magnet type

Consider the simplified two-phase stator with two-pole permanent magnet rotor shown in Fig. 7.8. The rotor is shown in a stable equilibrium position for the case when $i_1 > 0$ and $i_2 = 0$. Stepping action is produced by switching current i_1 and i_2, supplied from a bipolar source. Three possible sequences for producing clockwise rotation are given. The first energises only one phase

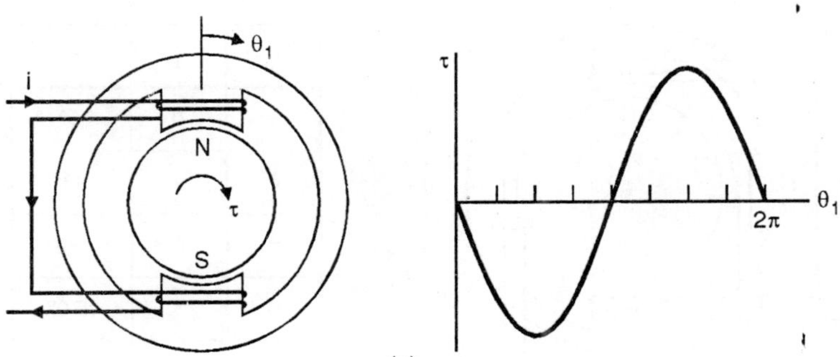

(a)

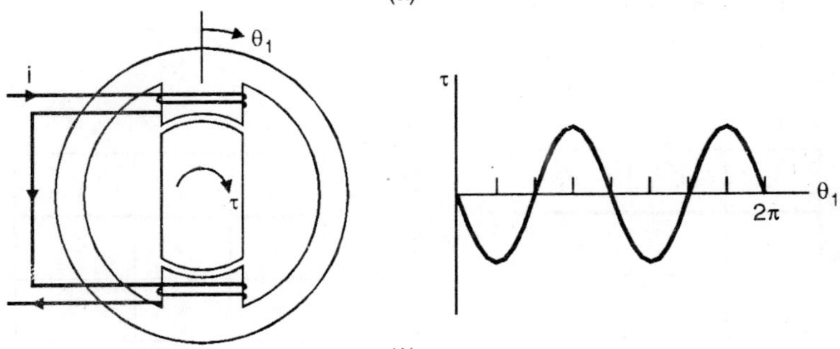

(b)

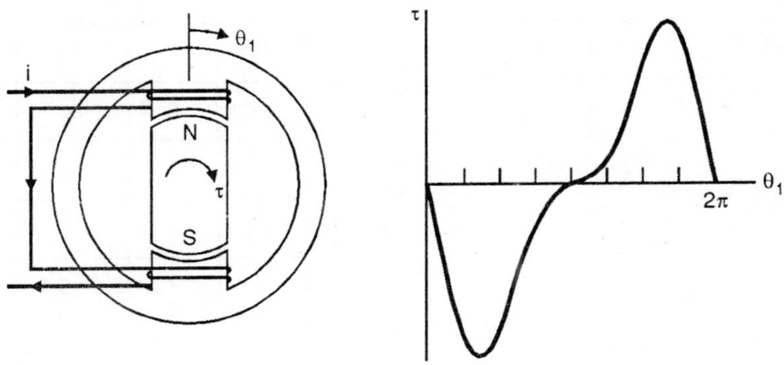

(c)

Fig. 7.7. Torque development in the three basic types of stepper motors. (*a*) Permanent-magnet rotor. (*b*) Variable reluctance rotor. (*c*) Hybrid rotor.

winding at a time. The resulting step size is one-half the pole pitch. The second method energises both windings simultaneously and would thus require larger power supply. The resulting steps are the same size, but the effective stator pole positions are located midway between the pole pieces. Advantages of this scheme are increased torque and damping. The final method, known as half-stepping, alternates between one and two phases. The resulting step size is one-fourth the pole.

Various techniques have been used to produce smaller step sizes. Perhaps the most obvious is to increase the number of poles or the number of phases or both.

Both permanent magnet and variable reluctance rotor can be used. However, a permanent magnet motor provides smooth response.

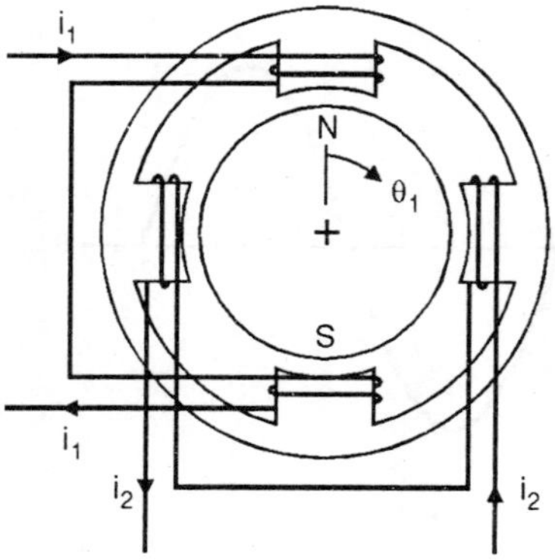

(a)

i_a	i_b	θ_1
+	0	0°
0	+	90°
–	0	180°
0	–	270°
+	0	0°

(b)

i_a	i_p	θ_1
+	+	45°
–	+	135°
–	–	225°
+	–	315°
+	+	45°

(c)

i_a	i_b	θ_1
+	0	0°
+	+	45°
0	+	90°
–	+	135°
–	0	180°
–	–	225°
0	–	270°
+	–	315°
+	0	0°

(d)

Fig. 7.8. Two-phase permanent-magnet stepper motor. (*a*) Cross-section (*b*) Full-step, single-winding sequence. (*c*) Full-step, two-winding sequence. (*d*) Half-step sequence.

An identical torque/speed characteristic for a stepper motor is shown in Fig. 7.9. Note that two modes of operations are possible. The normal mode could be called *locked step.* In this mode the combined load plus the rotor inertia is less than the developed torque, with the result that the rotor may come to rest (or at least experience a reversal of momentum) between steps. Starts, stops and instantaneous reversals are possible. The second mode called *slewing* maintains synchronism but does not permit stops or reversals. Some applications utilise both modes *e.g.* a magnetic tape drive may read/write at a step rate in the normal mode but rewind/fast forward in the slewing mode. The transition to and from slewing must be properly controlled to avoid loss of synchronism.

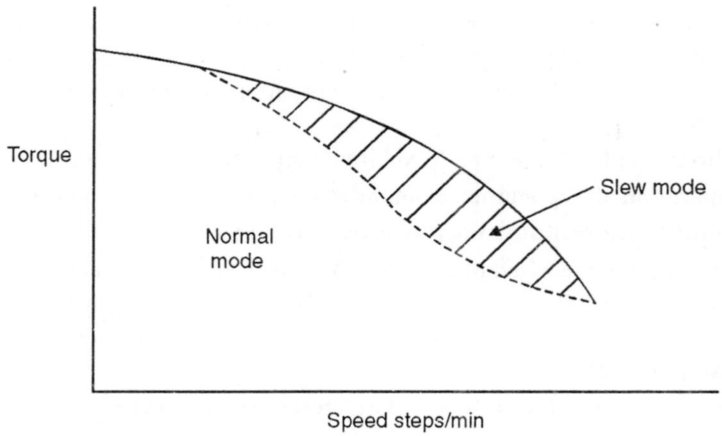

Fig. 7.9. Typical speed-torque characteristic of stepper motor.

7.9.2 Comparison of stepper motors

Stepper motors are available with step sizes ranging from 90° to 0.72°. Maximum step rates vary from about 100 to as many as 10,000 per sec. Torque capacities vary over a wide range. Selection of a stepper motor for a particular application must consider the overall performance of the motor with electronic drive.

Permanent magnet steppers are available in the larger step sizes, usually about 5°. They exhibit excellent torque/volume ratios but typically have high inertia rotors. The rotor provides braking and holding torque even with the stator windings deenergised. The variable reluctance motor is ideally suited to the smaller step sizes. By comparison, it develops less torque. However, since it may be constructed with a low inertia rotor, its torque/inertia ratio may be superior. The rotor is capable of "free-wheeling" since braking and detent torque must be provided by the stator. The hybrid stepper combines the smaller step size of the variable reluctance rotor with the higher torque of the permanent magnet. Some braking and holding torque is also present. As might be expected, it is the most expensive of the three types to manufacture.

7.9.3 Application

The characteristics of stepper motors make them ideally suited to many applications requiring incremental motion, especially where digital control is used. A few common ones are printer head and paper feed drives in computer printers, disc drives, digital plotters, medical equipment that dispenses precise dosages, computer controlled tools and process control valves.

PROBLEMS

7.1. Explain why a synchronous motor is not self-starting.

7.2. What is a damper winding ? Explain its application for a synchronous motor.

7.3. Describe briefly the construction, principle of operation and application of a synchronous motor.

7.4. Explain why a synchronous motor develops torque at synchronous speed whereas an induction motor develops torque at all speeds except the synchronous speed.

7.5. Derive expressions for active power and reactive power input to a synchronous motor when connected to a large system. Neglect resistance.

7.6. What is a synchronous condenser ? Explain its operation for improvement of power factor of the electric installation.

7.7. What is synchronous phase modifier ? Explain its operation.

7.8. Explain the operation of a synchronous motor when a change in load on the motor takes place.

7.9. Explain different methods of starting a synchronous motor.

7.10. A synchronous motor overexcited to take 600 kVA gives 360 hp at an efficiency of 0.9. It runs in parallel with a load of 1000 kVA at p.f. 0.8 lag. Find the resultant power factor. ((0.99) lag)

7.11. Name different types of stepper motors and explain the construction and operation of permanent magnet type stepper motor.

7.12. Draw and explain a typical torque/speed characteristic of a stepper motor and compare the performance of the three types of motors.

CHAPTER

8 Three-Phase Induction Motors

8.1 INTRODUCTION

The three-phase induction motor is the most commonly used motor in any power system. It forms about 50 to 70% of the total load in the system. Ordinarily, it is more or less a constant speed motor and thus has a disadvantage as compared to d.c. motors. However, with the advent of thyristorised control, using variable voltage variable frequency control, induction motors which are inherently simple in construction and robust, have replaced the d.c. series motors used for traction purposes.

8.2 CONSTRUCTION

In the most commonly form a polyphase induction motor consists essentially of a stator and a rotor. The stator supports windings which receive energy from the source and the rotor carries windings in which the working current is induced.

The core of stator is laminated and is made of sheet steel punchings and is slotted in its inner cylindrical surface and the winding consists of embedded coils disposed in exactly the same manner as those of a revolving field synchronous generator for the same number of poles. In motors of large ratings, the stator slots are of the open type *i.e.*, the side walls of the slots are parallel to facilitate the insertion of form-wound coils which are completely insulated before they are slipped into place ; but in smaller size motors, the slots are partially closed in order to reduce the effective length of the air gap between stator and rotor.

For small and medium size motors the entire cage winding, bars and end rings together is made of aluminium cast in place as an integral unit. This construction has the advantage of economy in manufacture.

Basically there are two types of poly phase induction motors (*i*) squirrel cage rotor type (*ii*) slip ring induction type. The difference in the two types is because of difference in construction of rotor, the stator design is same in both the cases.

The squirrel cage rotor has winding consisting of conducting copper or aluminium bars in slots in the rotor iron and short-circuited at each end by conducting end rings. The extreme simplicity and ruggedness of the squirrel cage construction are outstanding advantages of this induction motor. Thus the rotor in case of squirrel cage construction acts as the short-circuited secondary of a transformer of which the stator acts as the primary winding.

The slip ring rotor or the wound rotor carries a polyphase winding similar to and wound for the same number of poles as the stator. The terminals of the rotor winding are connected to insulated slip rings mounted on the shaft. Carbon brushes, bearing on these rings make the rotor terminals available external to the motor. The terminals of the rotor winding are brought out through slip ring to a suitably controlled and balanced rheostat.

8.3 PRINCIPLE OF OPERATION

When a 3-phase supply is connected to the stator terminals (normally delta connected as per the same voltage the torque developed is higher as compared to when it is star connected as

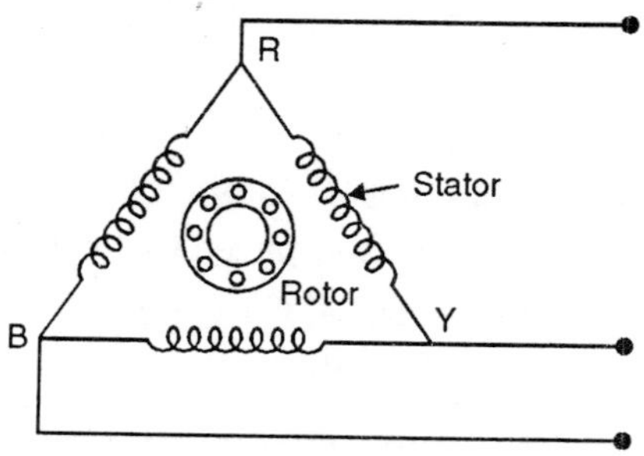

Fig. 8.1. Schematic diagram of squirrel cage motor.

voltage per phase is line voltage in delta connection) of the squirrel cage motor, the currents in the stator winding give rise to a constant magnitude rotating magnetic field which induces voltage in the stationary rotor conductors which is short circuited and hence circulates currents in the rotor conductors which produces certain flux and the interaction of stator and rotor fluxes give rise to torque and the rotor starts rotating in the direction of the magnetic field due to stator winding. Initially the frequency of voltages induced in rotor conductors corresponds to supply frequency and as the relative speed between the stator magnetic field and rotor decreases the frequency of the voltage induced in rotor also decreases. Under no load condition the rotor tries to catch up with the stator magnetic field but it can't run at the same speed as the stator magnetic field as otherwise the relative speed will be zero and no voltage would be induced in the rotor conductors and hence no torque will be developed in the rotor. Therefore, the rotor always rotates at a speed less than the synchronous speed of the rotating stator magnetic field. As the motor is loaded the slip between the stator magnetic field and rotor increases. Under no load condition since the rotor has to meet its copper and iron losses only, the slip between the rotor and stator magnetic field is around 1% and as the motor is loaded the slip increases *i.e.* the speed of the rotor decreases. The action is similar to dc shunt motor. The full load slip is normally 5%. The slip is defined as

$$s = \frac{n_s - n_r}{n_s} \quad ...(8.1)$$

where s is the slip, n_s the synchronous speed of the stator rotating magnetic field, n_r the actual speed of the rotor. The rotor moves in the same direction as the stator field because according to Lenz's law the rotor should move in a direction such that the emf induced in its conductors (or current flow through its conductors) decreases which is possible only if the relative speed between the rotor and the stator magnetic field decreases which means the rotor must rotate in the same direction as the stator magnetic field.

The starting torque of an ordinary (the other one is double cage) squirrel cage motor is limited to approximately double the full load torque when full voltage is applied to the stator winding and under these conditions the starting current is five to eight times its full load value. It, therefore, follows that in cases where high starting torque is required a different type of rotor is necessary. In fact we will see that a double cage rotor or slip ring induction motor with rotor resistances provides high starting torque without drawing large starting current.

In both types of induction motors described the stator carries the primary winding and the rotor the secondary. There may be situations when it is desirable to have the primary on the rotor and the secondary on the stator *e.g.* if the line voltage is high and the secondary current large it would be inconvenient to conduct large current to the rheostat through slip rings. In such a case, the primary wound on the rotor can be supplied with its relatively small current through slip rings, while the stationary secondary terminals can be connected directly to the rheostat. In either case, the revolving magnetic field will travel at synchronous speed with respect to the primary winding. In the first case, it will move at synchronous speed in space, since the primary itself is stationary; in the second case, it will move in space at a speed corresponding to slip frequency.

The number of rotor slots should not be equal to stator, for if they were, the reluctance of the magnetic circuit would vary from maximum when teeth are opposite slots to minimum when teeth are opposite teeth. Such a pulsation would result in additional core losses and in fact more important is that the rotor would tend to lock with the stator especially at starting with teeth opposite teeth. This can be avoided by making the number of stator and rotor teeth prime to each other, thereby providing a sort of vernier action between the two sets of teeth.

It is desirable to build the stator laminations so that the teeth and slots are parallel to the axis of the shaft but in the case of the rotor the laminations are slightly skewed to eliminate the locking action which is most pronounced if the air gap flux is radially disposed along the whole length of the opposite teeth, the skewing introduces a tangential component into the pull between opposite teeth and so tends to minimise locking action.

The air gap or more accurately the gap between the stator and rotor teeth should be as small as possible so as to have perfect coupling between the stator and rotor magnetic field and thus reducing the leakage flux between the stator (primary) and rotor (secondary) winding. This is desirable as the secondary (rotor) current upon which the torque and power developed depend is supplied inductively from the primary the stator winding and hence good coupling between the two windings is essential. From this point of view the induction motor has similar properties as the transformer and any air gap, however small, will cause the leakage reactance and the magnetising current to be greater than in an equivalent stationary transformer having a closed magnetic circuit. The effect of the air gap is to reduce the p.f. especially at light loads. The length of air gap is, therefore, determined mainly by mechanical consideration to avoid vibration of the shaft and wearing of the bearings and this may be smaller in low speed than in high speed machines.

8.4 EMF AND CURRENT RELATIONS (EQUIVALENT CIRCUIT)

In order to simplify the analysis the assumptions made are

(*i*) that the voltage V_1 applied across each phase of primary is purely sinusoidal and alternating at frequency f_1;

(*ii*) that the flux per pole is sinusoidally distributed in space around the air gap and that it is rotating at synchronous speed and (*iii*) that the rotor is equipped with a coil winding arranged for the same number of poles as the primary.

When 3-phase supply is fed to the stator winding it gives rise to a rotating magnetic field of constant magnitude. The voltage induced in the stator winding is

$$E_1 = 4.44\ K_1 f_1 N_1 \phi \qquad ...(8.2)$$

where K_1 is a constant which depends upon the distribution of the winding. Similarly the voltage induced in the rotor under stationary condition is

$$E_2 = 4.44\ K_2 f_1 N_2 \phi \qquad ...(8.3)$$

While under running condition when the slip is s, the secondary emf is given by

$$sE_2 = 4.44\ K_2 s f_1 N_2 \phi \qquad ...(8.4)$$

Each of these two emfs will lag behind ϕ by 90 electrical degrees as in case of a transformer as the currents induced in the secondary are exactly on the same line as in case of transformer. The only difference here is that the frequency of the secondary currents when the rotor is in motion is not same as that of primary (stator) and keeps on varying depending upon the speed of the rotor. Since the secondary (rotor) winding constitutes a closed circuit a current will be produced which is given as

$$I_2 = \frac{sE_2}{\sqrt{R_2^2 + s^2 X_2^2}} = \frac{E_2}{\sqrt{(R_2/s)^2 + X_2^2}} \qquad ...(8.5)$$

where R_2 is the rotor resistance per phase and X_2 is the rotor reactance per phase at frequency f_1. The two equivalent forms of equation (8.5) suggest two different interpretation.

The first form suggests that the voltage induced in rotor conductors is of frequency sf_1 and is of magnitude sE_2 and this emf. acts on a circuit of resistance R_2 and reactance sX_2 and this is in accordance with the actual operating condition. However, the second form suggests that the rotor is stationary and the emf induced is E_2 and this is connected to a variable resistance R_2/s and fixed reactance X_2 at f_1 frequency. Thus the current and emf being of supply frequency. This condition is similar to the transformer and, therefore, the second form of interpretator can be used to draw the equivalent circuit of the induction motor. There is, however, one variation with regard to a transformer. The no load current I_0 in case of induction motor is more in magnitude and leads the flux ϕ more because of the presence of the component of current to supply the friction and windage losses in addition to the core loss. Also I_m the magnetising component of current is larger in case of induction motor due to airgap as compared to the transformer.

Using the second form of equation (8.5) the equivalent circuit of the induction motor is given in Fig. 8.2.

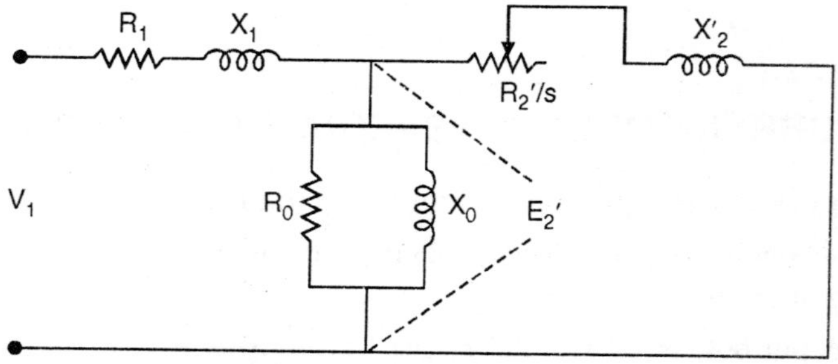

Fig. 8.2. Equivalent circuit of induction motor.

Here R_2' is the equivalent rotor resistance referred to stator side and X_2' is the equivalent rotor reactance at line frequency and referred to the stator side. As in case of transformer, the shunt branch consisting of parallel combination of R_0, X_0 can be shifted to the left of the circuit of Fig. 8.2 as I_0 is negligibly small as compared to I_2' and hence the approximate equivalent circuit is given in Fig. 8.3.

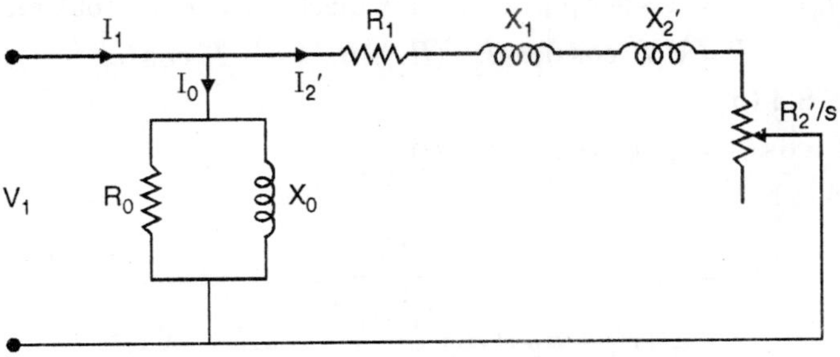

Fig. 8.3. Approximate equivalent circuit of induction motor.

From Fig. 8.3 the secondary current referred to primary side (stator) I_2' is given by

$$I_2' = \frac{V_1}{\sqrt{(R_1 + R_2'/s)^2 + (X_1 + X_2')^2}} \quad \text{...(8.6)}$$

In the equivalent circuit of Fig. 8.3, the effect of motor load is included in the variable resistance R_2'/s but since

$$\frac{R_2'}{s} = R_2' + \frac{1-s}{s} R_2' \quad \text{...(8.7)}$$

the approximate equivalent circuit can be redrawn as in Fig. 8.4 (*a*). Here the resistance R_2' is an inherent part of the internal impedance of the machine but the variable resistance $R_2' \frac{1-s}{s}$ is the electrical analog of the mechanical load. It is this fact which accounts for the concept that the induction motor can be thought of as a generator feeding a fictitious resistance.

8.5 MECHANICAL POWER DEVELOPED

From Fig. 8.3 it is seen that the power input (per phase) to the motor is

$$P_1 = V_1 I_1 \cos\theta_1 \quad \text{...(8.8)}$$

where the meaning of the symbols is indicated in the phasor diagram of Fig. 8.4 (*b*) constructed to agree with the circuit of Fig. 8.3.

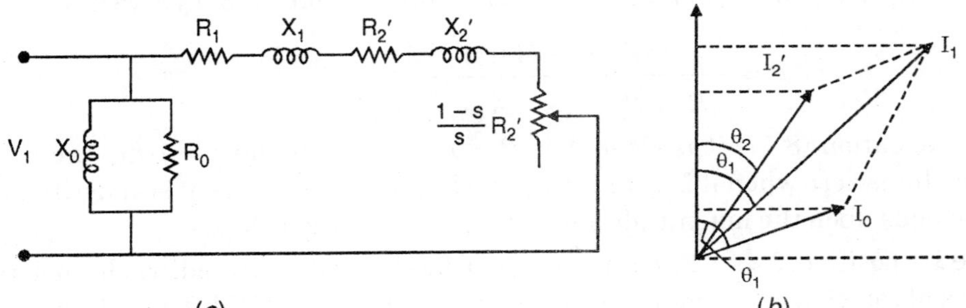

Fig. 8.4. (*a*) Approximate equivalent circuit of 3-phase induction motor (*b*) Power component of current of Fig. 8.3.

A part of the power input to the stator, amounting to $I_2'^2\,(R_1 + R_2')$ watts per phase is consumed as copper loss in the primary (stator) and secondary (rotor) windings and the remainder is converted into mechanical power. Not all the mechanical power thus developed is available at the shaft, as there is further loss due to friction windage and core loss. These losses have been taken into account in the equivalent circuit by the shunt branch R_0 and it is seen from phasor diagram 8.4 (*b*) that it amounts to $V_1 I_0 \cos\theta_0$ watts per phase. The net mechanical power output per phase is, therefore,

$$P = V_1 I_1 \cos\theta_1 - I_2'^2 (R_1 + R_2') - V_1 I_0 \cos\theta_0 \qquad ...(8.9)$$

But from Fig. 8.4 (*b*)

$$I_1 \cos\theta_1 = I_2' \cos\theta_2 + I_0 \cos\theta_0 \qquad ...(8.10)$$

From Fig. 8.4 (*a*)

$$I_2' = \frac{V_1}{\sqrt{\left\{R_1 + R_2' + R_2' \dfrac{(1-s)}{s}\right\}^2 + (X_1 + X_2)^2}} = \frac{V_1}{Z'} \qquad ...(8.11)$$

and

$$\cos\theta_2 = \frac{R_1 + R_2' + R_2'(1-s)/s}{Z'} \qquad ...(8.12)$$

Substitution of equations (8.10) through (8.12) in equation (8.9) gives

$$P = \frac{V_1^2\, R_2'}{(R_1 + R_2'/s)^2 + (X_1 + X_2)^2} \cdot \frac{1-s}{s} \qquad ...(8.13)$$

which gives the mechanical power developed per phase in terms of the slip s, all other terms in the expression being constants of the machine.

Further using equation 8.6 and substituting in equation 8.13 we have

$$P = I_2'^2 \,.\, R_2' \frac{1-s}{s} \qquad ...(8.14)$$

which means that the mechanical power developed per phase may be regarded as the ohmic loss in a fictitious secondary resistance of $R_2' \dfrac{1-s}{s}$ ohms per phase

Now total power input to the secondary (rotor)

= Mechanical power developed + rotor copper loss

$$= I_2'^2 R_2' \frac{1-s}{s} + I_2'^2 R_2' = \frac{I_2'^2 R_2'}{s} \qquad ...(8.15)$$

$$= \frac{\text{Secondary copper loss per phase}}{\text{Slip}} \qquad ...(8.16)$$

On multiplying numerator and denominator of equation (8.13) by s we have

$$P = \frac{V_1^2\, R_2'\, s\,(1-s)}{(sR_1 + R_2')^2 + s^2\,(X_1 + X_2)^2} \qquad ...(8.17)$$

From equation (8.17) it is clear that $P = 0$ when $s = 0$ and $s = 1$ *i.e.* the output of an induction motor is zero when it is running at synchronous speed or is at standstill. Further the sign of P depends upon the magnitude and sign of s as outlined below :

Case I. s positive, $0 < s < 1$: The motor speed lies between zero and synchronous speed and the direction of rotation of the rotor is same as that of the stator field. The power P is therefore positive and indicates the motor action of the machine.

Case II. s negative, $s < 0$: The rotor rotates in the same direction as the stator field but at a speed higher than the synchronous speed. The relative direction of rotation of the rotor and magnetic field are reversed which induces currents and voltage in the rotor circuit in the reverse direction which means the machine acts as a generator.

Case III. s positive, $s > 1$: The rotor is physically rotating in a direction opposite to the direction of rotation of the magnetic field. The rotor currents and voltages induced are in the same direction as in case I but become greater in magnitude because of the higher relative speed. The machine acts as an electric brake and thus power P is negative.

From equation (8.17)

$$2\pi\, nT = 2\pi\, n_s(1-s)T = \frac{3V_1^2\, R_2\, s\,(1-s)}{(sR_1 + R_2)^2 + s^2\,(x_1 + x_2)^2} \qquad \text{...(8.18)}$$

where n = speed of machine

n_s = synchronous speed of machine

Typical torque-speed curves have been drawn in Fig. 8.5, one for the forward direction of rotation and the other for reverse.

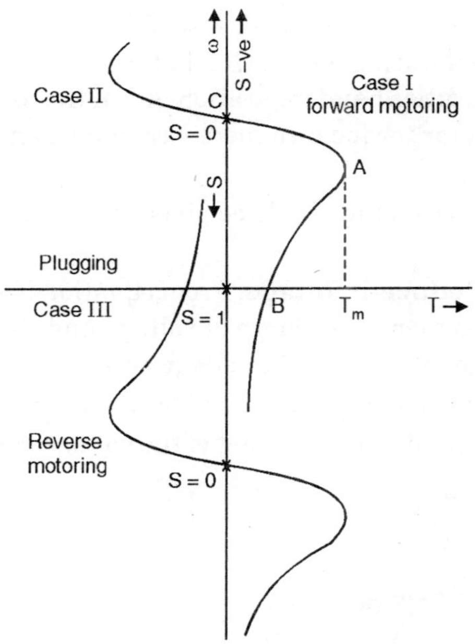

Fig. 8.5. Torque-slip characteristic of induction motor.

The regenerative action of the machine when the rotor rotates in the forward direction, at a speed greater than synchronous speed, is shown by the curve in second quadrant (forward motor curve). However, it is so be noted that this action of the machine is possible only if the system to which the motor is connected supplies reactive power for its excitation.

The maximum torque which the motor can develop is determined by the condition dT/ds = 0. On differentiating equation (8.18) with respect to s the condition for maximum torque is

$$s = \pm \frac{R_2}{\sqrt{R_1^2 + (X_1 + X_2)^2}} \qquad \text{...(8.19)}$$

The positive sign stands for motoring action whereas negative sign for generating action.

Substituting equation (8.19) for s in equation (8.18)

$$T_{max} = \frac{3 V_1^2}{\omega_s \times 2 [R_1^2 + (x_1 + x_2)^2]^{1/2}} \quad ...(8.20)$$

From equations (8.19) and (8.20), it is clear that the slip at which maximum torque occurs depends upon the rotor resistance but the magnitude of maximum torque is independent of rotor resistance. Neglecting R_1 the maximum torque under standstill occurs when

$$R_2 \simeq X_1 + X_2$$

i.e. rotor resistance equals the total leakage reactance. R_2 can be made equal to the leakage reactance only in case of the slip ring induction motor by inserting additional resistance between the slip rings. Corresponding to this operation the p.f. is 0.707. It is to be noted that higher starting torques can be obtained by inserting rotor resistance, but a permanent resistance in the rotor will reduce the efficiency under normal running condition.

In Fig. 8.5, the portion *AB* of the speed torque Characteristic is the unstable operation of the motor and *AC* the stable. During the portion *AB* if the torque developed by the motor is greater than the resisting torque (load torque) the speed continues to increase until the point *A* (corresponding to maximum torque) is reached. Also if the load torque is greater than the motor torque, the motor starts decelerating along *AB* till it reaches the point *B* where the machine comes to a standstill. On the other hand in position *AC*, if the load torque increases the motor slow down and hence the motor torque also increases and when the two are equal, the motor comes to steady operation.

The torque speed characteristic of a three phase induction motor can be modified by the following methods :

(*i*) **Application of Reduced Voltage.** From equations (8.18) and (8.19) it is clear that the torque is proportional to square of voltage and the slip at which the maximum torque occurs is independent of the voltage applied as shown in Fig. 8.6.

From Fig. 8.6, it is clear that the difference between the speed corresponding to maximum torque and the synchronous speed is independent of the voltage applied.

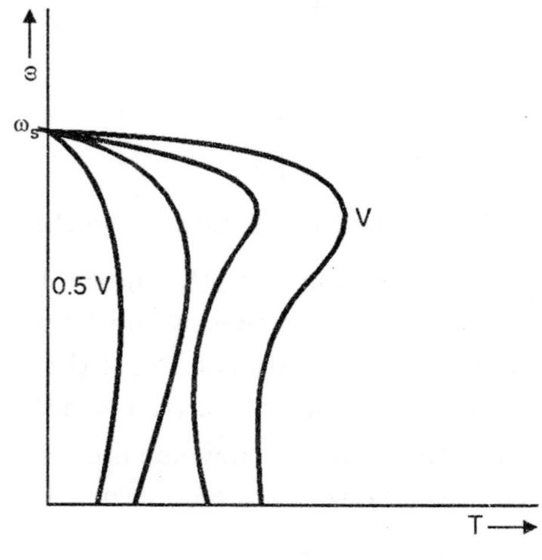

Fig. 8.6. Speed control using reduced voltage.

(*ii*) **Application of Variable Frequency.** The speed is a function of frequency as given by the synchronous speed equation $f_1 = \frac{np}{120}$. The variable frequency can be supplied by a special generator or a frequency changer. While varying frequency, it is necessary to see that the motor retains its shunt characteristics and has adequate overload torque capacity. Both these objectives are achieved by maintaining air gap flux constant. For an induction motor $V_1 \propto f\phi$ or $\frac{V_1}{f} = K\phi$. Therefore, in order to maintain constant air gap flux $\frac{V_1}{f}$ should be kept constant. A set of torque speed characteristics are shown in Fig. 8.7 using variable frequency control.

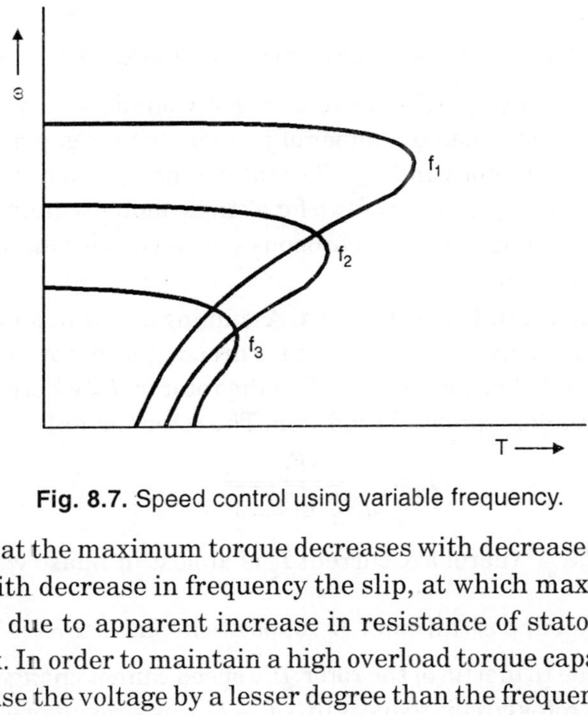

Fig. 8.7. Speed control using variable frequency.

It can be seen that the maximum torque decreases with decrease in frequency below normal frequency. Also with decrease in frequency the slip, at which maximum torque occurs increases. This is partly due to apparent increase in resistance of stator and partly due to the decrease in air gap flux. In order to maintain a high overload torque capacity at low frequencies, it is desirable to decrease the voltage by a lesser degree than the frequency *i.e.* have higher ratio of *V/f* as frequency is decreased. Variable frequency control provides speed control over the range 10 : 1 to 12 : 1 and high smoothness of control.

(*iii*) **Addition of Rotor Resistance.** The torque speed characteristic can be modified by inserting resistance in the rotor circuit. Obviously this method is applicable to wound rotor motor. The characteristics are shown in Fig. 8.8 for different values of rotor resistance. It can be seen from Fig. 8.8 that the magnitude of maximum torque is independent of rotor resistance but the slip at which maximum torque occurs is directly proportional to the rotor resistance.

Further, with increase in rotor resistance the starting torque increases upto certain value of resistance, beyond which the starting torque decreases which is due to the enormous decrease in the value of the starting current.

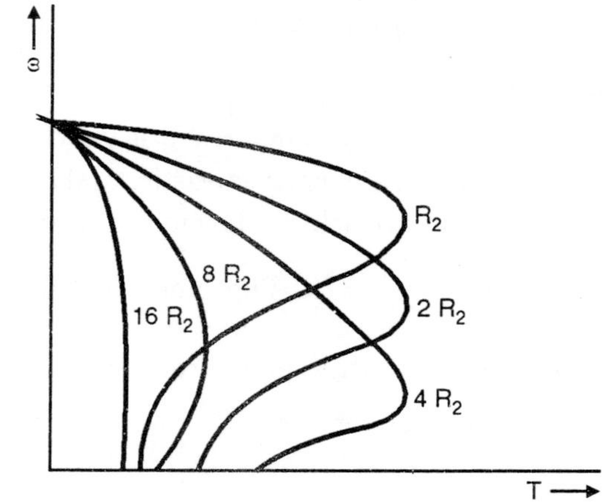

Fig. 8.8. Speed torque curves with rotor resistance control.

(*iv*) **Pole Changing.** A squirrel cage rotor is not wound for a specific number of poles, rather it adapts to the same number of poles as the air gap magnetic field, which is determined by the stator winding. The stator winding can be reconnected to multiply the number of poles by a factor of two for a single motor winding. If the motor has two distinct stator windings, then synchronous speeds can be obtained again by reconnecting the windings suitably.

(*v*) **Voltage Injection in Rotor Circuit.** Assuming constant supply voltage frequency and load torque and neglecting the voltage drop in the stator, the applied voltage $V_1 \simeq E_1 = K\phi$ = constant. The induced e.m.f. in the rotor is sE_2 where s is the slip and E_2 the induced e.m.f. under standstill condition. The rotor current

$$I_2 = \frac{sE_2}{\sqrt{R_2^2 \pm (s\,x_2)^2}}$$

Normally, $R_2^2 >> (sx_2)^2$ therefore, current I_2 is almost in phase with E_2 and is approximately equal to $\frac{s\,E_2}{R_2}$. Now suppose an e.m.f. E_1, opposite in phase to E_2 is injected into the rotor (wound rotor). Initially, due to inertia of the rotor, the speed cannot change instantaneously and hence the net e.m.f. in the rotor circuit reduces to $(sE_2 - E_1)$, as a result of which the rotor current and hence the torque developed decrease. With load torque remaining constant the speed decreases (slip increases). This process continues till a new slip $s_i > s$ is achieved, when the rotor induced e.m.f. increases to circulate enough current in the rotor to develop the desired torque. Similarly, if the injected e.m.f. is in phase with the rotor induced e.m.f., the speed of the motor will increase. It should be noted that injected e.m.f. has to be at slip frequency.

Example 8.1. A 12 pole, 3-phase alternator is coupled to an engine running at 500 rpm. It supplies an induction motor which has a full load speed of 1440 rpm. Find the percentage slip and the no. of poles of the motor.

Solution. The frequency of the output of the alternate.

$$f = \frac{pn}{120} = \frac{12 \times 500}{120} = 50 \text{ Hz}$$

The minimum no. of poles can be 2, for which at 50 Hz the synchronous speed is 3000 rpm. Since the full load slip usually is of the order of 5% a speed of 3000 is not possible as the rotor speed given is 1440 rpm.

Next the no. of poles could be 4, for which at 50 Hz the synchronous speed is 1500 rpm and this is close to 1440 rpm. Hence no. of poles is 4 and the percent slip = $\frac{1500-1440}{1500} \times 100 = 4\%$. **Ans.**

Example 8.2. If the e.m.f. in the stator of an 8-pole induction motor has a frequency of 50 Hz and that in the rotor $1\frac{1}{2}$ Hz, at what speed is the motor running and what is the slip.

Solution. The supply frequency is 50 Hz whereas the rotor emf frequency is $1\frac{1}{2}$ which means

$$sf = \frac{3}{2} \quad \text{or} \quad s = \frac{3}{2 \times 50} = 0.03$$

Now for 8-pole at 50 Hz the speed comes out to be

$$\frac{120f}{p} = \frac{120 \times 50}{8} = 750 \text{ rpm}$$

Since slip is 0.03, therefore, the actual speed of the rotor is

$$0.03 = \frac{750 - n_r}{750}$$

or $$n_r = 750(1 - 0.03) = 750 \times 0.97$$

$$= 727.5 \simeq 728 \text{ rpm.}$$ **Ans.**

Example 8.3. A 3-phase induction motor has a 4-pole star-connected stator winding. The motor runs on a 50 Hz supply with 200 V between lines. The rotor resistance and standstill reactance per phase are 0.1 ohm and 0.9 ohm respectively. The ratio of rotor to stator turns is 0.67. Calculate (*a*) total torque at 4% slip (*b*) total mechanical power at 4% slip (*c*) maximum torque (*d*) speed at maximum torque (*e*) maximum mechanical power. Neglect stator impedance.

Solution. (*a*) Neglecting stator impedance, the power developed using equation 8.17 is given as

$$P = \frac{V_1^2 R_2' (1-s) s}{R_2'^2 + s^2 X_2'^2}$$

Referring the quantities to rotor side the equation becomes

$$\frac{V_2^2 R_2 (1-s) s}{R_2^2 + s^2 X_2^2} = \frac{0.67^2 \times 200^2 \times 0.1 \times 0.96 \times 0.04}{0.1^2 + 0.04^2 \times 0.9^2}$$

$$\simeq \frac{68.95}{0.01} = 6895 \text{ watts}$$

or $$\frac{6895}{735} = 9.38 \text{ metric hp.}$$ **Ans.**

(*b*) The torque at this slip $= 2\pi n T = 6895$

The speed per second $= \frac{1440}{60}$

Hence torque in N-m $= \frac{6895 \times 60}{2\pi \times 1440} = 45.7$ Nm. **Ans.**

It is to be noted that the motor is star connected and hence the voltage per phase would be $\frac{200}{\sqrt{3}}$ and since the power is to be obtained for all the three phases and the equation 8.17 gives power per phase, therefore we have taken voltage as 200 volts rather than $\frac{200}{\sqrt{3}}$.

(*c*) The slip at which maximum torque occurs is given by

$$s = \frac{R_2}{X_2} = \frac{0.1}{0.9} = 0.111$$

and
$$P_{max} = \frac{V_1^2\, s\,(1-s)}{R_2^2 + s^2 X_2^2} = \frac{0.67^2 \times 200^2 \times 0.111 \times 0.889 \times 0.1}{0.1^2 + 0.11^2 \times 0.9^2}$$

$$= \frac{177.188}{0.01 + 0.0098} = 8948.8 \text{ watts}$$

The speed at this slip 1500(1 – 0.111) = 1334 rpm

Hence $T_{max} = 2459$ Nm

Example 8.4. An induction motor has an efficiency of 0.9 when the load is 50 hp. At this load, the stator and rotor copper loss each equals the iron loss. The mechanical losses are one-third of the no load loss. Calculate the slip.

Solution. Let the rotor copper loss be x watts

The total losses are $(x + x + x + x/3) = \frac{10x}{3}$ watts

$$0.9 = \frac{50 \times 735}{50 \times 735 + 10\,x/3}$$

$$36750 + \frac{10x}{3} = 40833$$

$$\frac{10x}{3} = 4083 \quad \text{or} \quad x = 1225 \text{ watts}$$

Now rotor copper loss $I_2'^2 R_2'' = 1225$ watts

The load is 36750 watts $\doteq I_2'^2 R_2 \frac{1-s}{s} = 1225 \cdot \frac{1-s}{s}$

or
$$\frac{1-s}{s} = 30 \text{ or } 31s = 1 \quad \text{or} \quad s = \frac{1}{31} = 0.032 \quad \textbf{Ans.}$$

8.6. STARTING OF 3-PHASE INDUCTION MOTORS

Using equation (8.13) the expression for torque can be obtained as follows

$$P = \frac{V_1^2 R_2'}{(R_1 + R_2'/s)^2 + (X_1 + X_2')^2} \cdot \frac{1-s}{s}$$

Now $P = 2\pi n T$ where n is the rotor speed

or $P = 2\pi\, n_s(1-s)T$ where n_s is the synchronous speed in rps.

Hence
$$T = \frac{V_1^2 R_2'/s}{2\pi n_s\, [(R_1 + R_2'/s)^2 + (X_1 + X_2')^2]} \quad ...(8.21)$$

and rewriting equation (8.6) for current we have

$$I = \frac{V_1}{\sqrt{(R_1 + R_2'/s)^2 + (X_1 + X_2')^2}}$$

It is thus seen that contrary to condition with the dc motor, the torque depends on the speed, a feature which gives rise to the well-known torque slip curves of the induction motor. For starting purposes the torque and current at standstill ($s = 1$) are important and these are given by

$$T_s = \frac{V_1^2 R_2'}{2\pi n_s \, [(R_1 + R_2')^2 + (X_1 + X_2')^2]} \quad \text{...(8.22)}$$

and

$$I_s = \frac{V_1}{\sqrt{(R_1 + R_2')^2 + (X_1 + X_2')^2}} \quad \text{...(8.23)}$$

For given value of reactance, therefore, the starting torque and the starting current vary with rotor resistance in the manner shown in Fig. 8.9, the torque reaching a maximum when R_2' is

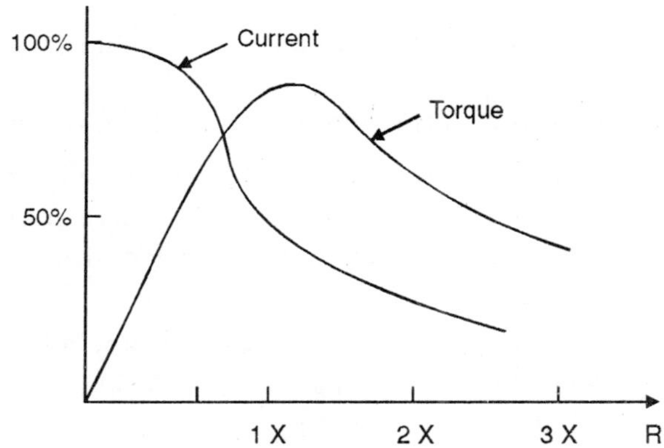

Fig. 8.9. Starting torque and current of induction motor with various values of rotor resistance.

approximately equal to $X_1 + X_2'$. For normal running the value of R_2' should not exceed 1 or 2 per cent of X_2' or the efficiency will be low as there will be lot of ohmic losses in R_2'. Therefore, if starting torque required is high, resistance must be added to the secondary circuit and cut out gradually as the motor speeds up. The resistance also keeps the starting current within allowable limits. This kind of control on rotor resistance is possible in case of slip ring or wound rotor induction motors. However, in case of squirrel cage induction motor, it is not possible to insert resistance in series with the rotor circuit and hence it is inherently a low starting torque motor. As mentioned earlier, the induction motors when running are connected in delta as the torque is proportional to square of the voltage and per phase voltage is the line voltage when the stator is delta connected. Usually motors of 15 to 20 hp are normally connected directly on to the supply without reducing the voltage at starting. However, following methods are used to start induction motors :

(*a*) Direct on-line starter (*b*) Star delta starter (*c*) Auto transformer starting (*d*) Stator resistance starting. We will now discuss salient features of various starting methods.

(*a*) **Direct on-line starter.** In this method it is presumed that the motor is already connected in delta for the reason mentioned earlier. Here the motor is connected direct on to the supply without having any voltage reducing device in between the motor and the supply. The starter has a set of contactor, a no-volt coil and an overcurrent release. Initially as the supply is given to the starter, the no-volt coil is energised and operates the set of contactors which connect supply to the motor. For any reason, if the supply voltage is less than the pick-up voltage of the no-volt coil the contactors will not be operated and the supply will not be connected to the motor. Or during operation if the voltage goes below the drop down voltage of the no-volt coil, the no-volt coil gets deenergized and releases the contactor thereby the motor is disconnected from the supply. This is desirable as for the same load if the voltage goes below certain value the motor will draw large current and hence, it should be instantaneously disconnected. Whenever, the voltage of the system is more than the normal voltage or there is overloading due to any reason the overcurrent release will operate the contactors and disconnect the supply to the motor. This type of starter is normally used for motors rated upto 15 to 20 hp.

(*b*) **Star delta starter.** In this starter, it is required that all the six terminals of the motor stator winding should be available on the body of the motor as these are connected to the respective terminals of the star delta starter. Here when the starter is operated, initially the winding of motor is connected in star so that the voltage per phase is $\frac{V}{\sqrt{3}}$ where V is the supply line voltage and therefore the starting torque is $\frac{1}{3}$ of the final torque and the starting current is $\frac{1}{\sqrt{3}}$ of the current when V is applied. This is the additional feature of star-delta starter over the direct on line starter. Once the motor runs upto speed, the starter is put on to delta position and the motor stator winding is finally connected in delta under running condition. This method is used for more than 20 hp motors.

(*c*) **Auto transformer starter.** Here reduced voltage is applied. The reduction of voltage may be carried out by an auto transformer tapping giving 50%, 60% and 75% of normal voltages usually being provided and giving therefore starting torques of 25 per cent, 36 per cent and 56 per cent and starting currents of 50%, 60% and 75% respectively of that obtainable by direct switching.

(*d*) **Stator resistance starting.** A resistance in series with the stator can be used to limit the starting current but the starting torque available with a given current is less than with an auto transformer and even much less as compared to other methods.

The cost of the stator resistance is, however, less than that of an auto transformer, and it finds application in certain cases where only low starting torques are required.

Before we compare various starting methods through typical example, following relation is derived.

Using equation (8.14) we have

$$P = I_2'^2 R_2' \frac{1-s}{s}$$

Also $$P = 2\pi n_f T = 2\pi n_s(1-s)T$$

Hence $$T = I_{2s}'^2 \frac{R_2'}{s} \cdot \frac{1}{2\pi n_s}$$

Starting torque $$T_s = \frac{1}{2\pi n_s} \cdot I_2'^2 R_2'$$

and Full load torque $$T_f = \frac{1}{2\pi n_s} \cdot I_{2f}'^2 \frac{R_2'}{s_f}$$

Hence $$\frac{T_s}{T_f} = \left(\frac{I_{2s}'^2}{I_{2f}^2}\right) s_f \qquad ...(8.24)$$

By making use of this equation, the starting currents and torques for any particular motor condition can be calculated.

Example 8.5. A 400 V 3-phase squirrel cage induction motor has a full load slip of 4 per cent, a standstill impedance of 1.54 ohms and a full load current of 30 A. The maximum starting current which may be taken from the line is 75 A. What tapping must be provided on an auto transformer starter to limit the current to this value and what would be the starting torque available in terms of the full load torque ?

Solution. Suppose the tapping is t per cent of normal voltage

Normal voltage $= \frac{400}{\sqrt{3}} = 231$ volt

Starting current in the motor $\dfrac{\frac{t}{100} \times 231}{1.54}$ amp

Current on supply side of the auto transformer $= \frac{t}{100}$ of current in the motor

Therefore, starting current taken from supply $= \left(\frac{t}{100}\right)^2 \cdot \frac{231}{1.54}$ A

and this is to be limited to 75 A

Hence $$\left(\frac{t}{100}\right)^2 \cdot \frac{231}{1.54} = 75$$

or $$t^2 = \frac{75 \times 1.54 \times 100^2}{231} \quad \text{or} \quad t = 70.7$$

i.e. a 70.7 per cent tapping should be provided.

The starting current in the motor

$$= 75 \times \frac{100}{70.7} = 106 \text{ A}$$

Therefore, from equation (8.24) $\frac{\text{Starting torque}}{\text{Full load torque}} = \left(\frac{106}{30}\right)^2 . 0.04 = 0.5$

i.e. the starting torque available would be only half of the full load torque.

Example 8.6. The power input to a 3-phase induction motor is 60 kW. The stator losses total 1 kW. Find the total mechanical power developed and the rotor copper loss per phase if the motor is running with a slip of 3%.

Solution. Total power input to the stator 60 kW

Total stator losses = 1 kW

Therefore total input to the rotor = 59 kW

Using equation 8.15 we have $\frac{3I_2'^2 R_2'}{s} = 59 \times 1000$ watts

Hence $\frac{3I_2'^2 R_2'}{0.03} = 59000$

Hence rotor copper loss $= I_2'^2 R_2' = \frac{59000 \times 0.03}{3} = 590$ watt per phase

and mechanical power developed = rotor input – rotor copper loss

$= 59000 - 590 \times 3 = 57.23$ kW. **Ans.**

It is to be noted that the parameters of the induction motor $(R_1 R_2', X_1 X_2', R_0, X_0)$ can be obtained on the similar lines as in case of a transformer by performing short circuit (blocked rotor) and open circuit (light running) tests.

Example 8.7. A delta connected, 400 V, 50 h.p. 750 r.p.m. squirrel cage motor takes a full load current of 50 A and has a full load slip of 4.5 per cent. The impedance per phase is 2.5 ohm. Determine the starting torque and the starting current taken from the supply if the motor is started by (*i*) D.O.L. starter ; (*ii*) Star-delta starter ; (*iii*) an auto transformer starter with 70% tapping.

Solution.

(*i*) D.O.L. Starter

The starting current per phase $= \frac{400}{2.5} = 160$ A

Therefore Line current $= \sqrt{3} I_{ph} = 277.12$ **Ans.**

Full load torque $= \frac{50 \times 735.5 \times 60}{2\pi \times 750} = 468.47$ Nm.

Now $$\frac{T_{st}}{T_{ft}} = \left(\frac{I_{st}}{I_{fl}}\right)^2 s_f$$

$$T_{st} = \left(\frac{277.12}{50}\right)^2 \times 0.045 \times 468.47 = 647.57 \text{ Nm} \quad \textbf{Ans.}$$

(*ii*) When started by star-delta starter

$$T_{st} = T_f \left(\frac{1}{\sqrt{3}}\right)^2 = \frac{468.47}{3} = 156.15 \text{ Nm} \quad \textbf{Ans.}$$

8.7. INDUSTRIAL APPLICATION

We have so far studied operating characteristics of induction motor. Based on these characteristics following applications are suggested. In agriculture induction motors are invariably used. The other applications are as follows :

Textile mills. In a textile mill, following operations are carried out, cotton to slivers, spinning, weaving and finishing. The process of obtaining slivers from cotton involves, removal of seeds and other impurities from cotton, transformation into laps and finally into slivers. After this follows spinning where a continuous yarn of sufficient strength is produced which is wound on bobbins. The process of weaving consists of joining two sets of threads, one extends throughout the length of the fabric and the other one has its threads going across. This process is carried out in a loom. The process of finishing consists of bleaching, dyeing, printing, and packing.

Loom motors require high starting torque, frequent starting and stopping and low temperature rise under full load condition. Therefore, totally enclosed fan cooled, high torque squirrel cage induction motors are used.

Spinning motors require moderate starting torque and smooth acceleration, so as to have straight and unbroken yarns. Normally a 4 or 6 pole squirrel cage induction motor is used.

Cement mills. The raw material required consist of limestone, silica, alumina and ferric oxide. The lime stone is crushed and a required amount of iron ore is added and then they are grounded in a grinding mill. The powder from this grinding mill is homogenised in silos, by passage of air from bottom through the medium and is then fed into the kiln, from where cement clinker at high temperature is obtained. This clinker is then air cooled in special type of coolers. The motors used for crushers is the *slip ring* induction motor having 200 to 250% pull out torque.

Coal mining. For drilling in coal mines, usually induction motors with flame proof enclosures are used. Sometimes high frequency (150 to 200 Hz) motors are used. This is done to reduce the size of the motor for the same power ratings. For raising coal or raising or lowering men and some equipments usually wound rotor induction motor is used. While applying brakes, initially d.c. rheostatic braking is used but finally for dead stops, mechanical brakes are used. For large size mines, usually d.c. motor with Ward-Leonard Ilgner system is employed for mine winders.

Lifts. For lifts normally slip ring induction motors having high starting torque, high overload capacity and pull out torques, are used.

Lathe, grinding and milling machine. Here the machine tools require constant operating speed and low starting torque and therefore, d.c. shunt motors or squirrel cage motors are usually used.

Cranes. For cranes or hoist, the requirements are high starting torque and frequent switching operations. For this d.c. series motor is used. In case of a.c. supply, slip ring induction motors are used where speed control is obtained by rotor resistance.

A typical but undesirable operation of a 3-phase induction motor is what is known as single phasing operation of a 3-phase induction motor. Suppose a three-phase induction motor is operating under normal condition of supply and suddenly one of the three phases of the lines blows off. The machine is put on the remaining two phases. This is known as single-phasing operation of a 3-phase induction motor. The machine continues to run but at a slightly reduced speed. A humming noise from the machine is heard. The current in the line becomes $\sqrt{3}$ times its normal value. It is not the quantity of current that is dangerous for the operation of the motor to continue but the quality of current which may overheat the stampings of the rotor which may behave a short-circuited secondary and the stator may draw large current resulting in complete damage of the motor. Therefore, moment humming noise is heard the supply should be switched off from the motor.

PROBLEMS

8.1. A delta connected 400 V, 30 hp, 750 rpm squirrel cage motor takes a full load current of 37A and has a full load slip of 4.5 per cent. The impedance per phase is 3.9 ohms. Determine the starting torque and the starting current taken from the supply if the motor is started by (*a*) Direct switching

(*b*) An auto-transformer starter with a 70% tapping

(*c*) A star-delta starter. [**Ans.** (*a*) 291.6 Nm, 177 A (*b*) 143 Nm, 86.8 A (*c*) 97 Nm, 59 A]

8.2. A 3-phase squirrel cage motor has maximum torque equal to twice full load torque $R_2' = 0.2\ \Omega$ and $X_2' = 2\ \Omega$. Determine the ratio of starting torque to full load torque, if it is started by (*a*) DOL starter (*b*) Star-delta starter (*c*) Auto-transformer starter with 70% tapping.

[**Ans.** (*a*) 0.396 FLT (*b*) 0.132 FLT (*c*) 0.194 FLT.]

8.3. Explain why the rotor of an induction motor rotates in the same direction as the stator magnetic field and why the speed of the motor is less than the synchronous speed.

8.4. Explain what you mean by single phasing operation of a 3-phase induction motor.

8.5. Comment on the statement

(*a*) The rotating fields due to stator and rotor are stationary with respect to each other.

(*b*) Induction motor can be regarded as a generator feeding a fictitious resistance load.

8.6. A 4-pole induction motor is running at 1440 rpm from a 50 Hz supply. Find the percentage slip and the frequency of rotor current. [**Ans.** 4% 2 Hz]

8.7. A 3-phase induction motor has 4-pole star-connected stator winding and runs on a 220 V 50 Hz supply. The rotor resistance is 0.1 ohm and reactance 0.9 ohm. The ratio of stator to rotor turns is 1.75. The full load slip is 5%. Determine for this load (*a*) the total torque (*b*) the horsepower. Also find (*c*) the maximum torque (*d*) the speed at maximum torque.

[**Ans.** 4.28 kgm (*b*) 8.4 hp (*c*) 5.73 kgm (*d*) 1330 rpm.]

8.8. A slip ring induction motor runs at 290 rpm on full load when connected to a 50 Hz supply. Calculate (*a*) the number of poles (*b*) the slip (*c*) the slip for full load torque if the total resistance of the rotor circuit is doubled. [**Ans.** (*a*) 20 (*b*) 3.3% (*c*) 6.6%.]

8.9. The power input to a 500 V, 50 Hz, 6 pole, 3-phase induction motor running at 975 rpm is 40 kW. The stator losses are 1 kW and the friction windage losses total 2 kW. Determine (*a*) the slip (*b*) the rotor copper loss (*c*) the brake horsepower (*d*) the efficiency.

[**Ans.** (*a*) 0.025 (*b*) 975 W (*c*) 48.4 hp (*d*) 90%.]

CHAPTER

9 Single Phase Induction Motors

9.1 INTRODUCTION

The characteristics of single phase induction motors are identical to 3-phase induction motors except that single phase induction motor has no inherent starting torque and some special arrangements have to be made for making it self starting. It follows that during starting period the single phase induction motor must be converted to a type which is not a single phase induction motor in the sense in which the term is ordinarily used and it becomes a true single phase induction motor when it is running and after the speed and torque have been raised to a point beyond which the additional device may be dispensed with. For these reasons, it is necessary to distinguish clearly between the starting period when the motor is not a single phase induction motor and the normal running condition when it is a single phase induction motor. The starting device adds to the cost of the motor and also requires more space. For the same output a 1-phase motor is about 30% larger than a corresponding 3-phase motor.

The single phase induction motor in its simplest form is structurally the same as a polyphase induction motor having a squirrel cage rotor, the only difference is that the single phase induction motor has single winding on the stator which produces mmf stationary in space but alternating in time, a polyphase stator winding carrying balanced currents produces mmf rotating in space around the air gap and constant in time with respect to an observer moving with the mmf. The stator winding of the single phase motor is disposed in slots around the inner periphery of a laminated ring similar to the 3-phase motor.

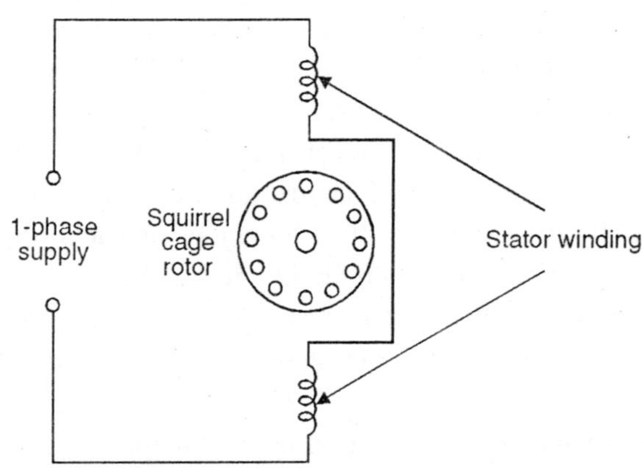

Fig. 9.1. Elementary single phase induction motor.

An induction motor with a cage rotor and single phase stator winding is shown schematically in Fig. 9.1. The actual stator winding as mentioned earlier is distributed in slots so as to produce an approximately sinusoidal space distribution of mmf.

9.2 PRINCIPLE OF OPERATION

Suppose the rotor is at rest and 1-phase supply is given to stator winding. The current flowing in the stator winding gives rise to an mmf whose axis is along the winding and it is a pulsating mmf, stationary in space and varying in magnitude, as a function of time, varying from positive maximum to zero to negative maximum and this pulsating mmf induces currents in the short-circuited rotor of the motor which gives rise to an mmf. The currents in the rotor are induced due to transformer action and the direction of the currents is such that the mmf so developed opposes the stator mmf. The axis of the rotor mmf is same as that of the stator mmf. Since the torque developed is proportional to sine of the angle between the two mmf and since the angle is zero, the net torque acting on the rotor is zero and hence the rotor remains stationary.

For analytical purposes a pulsating field can be resolved into two revolving fields of constant magnitude and rotating in opposite directions as shown in Fig. 9.2 and each field has a magnitude equal to half the maximum length of the original pulsating phasor.

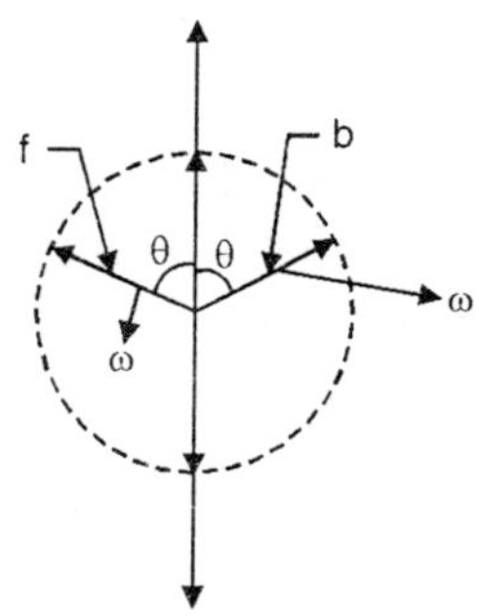

Fig. 9.2. Representation of the pulsating field by space phasors.

These component waves rotate in opposite direction at synchronous speed. The forward (anticlockwise) and backward-rotating (clockwise) mmf waves f and b are shown in Fig. 9.2. In case of 3-phase induction motor there is only one forward rotating magnetic field and hence torque is developed and the motor is self-starting. However, in single phase induction motor each of these component mmf waves produces induction motor action but the corresponding torques are in opposite direction. With the rotor at rest the forward and backward field produce equal torques but opposite in direction and hence no net torque is developed on the motor and the motor remains stationary. If the forward and backward air gap fields remained equal when the rotor is revolving, each of the component fields would produce a torque-speed characteristic similar to that of a polyphase induction motor with negligible leakage impedance as shown by the dashed curves f and b in Fig. 9.3.

The resultant torque-speed characteristic which is the algebraic sum of the two component curves shows that if the motor were started by auxiliary means it would produce torque in whatever direction it was started.

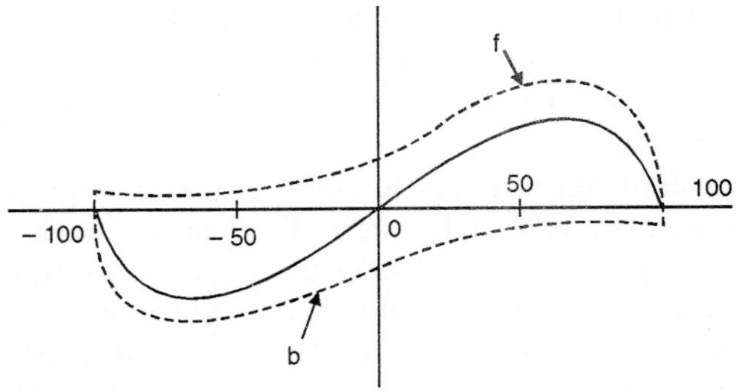

Fig. 9.3. Torque-speed characteristic of a 1-phase induction motor based on constant forward and backward flux waves.

In reality the two fields, forward and backward do not remain constant in the air gap and also the effect of stator leakage impedance can't be ignored. In the above qualitative analysis the effects of induced rotor currents have not been properly accounted for.

When single phase supply is connected to the stator and the rotor is given a push along the forward rotating field, the relative speed between the rotor and the forward rotating magnetic field goes on decreasing and hence the magnitude of induced currents also decreases and hence the mmf due to the induced current in the rotor decreases and its opposing effect to the forward rotating field decreases which means the forward rotating field becomes stronger as the rotor speeds up. However for the backward rotating field the relative speed between the rotor and the backward field increases as the rotor rotates and hence the rotor emf increases and hence the mmf due to this component of current increases and its opposing effect to the backward rotating field increases and the net backward rotating field weakens as the rotor rotates along the forward rotating field. However, the sum of the two fields remains constant since it must induce the stator counter emf which is approximately constant if the stator leakage impedance drop is negligible. Hence, with the rotor in motion the torque of the forward field is greater and that of the backward field is less than what is shown in Fig. 9.3. The true situation being as is shown in Fig. 9.4.

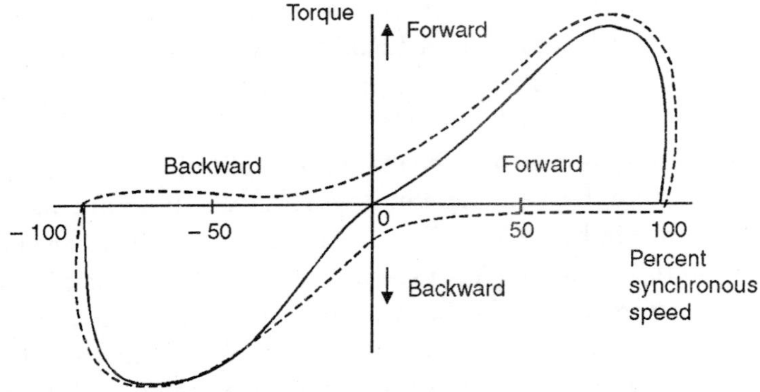

Fig. 9.4. Torque-speed characteristic of a 1-phase induction motor taking into account changes in the flux waves.

In the normal running region at a few per cent slip the forward field is several times stronger than the backward field and the flux wave does not differ materially from the constant amplitude revolving field in the air gap of a balanced polyphase motor. Therefore, in the normal running range of the motor, the torque-speed characteristic of a single phase motor is not very much different from that of a polyphase motor having the same rotor and operating with the same maximum air gap flux density.

In addition to the torque shown in Fig. 9.4, double-stator frequency torque pulsation are produced by the interaction of the oppositely rotating flux and mmf waves which move past each other at twice synchronous speed. These double frequency torques produce no average torque as these pulsations are sinusoidal and over the complete cycle the average torque is zero. However, sometimes these are additive to the main torque and for another half a cycle these are subtractive and therefore a variable torque acts on the shaft of the motor which makes the motor noisier as compared to a polyphase induction motor where the total torque is constant. Such torque pulsations are unavoidable in single phase circuits. Mathematically

$$T \alpha I^2 \quad \text{...(9.1)}$$

Let

$$I = I_m \sin \omega t$$

$$T = K I_m^2 \sin^2 \omega t$$

$$= KI_m^2 \frac{(1 - \cos \omega t)}{2} \quad \text{...(9.2)}$$

So the expression for torque contains a constant term superimposed over by a pulsating torque with pulsation frequency twice the supply frequency.

9.3 STARTING OF SINGLE PHASE INDUCTION MOTORS

The single phase induction motors are classified based on the method of starting method and in fact are known by the same name descriptive of the method. Appropriate selection of these motors depends upon the starting and running torque requirements of the load, the duty cycle and limitations on starting and running current drawn from the supply by these motors. The cost of single phase induction motor increases with the size of the motor and with the performance such as starting torque to current ratio (higher ratio is desirable), hence, the user will like to go in for a smaller size (hp) motor with minimum cost, of course, meeting all the operational requirements. However, if a very large no. of fractional horsepower motors are required, a specific design can always be worked out which might give minimum cost for a given performance requirements. Following are the starting methods.

(*a*) **Split-phase induction motor.** The stator of a split phase induction motor has two windings, the main winding and the auxiliary winding. These windings are displaced in space by 90 electrical degrees as shown in Fig. 9.5 (*a*). The auxiliary winding is made of thin wire (super enamel copper wire) so that it has a high R/X ratio as compared to the main winding which has thick super enamel copper wire. Since the two windings are connected across the supply the current I_m and I_a in the main winding and auxiliary winding lag behind the supply voltage V, I_a leading the current I_m Fig. 9.5(*b*). This means the current through auxiliary winding reaches maximum value first and the mmf or flux due to I_a lies along the axis of the auxiliary winding and after some time ($t = \theta/w$) the current I_m reaches maximum value and the mmf or flux due to I_m lies along the main winding axis. Thus the motor becomes a 2-phase unbalanced motor. It is unbalanced since the two currents are not exactly 90 degrees apart. Because of these two fields a starting torque is developed and the motor becomes a self-starting motor. After the motor starts, the auxiliary winding is disconnected usually by means of centrifugal switch that operates at about 75 per cent of synchronous speed. Finally the motor runs because of the main winding. Since this being single phase some level of humming noise is always associated with the motor during running. A typical torque speed characteristic is shown in Fig. 9.5 (*c*). It is to be noted that the direction of rotation of the motor can be reversed by reversing the connection to either the main winding or the auxiliary windings.

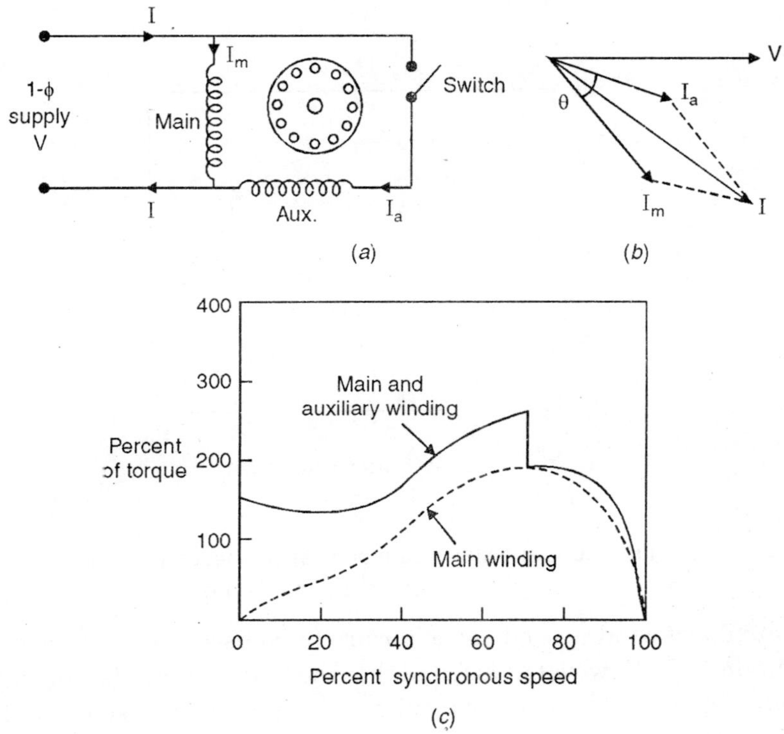

Fig. 9.5. Split phase induction motor (*a*) Connection (*b*) Phasor diagram at starting (*c*) Typical torque-speed characteristic.

(*b*) **Capacitor start induction motor.** Capacitors are used to improve the starting and running performance of the single phase inductions motors.

The capacitor start induction motor is also a split phase motor. The capacitor of suitable value is connected in series with the auxiliary coil through a switch such that I_a the current in the auxiliary coil leads the current I_m in the main coil by 90 electrical degrees in time phase so that the starting torque is maximum for certain values of I_a and I_m. This becomes a balanced 2-phase motor if the magnitude of I_a and I_m are equal and are displaced in time phase by 90° electrical degrees. Since the two windings are displaced in space by 90 electrical degrees as shown in Fig. 9.6 maximum torque is developed at start. However, the auxiliary winding and capacitor are disconnected after the motor has picked up 75 per cent of the synchronous speed. The motor will start without any humming noise. However, after the auxiliary winding is disconnected, there will be some humming noise.

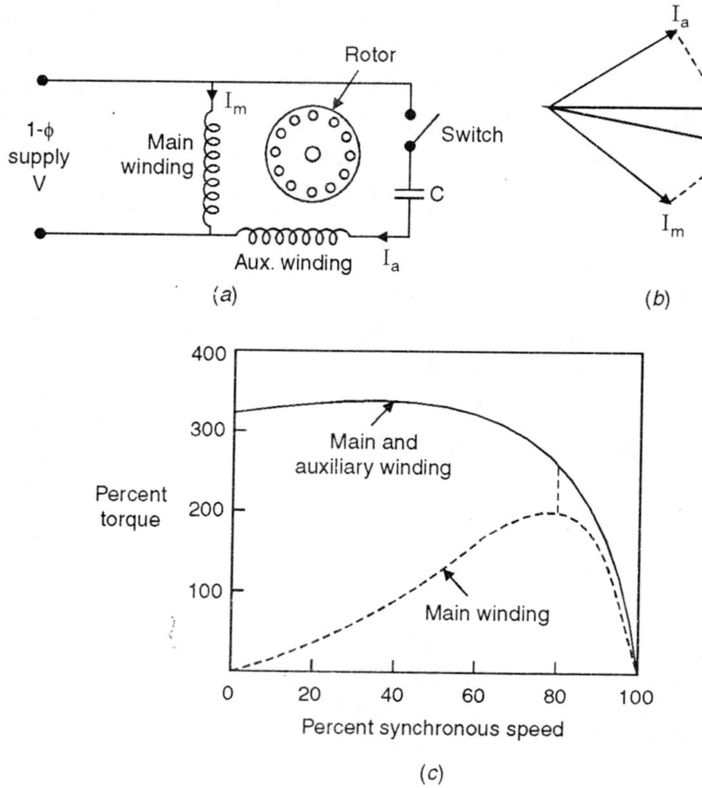

Fig. 9.6. Capacitor start motor (*a*) Connection
(*b*) Phasor diagram at start (*c*) Speed torque curve.

Since the auxiliary winding and capacitor are to be used intermittently, these can be designed for minimum cost. However, it is found that the best compromise among the factors of starting torque, starting current and costs results with a phase angle somewhat less than 90° between I_m and I_a. A typical torque-speed characteristic is shown in Fig. 9.6 (*c*) high starting torque being an outstanding feature.

(*c*) **Permanent-split capacitor motor.** In this motor the auxiliary winding and capacitor are not disconnected from the motor after starting, thus the construction is simplified by the omission of the switch as shown in Fig. 9.7 (*a*).

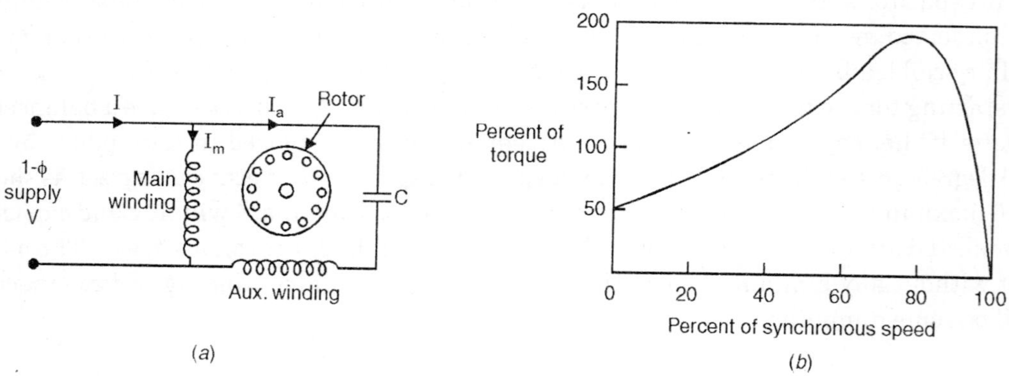

Fig. 9.7. Permanent split capacitor motor (*a*) Connection (*b*) Torque-speed characteristic.

Here the auxiliary winding and capacitor could be so designed that the motor works as a perfect 2-phase motor at anyone desired load. With this the backward rotating magnetic field would be completely eliminated. The double stator frequency torque pulsations would also be eliminated, thereby the motor starts and runs as a noise free motor. With this there is improvement in p.f. and efficiency of the motor. However, the starting torque must be sacrificed as the capacitance is necessarily a compromise between the best starting and running characteristics. The torque-speed characteristic of the motor is shown in Fig. 9.7 (*b*).

(*d*) **Capacitor start capacitor run motor.** If two capacitors are used with the auxiliary winding as shown in Fig. 9.8 (*a*), one for starting and other during the start and run, theoretically optimum starting and running performance can both be achieved.

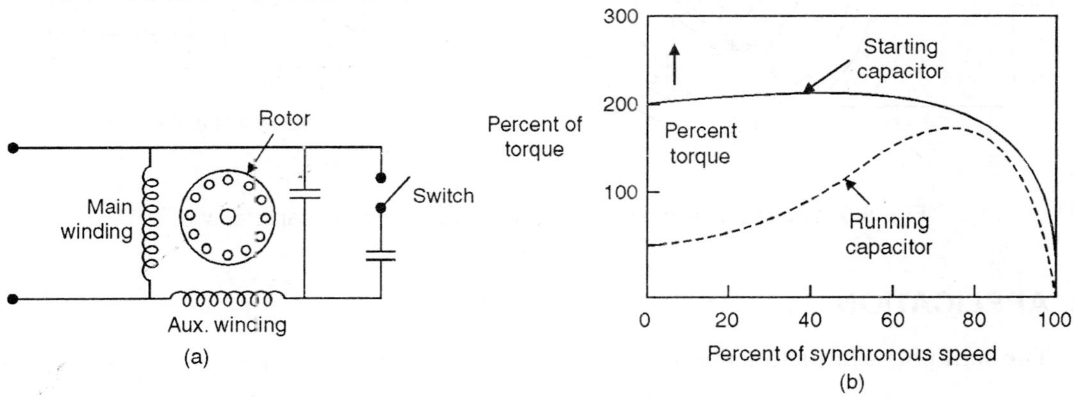

Fig. 9.8. (*a*) Capacitor start capacitor run motor (*b*) Torque-speed characteristic.

The small value capacitor required for optimum running conditions is permanently connected in series with the auxiliary winding and the much larger value required for starting is obtained by a capacitor connected in parallel with the running capacitor. The starting capacitor is disconnected after the motor starts.

The value of the capacitor for a capacitor start motor is about 300 μF for $\frac{1}{2}$ hp motor. Since this capacitor must carry current for a short starting period, the capacitor is a special compact ac electrolytic type made for motor starting duty. However, the capacitor permanently connected has a typical rating of 40 μF ; since it is connected permanently, the capacitor is an ac paper, foil and oil type. The cost of the motor is related to the performance ; the permanent capacitor motor is the lowest cost, the capacitor start motor next and the capacitor start capacitor run has the highest cost.

(*e*) **Shaded pole induction motor.** Fig. 9.9 (*a*) shows schematic diagram of shaded pole induction motor. The stator has salient poles with one portion of each pole surrounded by a short-circuited turn of copper called a shading coil. Induced currents in the shading coil (acts as an inductor) cause the flux in the shaded portion of the pole to lag the flux in the other portion. Hence the flux under the unshaded pole leads the flux under the shaded pole which results in a rotating field moving in the direction from unshaded to the shaded portion of the pole and a low starting torque is produced which rotates the rotor in the direction from unshaded to the shaded pole. A typical torque speed characteristic is shown in Fig. 9.9 (*b*). The efficiency is low. These motors are the least expensive type of fractional horse power motor and are built upto about $\frac{1}{20}$ hp. Since the rotation of the motor is in the direction from unshaded towards the shaded part of the pole, a shaded pole motor can be reversed only by providing two sets of shading coils which may be opened and closed or it may be reversed permanently by inverting the core.

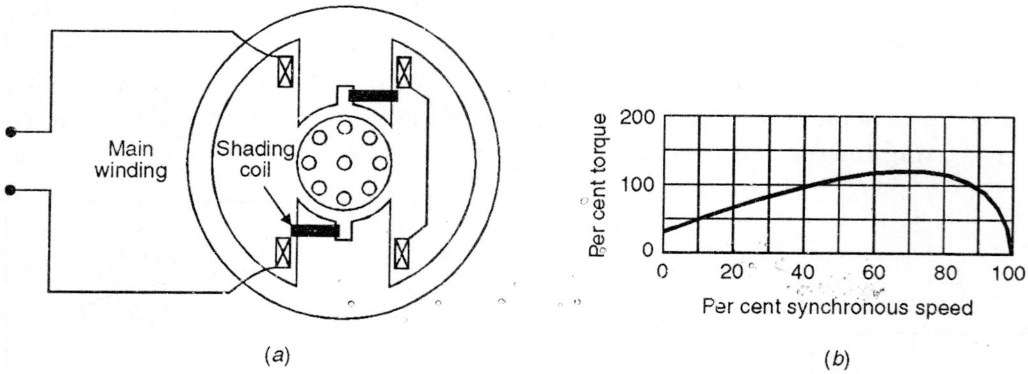

Fig. 9.9. Shaded-pole motor and typical torque-speed characteristic.

9.4 APPLICATION

The split phase induction motors are used for fans, blowers, centrifugal pumps and office equipments. Typical ratings are $\frac{1}{20}$ to $\frac{1}{2}$ hp; in this range they are the lowest cost motors available. The capacitor start motors are used for compressors, pumps, refrigeration and air-conditioning equipments and other hard to start-loads.

The capacitor start capacitor run motors are manufactured in a number of sizes from $\frac{1}{8}$ to $\frac{3}{4}$ hp and are used in compressors, conveyors, pumps and other high torque loads. The permanent split capacitor motors are manufactured in the range of $\frac{1}{20}$ hp to $\frac{3}{4}$ hp and are used for direct connected fans, blowers, centrifugal pumps and loads requiring low starting torque.

The shaded pole motors are used in toys, hair driers, deskfans etc.

PROBLEMS

9.1. Explain why 3-phase induction motor is self-starting and 1-phase is not.

9.2. Explain with the help of schematic diagram and phasor diagram the principle of operation of 1-phase split phase induction motor.

9.3. Describe how to reverse the following single phase induction motors (*a*) Resistance split phase (*b*) Capacitor start (*c*) Permanent slip capacitor (*d*) Two value capacitor (*e*) Shaded pole motor.

9.4. A single phase induction motor has poor performance as compared to a 3-phase induction motor. Discuss the reasons.

9.5. Discuss the differences between capacitor start, capacitor start capacitor run and permanent split capacitor motors.

CHAPTER

10 Power System

10.1 INTRODUCTION

Electric power system is the most capital intensive and the most complex system ever developed by man.

An electric power system consists of three major components; (*i*) Generation, (*ii*) Transmission and (*iii*) Distribution. A system may or may not have transmission component depending upon the distances of the generating system from the consumers of electric energy but distribution is an integral part of any electric power system. The intermediate phase between the generation and distribution has come into existence because of the long distances between the energy sources and the sinks (consumers). Electrical transmission is the most efficient method of transmitting power over long distances. Transmission system can further be classified as (*i*) primary transmission (275 KV and above) and (*ii*) secondary transmission (220 K V to 66 K V) depending upon the level of voltage used for transmission. Similarly distribution system can be further classified as (*i*) primary distribution (33 KV and above), (*ii*) secondary distribution (11 KV/6.6 KV/ 3.3 KV) and (*iii*) tertiary distribution (400 volts 3 phase) depending upon the voltage of distribution. A single line diagram of a typical electrical power system is shown in Fig. 10.1.

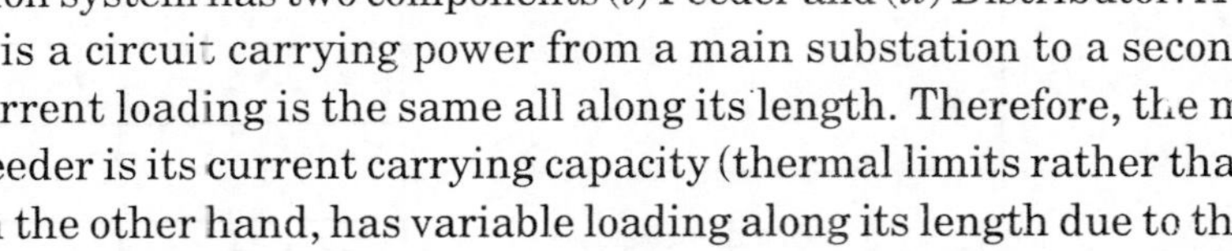

Fig. 10.1. A Typical Electric Power System.

Distribution system has two components (*i*) Feeder and (*ii*) Distributor. A feeder in a distribution network is a circuit carrying power from a main substation to a secondary substation, such that the current loading is the same all along its length. Therefore, the main criterion for the design of a feeder is its current carrying capacity (thermal limits rather than voltage drops). A distributor, on the other hand, has variable loading along its length due to the service connections, tapped off at intervals by the individual consumers. The voltage variation at consumer's terminals must be maintained within ± 5% in order to avoid unsatisfactory operation of equipment, *e.g.* a reduction in voltage will adversely affect the output of an induction motor or light from a filament lamp. Therefore, the main criterion for the design of a distributor is the voltage drop, which must be kept within certain limits. A typical distribution system in a city area is shown in Fig. 10.2.

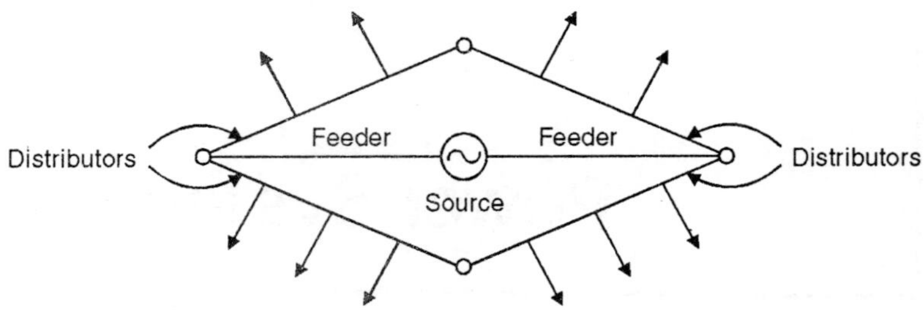

Fig. 10.2. A Typical Distribution System.

The distribution system occupies an important place in any electric power system. The effectiveness with which it achieves its objective of distributing electric energy to various consumers, is measured in terms of voltage regulation, flexibility, security of supply efficiency and cost.

10.2 TYPES OF DISTRIBUTION SYSTEM

Depending upon the type of supply, distribution system can be classified as follows :

(*i*) a.c. single phase (single phase loads only);

(*ii*) 3-phase, 3-wire (3 phase loads only); and

(*iii*) 3-phase, 4-wire (all types of loads).

The distribution system can also be classified depending upon the connections. These are the following two systems :

(*i*) Radial system and

(*ii*) Ring Mains system.

10.2.1 The Radial Systems

The electric energy distribution originally was through radial system. A typical radial distribution system is shown in Fig. 10.3.

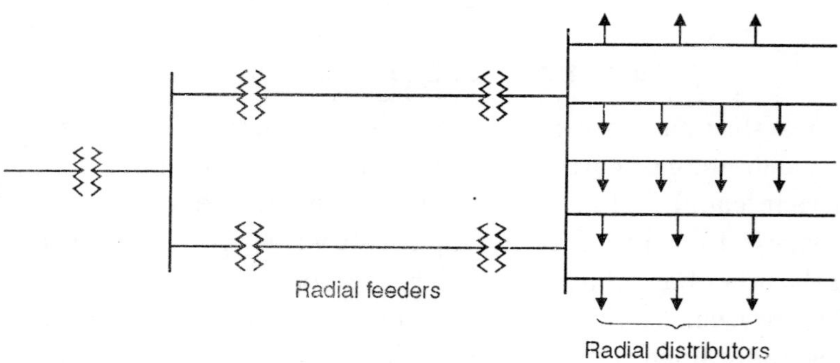

Fig. 10.3. Radial Distributors.

The advantages of radial system are its simplicity and low cost, which result from a straight forward circuit arrangement where a single or radial path is provided between the consumer and the source or bulk power supply. With such an arrangement, the amount of switching equipment required is small and the protective relaying is simple. The major disadvantage of radial system is its lack of security of supply. When a fault occurs on any section of the feeder; a number of consumers will be without supply for a considerable period. The radial system is normally used for rural distribution these days.

10.2.2 The Ring Mains Systems

The system is used most frequently to supply bulk loads such as small industrial loads and medium or large commercial buildings where continuity of supply is of considerable importance. A typical ring mains for feeder and distributors is shown in Fig. 10.4 (*a*).

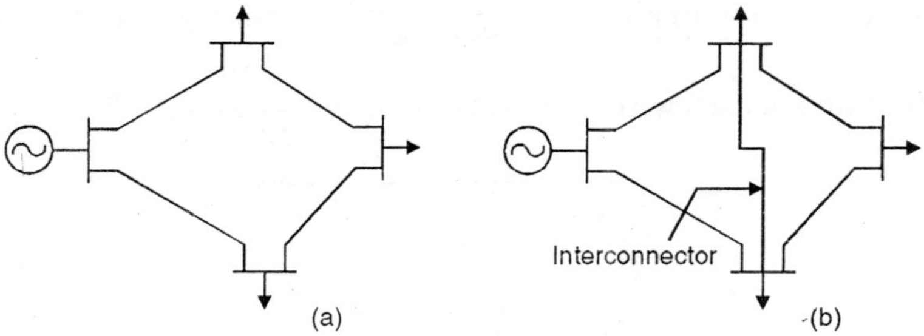

Fig. 10.4. Ring Mains (*a*) Ring (*b*) interconnected.

It can be shown that when an interconnector is used in a ring mains system [Fig. 10.4 (*b*)], it reduces the voltage drop between the points to which it is connected. The ring mains systems is used for urban distribution in contrast to the radial system.

10.3 COST COMPARISON OF SUPPLY SYSTEMS

The various systems considered for cost comparison are

(*i*) d.c. two wires

(*ii*) a.c. single phase two wire

(*iii*) a.c. three phase four wire.

The terms of comparison are (*a*) same maximum phase to ground voltage and (*b*) same efficiency or losses to transmit a given amount of power over a given distance. The quantity to be compared is the volume of copper required in each case.

D.C. two wire with midpoint earthed

Refer to Fig. 10.5 for d.c. two-wire midpoint earthed system. Let V be the voltage between either outer terminal and the earth so that voltage between the two outer terminals is 2V. Let P be the power transmitted. Therefore, the current in each outer conductor will be $I = \frac{P}{2V}$. If R_1 is the resistance of each conductor the copper loss will be

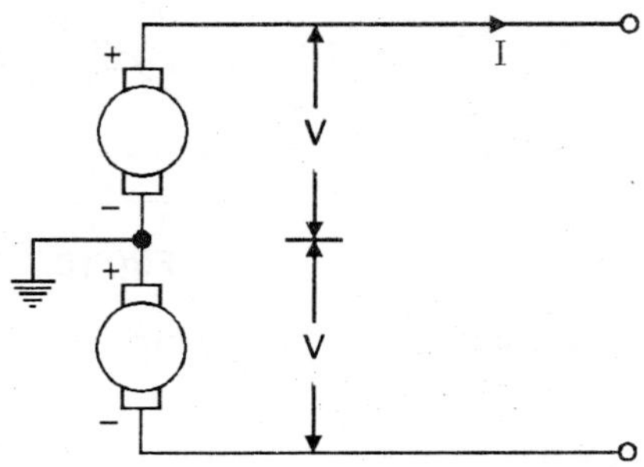

Fig. 10.5. D.C. Two-Wire System.

$$\text{Copper loss} = \frac{2P^2}{4V^2} R_1 = \frac{P^2}{2V^2} R_1$$

A.C. Single Phase Two-Wire

Refer to Fig. 10.6 for a.c. single phase two-wire system with middle point earthed. The maximum voltage between one of the outer wires and the earth is V volts. Therefore,

$$V_{rms} = \frac{V}{\sqrt{2}}$$

The voltage between the two outer conductors is $\frac{2V}{\sqrt{2}} = \sqrt{2}V$

Therefore current in the outer wire $I = \frac{P}{\sqrt{2}\, V \cos\phi}$ if ϕ is the power factor angle of the load.

Now if R_2 is the resistance of each wire, the copper loss $= 2 \cdot \frac{P^2}{2V^2 \cos^2\phi} R_2$

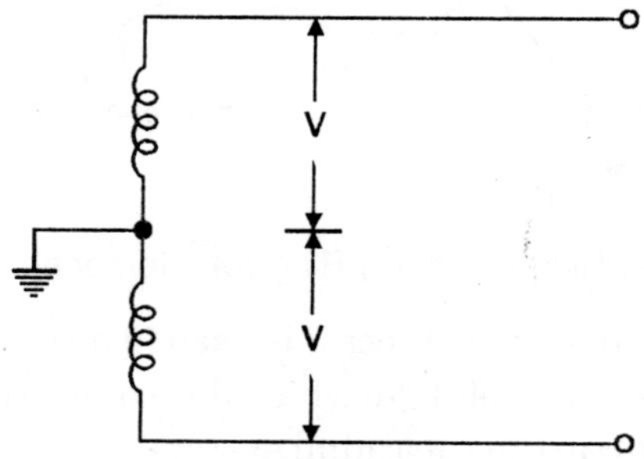

Fig. 10.6. A.C. Two Wire System.

$$\text{Copper loss} = \frac{P^2}{V^2 \cos^2\phi} R_2$$

A.C. 3-phase 4-wire system Refer to Fig. 10.7 for a 3-phase, 4-wire system

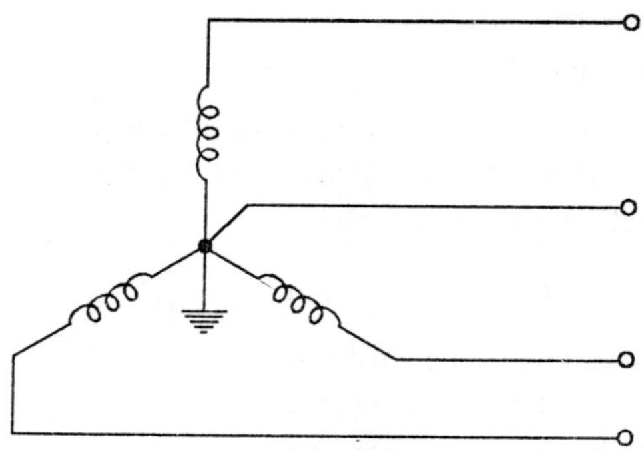

Fig. 10.7. Three Phase 4-Wire System.

The maximum voltage between one phase and grounded neutral is V volts.

$$V_{rms} = \frac{V}{\sqrt{2}}$$

$\therefore$ Current per phase under balanced condition $= \frac{P}{3 \times \frac{V}{\sqrt{2}} \cos\phi}$

where ϕ is the power factor angle of load. If R_3 is the resistance per conductor, copper loss is given by

$$\text{Copper loss} = 3\,\frac{2P^2}{9\,V^2\cos^2\phi}R_3 = \frac{2}{3}\frac{P^2}{V^2\,|\cos^2\phi}R_3$$

Equating the copper loss due to 3 phase 3-wires as calculated above with d.c. two wire system

$$\frac{P^2R_1}{2V^2} = \frac{2P^2}{3\,V^2\cos^2\phi}R_3$$

or

$$\frac{R_1}{R_3} = \frac{4}{3}\times\frac{1}{\cos^2\phi} \quad \text{or} \quad \frac{a_3}{a_1} = \frac{4}{3}\frac{1}{\cos^2\phi}$$

Since in case of three-phase three-wire system there are three wires and in case of d.c. two-wire system there are two wires, the volumes required for the same length l, are $3a_3\,l$ and $2a_1l$ respectively

$$\frac{3a_3l}{2a_1l} = \frac{3}{2}\frac{a_3}{a_1} = \frac{3}{2}\frac{4}{3}\frac{1}{\cos^2\theta}$$

In case the fourth wire, in case of 3-phase, 4-wire system is of the same size as the phase wires, the volume of copper required will be $4a_3l$ and the ratio

$$\frac{4a_3l}{2a_1l} = \frac{2a_3}{a_1} = 2\,\frac{4}{3}\frac{1}{\cos^2\theta} = \frac{8}{3}\frac{1}{\cos^2\theta}$$

Similarly, to compare the volume of copper required in case of a.c. single phase two wire system with the d.c. two-wire system we equate copper losses

$$\frac{P^2}{2V^2}R_1 = \frac{P^2}{V^2\cos^2\phi}R_2$$

or

$$\frac{R_1}{R_2} = \frac{2}{\cos^2\phi} \quad \text{or} \quad \frac{a_2}{a_1} = \frac{2}{\cos^2\phi}$$

Since the number of wires required is two in each case the ratio of copper volume is

$$1 : \frac{2}{\cos^2\phi}$$

Therefore, the ratio of copper volume required in the three cases under study when d.c. two wire system is taken as reference are

$$1 : \frac{2}{\cos^2\phi} : \frac{8}{3}\cdot\frac{1}{\cos^2\phi}$$

In case the system is single phase two wire with neutral grounded the reader will find that the ratio of volume of copper with reference to d.c. two wire will be

$$1 : \frac{8}{\cos^2\phi}$$

Example 10.1. A d.c. two-wire system is to be converted into a.c. 3-phase 3-wire system by the addition of a third conductor of the same cross-section as the existing conductors. If the rms voltage between wires and the percentage loss in the line remains unchanged, determine the additional load which can now be supplied. Assume balanced load of unity p.f.

Solution. (*a*) d.c. two-wire : Let the power transmitted in this case be P and V be the voltage between wires

$$I = \frac{P}{V}$$

Therefore $$\text{Loss} = \frac{2P^2}{V^2}R$$

where R is the resistance of each wire

Therefore,

$$\text{Per unit loss} = \frac{2P}{V^2}R$$

(*b*) 3-phase 3-wire : The r.m.s. voltage = V volts.

Let P' be the power transmitted then $I' = \dfrac{P'}{\sqrt{3}V}$ for unity p.f.

Therefore,

$$\text{Copper loss} = 3\left(\frac{P'}{\sqrt{3}V}\right)^2 R = \frac{P'^2}{V^2}R$$

Therefore $$\text{per unit loss} = \frac{P'}{V^2}R$$

Equating the per unit loss we have

$$\frac{2P}{V^2}R = \frac{P'}{V^2}R \quad \text{or} \quad P' = 2P$$

This shows that 100% additional power can be supplied by 3-phase 3-wire system.

10.4 POWER FACTOR IMPROVEMENT

Power factor of a circuit or installation is defined as the ratio of power consumed by the circuit (in watts or kW or MW) to the total complex power input to the circuit or the installation and is denoted by cos ϕ where ϕ is the power factor angle and is the angle between the circuit voltage and current through the circuit. Consider Fig. 10.8 (*a*) where the current lags

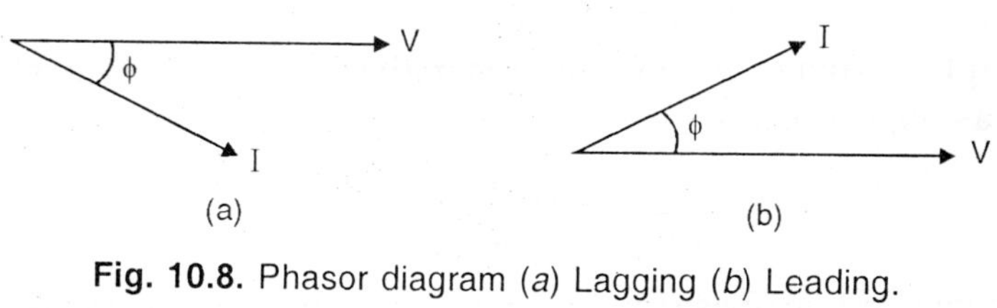

Fig. 10.8. Phasor diagram (*a*) Lagging (*b*) Leading.

the voltage by an angle ϕ whereas in (*b*) it leads the voltage by an angle ϕ. The active power consumed in both the circuit is $|V|\ |I|\cos\phi$ and the complex power is $|V|\ |I|$. The units are watts and volt-amperes respectively. However, the reactive power in (*a*) is being absorbed by the circuit as it is an inductive circuit whereas in (*b*) the reactive power is being supplied (lagging) or leading reactive power is taken from the supply due to the presence of capacitor in circuit (*b*) The p.f. in a circuit is decided by what the net reactive power is present in the circuit. So we first study the sources and sinks of reactive power. In brief the inductance wherever it be, whether the winding of a generator or a motor or the series inductance of a line or a simple coil absorb (sink) the reactive power whereas the capacitance of a line or static capacitor or synchronous motor when overexcited acts as a synchronous capacitor.

In a circuit if both the inductor and capacitor are present, the reactive power due to inductor (phasor) is in phase opposition to the reactive power due to capacitor. The net reactive power will decide whether the circuit as a whole is inductive or capacitive. However, if the two reactive powers are equal in the circuit, the net reactive power requirement of the circuit is zero and hence the p.f. of the circuit is unity. The three possible situations are shown in Fig. 10.9. Conventionally, the reactive power due to inductor is taken as positive and that due to capacitor as negative. This is because the reactive power supplied by the source is taken as positive and that supplied by the load to the source is taken as negative. This convention is used in Fig. 10.9.

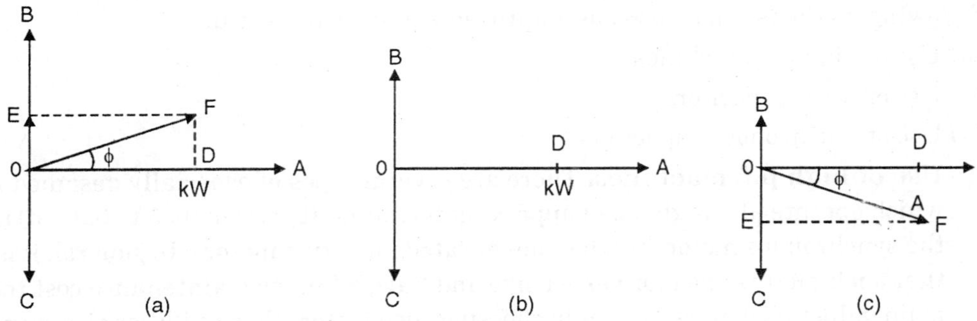

Fig. 10.9. Effect of reactive power on p.f.

In Fig. 10.9 (*a*) the inductive reactive power is greater than the capacitive and hence $OB - OC(BE) = OE$ is the net positive reactive power and OF the complex power in K VA and hence p.f.

$$\cos \phi = \frac{OD}{OF} \text{ lagging}$$

In Fig. 10.9 (*b*) since $OB = OC$ hence the net reactive power is zero and the p.f. of the circuit is unity *i.e.* the circuit behaves as resistive. In Fig. 10.9 (*c*) $OC > OB$, hence the net capacitive reactive power is $OE = OC - OB$ (CE) and the power factor is $\cos \phi = \dfrac{OD}{OF}$ lead. Reactive power in a circuit is defined as the wattless component of the complex power *i.e.* the component which does not do any effective work yet it is required in most of the electrical equipments and is sort of necessary evil just as friction is a necessary evil in our daily life activities. Reactive power is the time rate of change of electromagnetic energy stored in an inductor or the electrostatic energy stored in a capacitor. The presence of reactive power not only affects the p.f. of the circuit, it also affects the voltage levels at different points of the circuit. It is found that in a circuit if reactive power generated is more than the reactive power absorbed by the circuit, the voltage levels rise and voltage levels dip otherwise.

In an electric power system it is difficult to predict the load variation accurately. The load devices may vary from a few watt night lamps to multi megawatt induction motors. The following categories of loads are present in a system.

(*i*) Motor devices	70%
(*ii*) Heating and lighting equipment	25%
(*iii*) Electronic equipment	5%

10.4.1 Causes of low p.f.

Among the chief causes of poor p.f. may be mentioned, induction motors (which form almost 70% of the total load) particularly in small sizes (fractional hp motors) and large motors when running partially loaded, arc and induction furnaces, electric discharge lamps (associated with ballasts) and transformers. The transformers are also predominantly inductive circuits and in fact each megawatt generated, transmitted and distributed requires an equivalent of 3 to 4 MVA capacity transformer, at different voltage levels.

10.4.2 Methods to improve p.f.

Following methods can be used for improvement of p.f. of a circuit.

(*i*) Use of high p.f. machines

(*ii*) Use of static capacitors

(*iii*) Use of synchronous capacitors.

(*i*) **Use of high p.f. machines.** There are several types of especially designed motors which operate at a high p.f. of approximately unity (between 0.95 to 0.97 pf) notably the synchronous motor and the compensated induction motors. In general, it is found that such motors are more expensive and have a higher maintenance cost than the plain induction motors, so that in choosing such motors these additional costs must be balanced against the savings in energy cost due to their higher p.f. Of the above types, the synchronous motor is particularly useful as it can be very easily designed to operate at a leading p.f. and, therefore, compensate for the lagging p.f. of other machines.

(*ii*) **Use of static capacitor.** Consider Fig. 10.10 where load (inductive) is connected to the source.

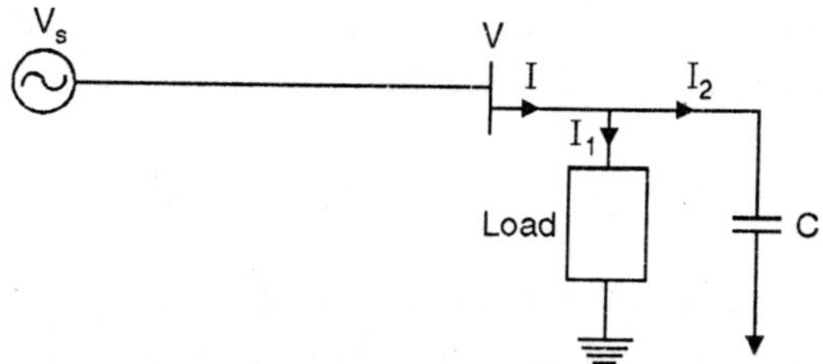

Fig. 10.10. Use of Static Capacitor.

For improvement of p.f. or improvement of voltage profile it is always desirable to connect a capacitor across the inductive load as shown in Fig. 10.10. As explained in article 7.7, a capacitor takes leading vars or it is equivalent to saying that it supplies lagging vars. Consider the phasor diagram in Fig. 10.11.

In Fig. 10.11 (*a*) the phasor diagram represents the kW, kVAr and kVA without the capacitor where the p.f. angle is ϕ and p.f. $\cos \phi$.

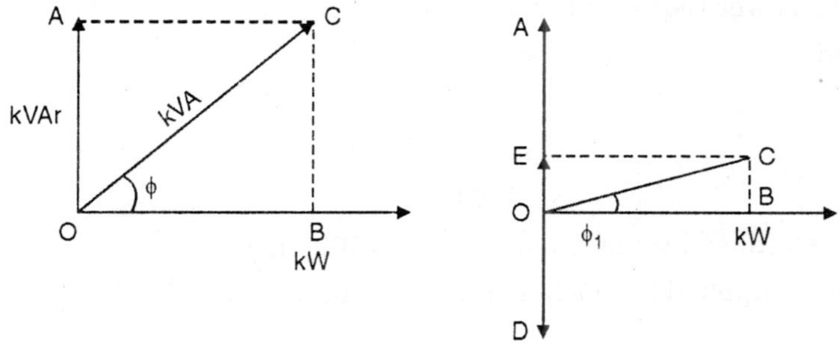

Fig. 10.11. p.f. improvement using static capacitor.

In Fig. 10.11 (*b*) when a capacitor *C* is connected across the load, it takes leading VArs or supplies the lagging VArs to the load and hence, the net lagging VArs to be supplied by the source are reduced by an amount *OD* and the source has to supply kVArs equivalent to $OA - OD = EO$ and the p.f. of the busbar improves to $\cos \phi_1$ as $\phi_1 < \phi$. It is to be noted very carefully that the p.f. of the load does not change here, it is a fixed quantity but it is the p.f. of the system which is improved with the use of a suitable value of capacitor across the load.

(*iii*) **Use of synchronous capacitor.** This has already been discussed at length in Article 7.7. The readers are advised to go through this article.

However, it may be mentioned here that to improve the p.f. above 0.95 requires a large amount of leading kVAr and the cost of the equipment to improve it to such high values may well out weigh any savings in energy costs.

Example 10.2. A slow speed alternator works in parallel with a turbo-alternator, the combined output being 2500 kW at 0.8 p.f. lag. If the turbo altenator provides 1000 kW at unity p.f., at what p.f. will the slow speed machine work ?

Solution. Total active power of the two alternators 2500 kW

Total reactive power 2500 tan $\cos^{-1} 0.8$

$$= 2500 \times 0.75$$
$$= 1875 \text{ kVAr}$$

Active power supplied by turbo alternator 1000 kW

Reactive power supplied by turbo alternator = zero

Hence active power supplied by the slow speed alternator = 2500 – 1000 = 1500 kW

And reactive power supplied by the slow speed alternator = 1875 kVAr

Hence p.f. of the slow speed alternator $\cos \tan^{-1} \dfrac{1875}{1500} = 0.624$ **Ans.**

Example 10.3. Two alternators working in parallel supply a lighting load of 3000 kW and a motor load aggregating 5000 kW at a p.f. of 0.71. One machine is loaded to 5000 kW at p.f. 0.8 lag. Determine the load and p.f. of the other machine.

Solution. Total load active power requirement

Lighting load		3000 kW
Motor load		5000 kW
	Total	8000 kW

Total reactive power requirement of the load

Lighting load $= \text{zero}$

Motor load $= 5000 \times \tan \cos^{-1} 0.71$

$= 5000 \times 0.9918$

$= 4959$ kVAr

Active power supplied by one machine $= 5000$ kW

Reactive power supplied by one machine $= 5000 \tan \cos^{-1} 0.8$

$= 5000 \times \frac{3}{4}$

$= 3750$ kVAr.

Active power supplied by the other machine = 8000 – 5000 = 3000 kW

Reactive power supplied by the other machine = 4959 – 3750 = 1209 KVAr

p.f. of the other machine $\cos \tan^{-1} \frac{1209}{3000} = 1209$ kVAr

$= 0.927$ **Ans.**

Example 10.4. A single phase motor takes a current of 40A at p.f. 0.7 lag from a 440 V 3-phase 50 Hz supply. Determine the value of a shunt capacitor so as to raise the p.f. to 0.9 lag, the load remaining the same.

Solution. The power active taken by the motor

$$\frac{440}{\sqrt{3}} \times 40 \times 0.7 = 7112 \text{ watt}$$

and

$$\text{reactive power} = \frac{440}{\sqrt{3}} \times 40 \sin \cos^{-1} 0.7$$

$$= 254 \times 40 \times 0.714$$

$$= 7254 \text{ VArs.}$$

The new p.f. is 0.9 lag, power remaining same reactive power requirement for this p.f.

$$7112 \tan \cos^{-1} 0.9 = 3444 \text{ VArs}$$

The VArs to be supplied by the shunt capacitor are 7254 – 3444 = 3810 VArs

The value of capacitor is given by

$$V^2wc = 254^2 \times 314\, c = 3810$$

or

$$c = \frac{3810}{254^2 \times 314} = 188\ \mu\text{F} \quad \textbf{Ans.}$$

Example 10.5. A 500 W discharge lamp takes a current of 4A at unity p.f. Calculate the inductance of a choke required to enable the lamp to work on 250 V 50 Hz mains. Determine also the capacitance to be connected across the mains to bring the resultant p.f. to unity.

Solution. Since the lamp takes 4A at unity p.f. the supply voltage should be $\frac{500}{4} = 125$ volts.

However we want to connect, the same lamp across 250 volt. The circuit is as follows

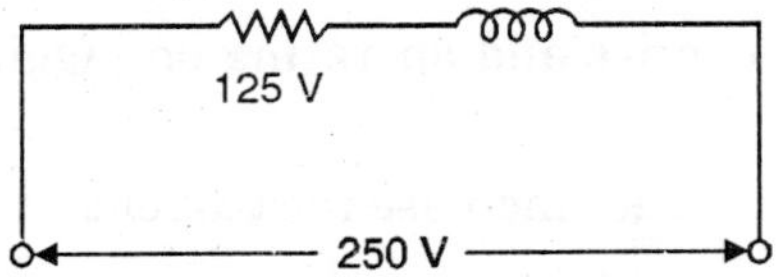

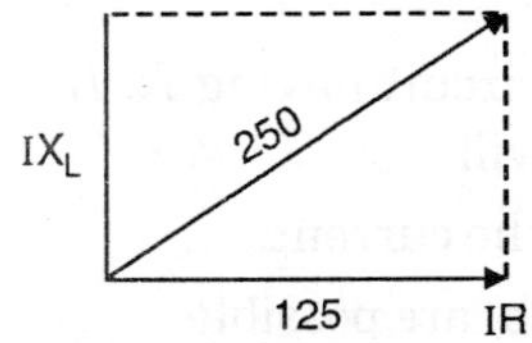

From the phasor diagram, the voltage across the choke would be

$$\sqrt{250^2 - 125^2} = 216.5 \text{ volt}$$

The reactance of the choke $= \dfrac{216.5}{4} = 54.12$

$$wL = 54.12$$

$$314 \text{ L} = 54.12. \quad \text{Hence L} = 0.172 \text{ H}$$

The reactive power requirement of the choke $I^2wL = 16 \times 0.172\,w$

and if C is the capacitance required across the supply, the reactive power supplied by the capacitor is

$$V^2wC = 250^2wC = 16 \times 0.172\,w$$

or $$250^2C = 16 \times 0.172$$

or $$C = 44\ \mu\text{F}.$$ **Ans.**

PROBLEMS

10.1. Draw single line diagram of a power system. Label all the major components of the system.

10.2. Discuss the advantages of ring main distribution system over the radial system.

10.3. Compare a 3-phase 4-wire system with the single phase system.

10.4. Describe briefly various supply systems in power system.

10.5. What are the causes of low p.f. in power system and discuss various methods of improving the power factor.

10.6. Compare the performance of static capacitor with a synchronous condenser as a p.f. improvement device.

10.7. A synchronous generator supplying 2000 kW operates in parallel with an induction generator supplying 1000 kW. The load has a lagging p.f. of 0.8. The induction generator has a p.f. of 0.95. Determine the p.f. at which the synchronous machine is working. **Ans.** 0.612

10.8. A synchronous motor overexcited to take 600 kVA gives 360 hp at an efficiency of 0.9. It runs in parallel with a load of 1000 kVA at p.f. 0.8 lag. Determine the resultant p.f. **Ans.** 0.99 lag

MULTIPLE CHOICE QUESTIONS

1. In an a.c. circuit having R, L and C in series and operating on lagging $p\ f$, increase in frequency will

(*a*) reduce the current. (*b*) increase the current.

(*c*) (a) and (b) are possible. (*d*) have no effect on current drawn.

2. In a circuit, voltage and current are given by

$$V = 10 \sin (\omega t + 30°)$$

and $$i = 10 \sin (\omega t - 30°)$$

The power consumed in the circuit is

(*a*) 100 watts (*b*) 50 watts

(*c*) 25 watts (*d*) 12.5 watts.

3. A series circuit is given by $z = 5 - j12$, its susceptance is

(*a*) 5/13 (*b*) 5/169

(*c*) 12/13 (*d*) 12/169.

4. In a pure LC circuit (parallel) under resonance condition, current drawn from the supply main is

(*a*) Very large (*b*) $V\sqrt{LC}$

(*c*) $V/\sqrt{LC}$ (*d*) Zero.

5. The current response as a function of frequency in a series resonant circuit is given by

(*a*) 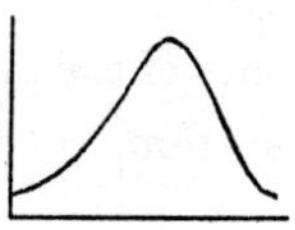 (*b*) 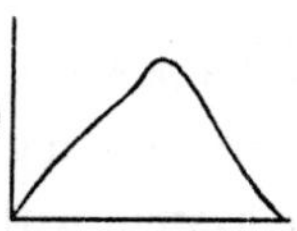

(*c*) 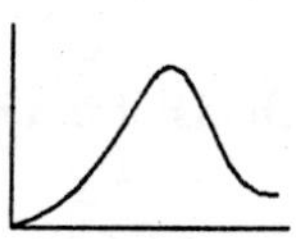 (*d*) 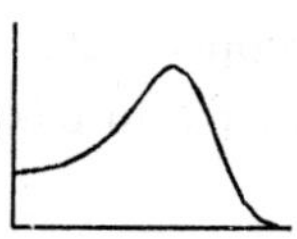

6. A drawn wire of resistance 5 Ω is further drawn so that its diameter becomes one fifth. Its resistance will now be (volume remaining same).

(*a*) 625 (*b*) 125

(*c*) 25 (*d*) None of these.

7. A 500 watt 220 V bulb is supplied with 110 volts. Power consumption by the bulb will be

(*a*) 250 watt (*b*) 125 watt

(*c*) slightly more than 125 watt (*d*) slightly less than 125 watt.

8. In series resonance circuit, increasing inductance to twice its value and reducing capacitance to half its value

(*a*) will change the resonance frequency.

(*b*) will change the impedance at resonance frequency.

(c) will change the maximum value of current.

(d) will increase the selectivity of the circuit.

9. For the circuit shown in figure the value of current I is

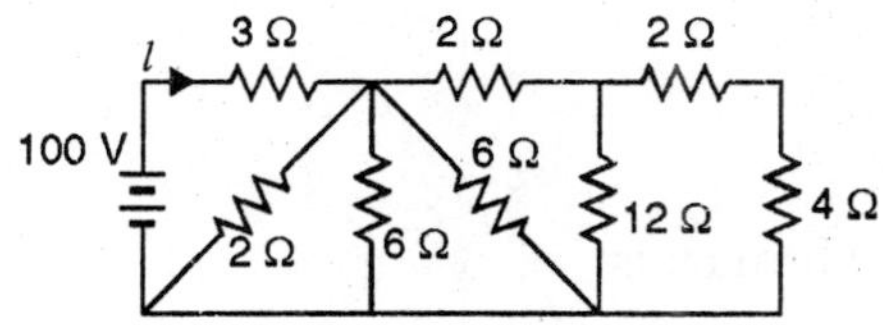

Fig. 1

(a) 10 A (b) 15 A

(c) 20 A (c) 25 A.

10. The maximum power that can be distributed in the load in the circuit shown is

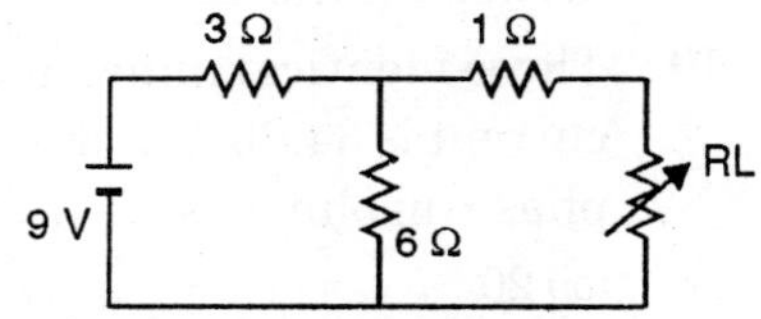

Fig. 2

(a) 3 watts (b) 6 watts

(c) 6.75 w (d) 13.5 watt

11. In delta connected circuit when one resistor is open, the power will be

(a) zero (b) reduced to 1/3

(c) reduced by 1/3 (d) unchanged.

12. A resistance of 10 ohm is connected in one branch of a network. The current in this branch is 2 A. If this 10 Ω resistor is replaced by a 20 ohms resistance, the current in this branch

(a) may be more or less than 2 A (b) will be more than 2 A

(c) will be less than 2 A (d) none of these.

13. If $V = a + jb$ and $I = (c + jd)$, the power is given by

(a) $ac - bd$ (b) $ac + bd$

(c) $bc - ad$ (d) $bc + ad$.

14. KCL is a consequences of law of conservation of

(a) energy (b) charge

(c) flux (d) all the above.

15. The current in the 1 Ω resistor is

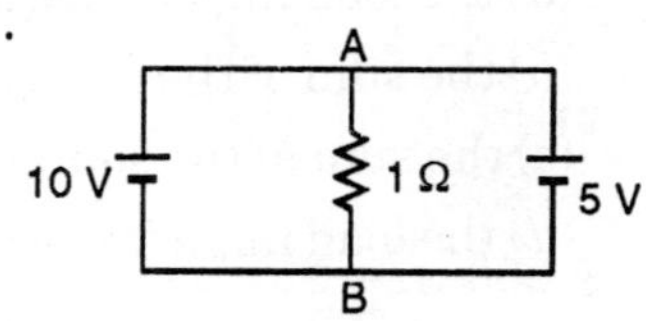

Fig. 3

(a) 5 A (b) 10 A

(c) 15 A (d) zero.

16. The current in a 5 Ω resistor branch in a linear network is 5A. If this branch is replaced by a resistor of 10 Ω, the current in this branch will be

(a) 5 A (b) 10 A

(c) less than 5 A (d) none of the above.

17. The potential of the point A in the given network

(a) 6 V

(b) 7 V

(c) 8 V

(d) none of the above.

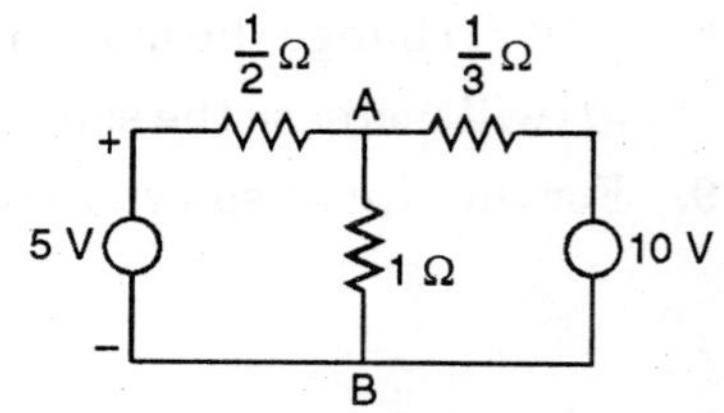

Fig. 4

18. The current through 8 Ω branch is

(a) 1 A

(b) 0.5 A

(c) 1.5 A

(d) none of the above.

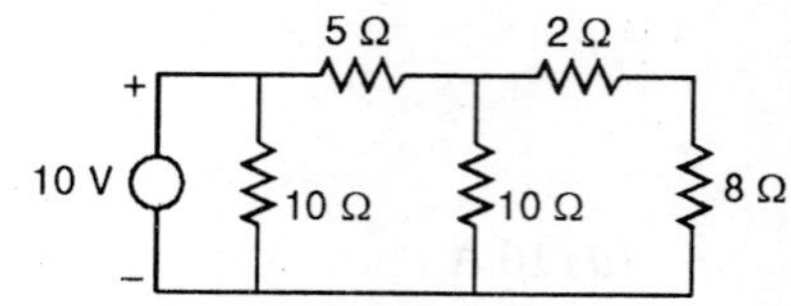

Fig. 5

19. Three identical impedances are connected in delta to a 3-phase supply of 400 V. The line current is 34.65 A and the total power taken is 14.4 kW. The resistance of the load in each phase (in ohms) is

(a) 20 (b) 16

(c) 12 (d) 10.

20. A coil having a resistance of 5 ohms and inductance of 0.1 H is connected in series with a condenser of capacitance 50μ F. A constant alternating voltage of 200 volts is applied to the circuit. The voltage across the coil at resonance is

(a) 200 volts (b) 1788 volts

(c) 1800 volts (d) 2000 volts.

21. If R_g in the circuit shown in the given figure is variable between 20 Ω and 80 Ω, then the maximum power transferred to the load R_L will be

(a) 15 W (b) 13.33 W

(c) 6.67 W (d) 2.4 W.

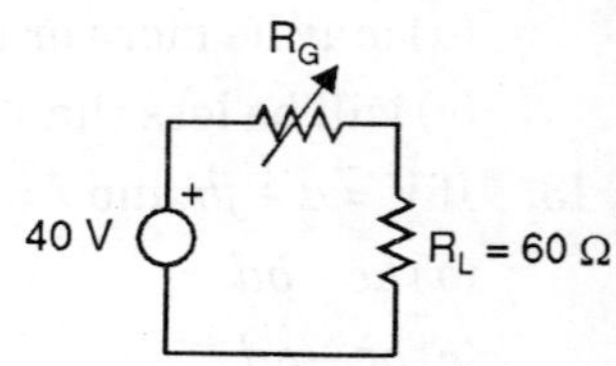

Fig. 6

22. Under the conditions of maximum power transfer from an a.c. source to a variable load

(a) the load impedance must also be inductive, if the generator impedance is inductive

(b) the sum of the source and load impedances is zero

(c) the sum of the source reactance and load reactance is zero

(d) the load impedance has the same phase angle as the generator impedance.

23. In the delta equivalent of the given star-connected impedance Z_{QR} is equal to

(a) 40 Ω (b) $(20 + j10)$ Ω

(c) $\left(5 + j\dfrac{10}{3}\right)\Omega$ (d) $(10 + j\,30)$ Ω.

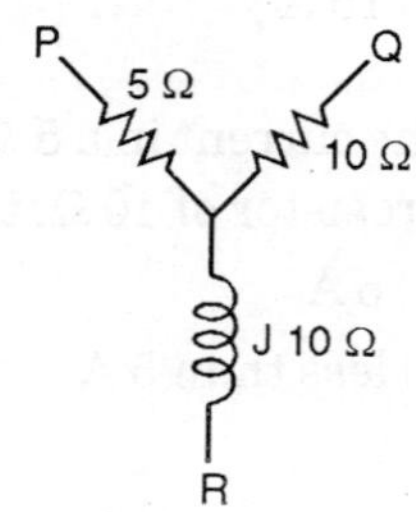

Fig. 7

24. The power delivered to a three-phase load can be measured by the use of 2 wattmeters only when the

(*a*) load is balanced
(*b*) load is unbalanced
(*c*) 3-phase load is connected to the source through 3 wires
(*d*) 3-phase load is connected to the source through 4 wires.

25. A 3-phase star-connected symmetrical load consumes P watts of power from a balanced supply. If the same load is connected in delta to the same supply, the power consumption will be

(*a*) P
(*b*) $\sqrt{3}\,P$
(*c*) $3P$
(*d*) not determinable from the given data.

26. Consider the circuit shown in the given figure. For maximum power transfer to the load, the primary to secondary turns ratios must be :

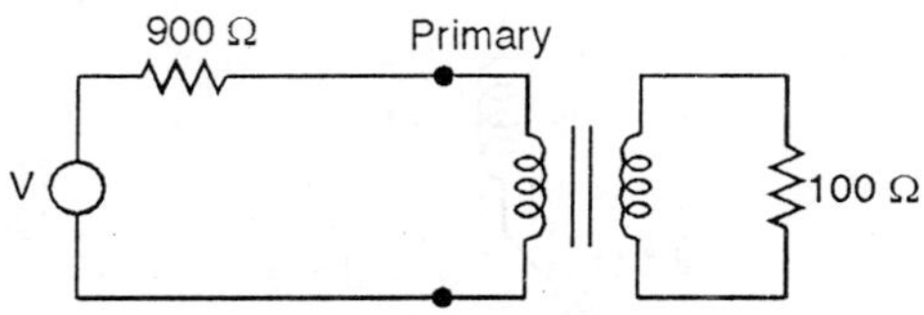

Fig. 8

(*a*) 9 : 1
(*b*) 3 : 1
(*c*) 1 : 3
(*d*) 1 : 9.

27. In two-wattmeter method of power measurement, one of the wattmeters will show negative reading when the load *pf* angle is strictly

(*a*) less than 30°
(*b*) less than 60°
(*c*) greater than 30°
(*d*) greater than 60°.

28. A 3-phase, 3 wire supply feeds a load consisting of three equal resistors connected in star. If one of the resistors is open circuited, then the percentage reduction in the load will be

(*a*) 75
(*b*) 66.66
(*c*) 50
(*d*) 33.33.

29. An a.c. source of 200 V rms supplies active power of 600 W and reactive power of 800 VAR. The rms current drawn from the source is

(*a*) 10*A*
(*b*) 5*A*
(*c*) 3.75*A*
(*d*) 2.5*A*.

30. Which one of the following networks is the *Y* equivalent of the circuit shown in Fig. 9 ?

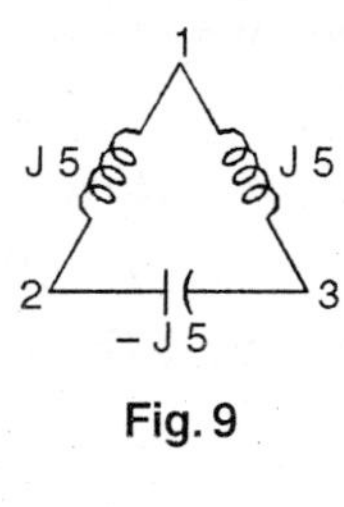

Fig. 9

(a) 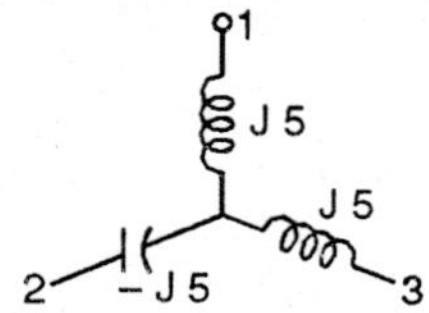

(b) 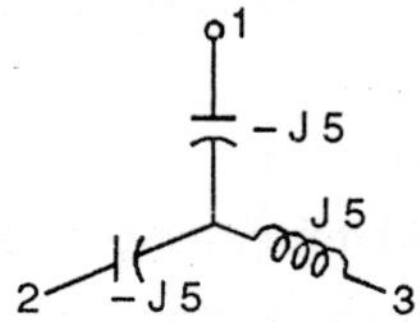

(c) 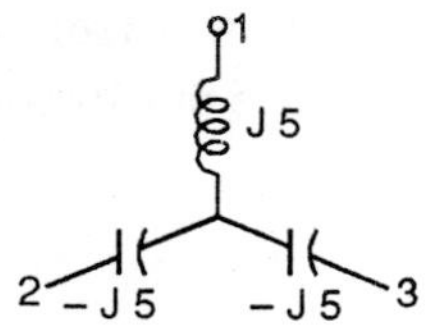

(d) 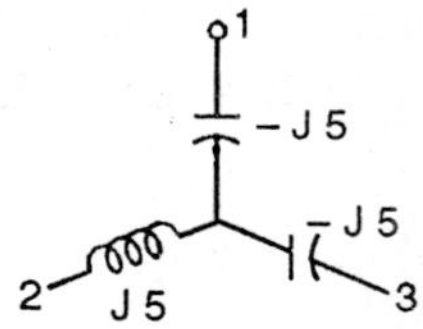

31. Which one of the following circuits is the delta equivalent of the star circuit given in Fig. 10 ?

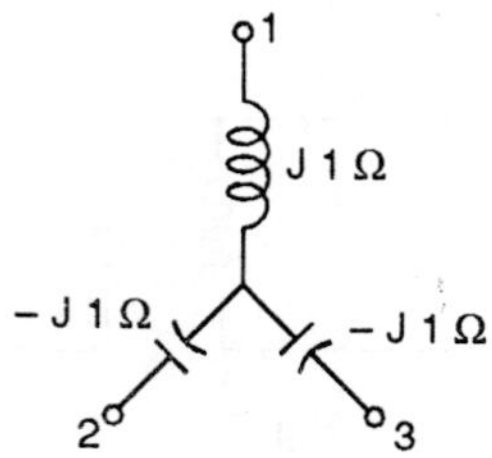

Fig. 10

(a) 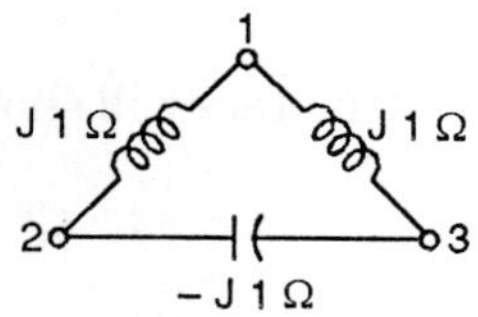

(b) 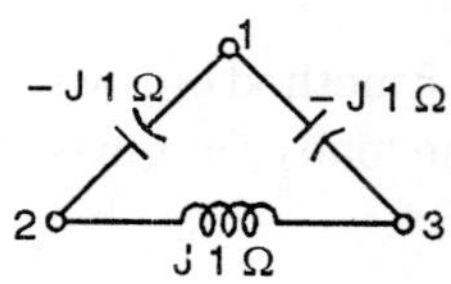

(c) 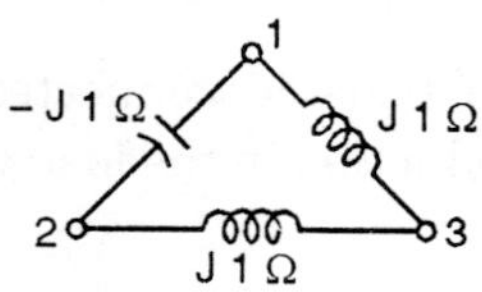

(d) 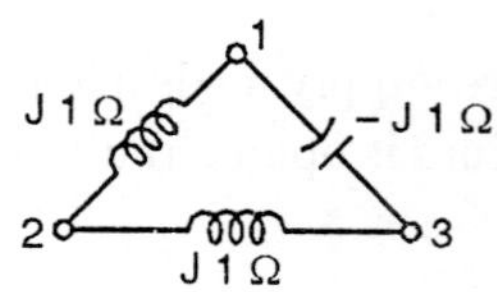

32. For the circuit shown in the adjoining figure, the voltage V_{AB} is

(a) 6 V (b) 10 V

(c) 25 V (d) 40 V.

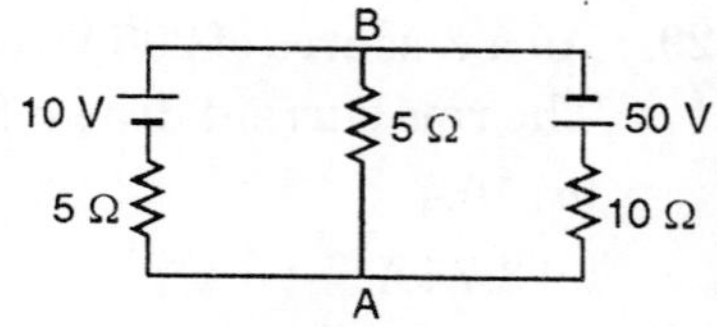

Fig. 11

33. If permanent brake magnet is moved away from the spindle of the moving disc, energy meter

(a) will run fast (b) will run slow

(c) no effect (d) will come to a stop.

34. If voltage supply to the energy meter is more than the rated and no load is connected, the energy meter may run

(*a*) slow (*b*) fast
(*c*) could be (*a*) and (*b*) (*d*) not run at all.

35. Which of the following materials will be used as a resistance for extending the range of measurement of a voltmeter ?
(*a*) Copper (*b*) Steel
(*c*) Aluminium (*d*) Manganin.

36. The advantage of PMMC type instrument is
(*a*) low power consumption (*b*) no hysteresis loss
(*c*) efficient current damping (*d*) all of the above.

37. Which of the following types of instruments can be used for d.c. only ?
(*a*) Moving iron-Attraction (*b*) Moving iron-Repulsion
(*c*) Permanent magnetic moving coil (*d*) Hot wire.

38. To measure power in a 3-phase circuit 2-wattmeters are used. It is applicable only if the load is
(*a*) Balanced (*b*) Unbalanced
(*c*) (*a*) and (*b*)
(*d*) Three wattmeters are required for 3-phase circuits.

39. By mistake voltmeter and ammeter are connected as shown in adjoining figure, then
(*a*) both voltmeter and ammeter will burn
(*b*) only ammeter will burn
(*c*) only voltmeter will burn
(*d*) none will burn.

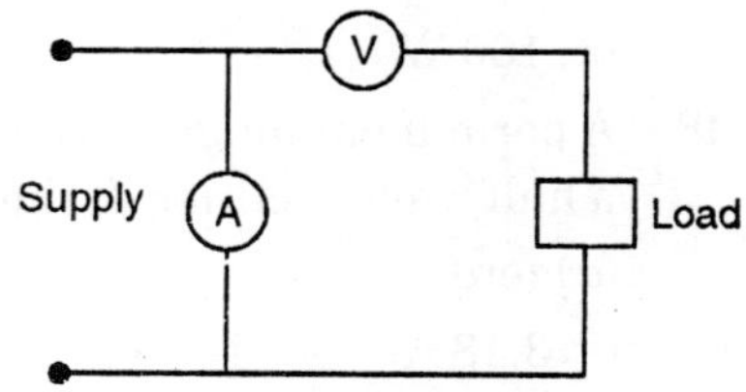

Fig. 12

40. An ordinary electrodynamometer wattmeter is not suitable for low p.f. power measurement because of
(*a*) power loss in voltage coil (*b*) power loss in current coil
(*c*) inductance of voltage coil (*d*) All of the above.

41. In the power measurement by two-wattmeter method, reading of one wattmeter will be negative, if the p.f. of the load is
(*a*) 0.5 lag (*b*) 0.8 lag
(*c*) 0.4 lead (*c*) none of the above.

42. A moving coil galvanometer has a resistance of 4 ohms and gives full scale deflection when carrying 30 mA. The instrument can be used to measure 150 volts by connecting in series with the instrument a resistance of
(*a*) 9996 Ω (*b*) 5004 Ω
(*c*) 5000 Ω (*d*) 4996 Ω.

43. A 15 volt moving iron voltmeter has a resistance of 300 ohms and an inductance of 0.12 henry. The instrument reads correctly on d.c. and on a.c. at 35 Hz it shows a voltage of 14.75 V. What will be its voltage reading at 100 Hz ?

(*a*) 15.4 V (*b*) 15 V
(*c*) 14.85 V (*d*) 14.5 V.

44. If torque to weight ratio of an instrument is low, it can be concluded that
(*a*) the meter will have uniform scale (*b*) the meter will have non-uniform scale
(*c*) the sensitivity of the meter will be high (*d*) the sensitivity of the meter will be low.

45. A moving coil instrument gives full deflection with 15 mA. The instrument has resistance of 5 Ω. If a resistance of 0.08 ohm is connected in parallel with the instrument, the instrument will be capable of reading upto
(*a*) 150 mA (*b*) 600 mA
(*c*) 750 mA (*d*) 950 mA.

46. In which instrument the deflecting torque depends upon frequency
(*a*) Hot wire (*b*) Moving coil
(*c*) Moving iron (*d*) Induction type.

47. A dynamometer type wattmeter with its voltage coil connected across the load side of the instrument reads 180 W. If the load voltage be 200 V and voltage coil branches a resistance of 2000 Ω, the power being taken by the load is
(*a*) 200 W (*b*) 180 W
(*c*) 160 W (*d*) 150 W.

48. A permanent magnet MC type ammeter and a MI are connected in rises with the output of a half wave rectifier. If the moving iron type instrument reads 5A, moving coil would read
(*a*) zero (*b*) 5 A
(*c*) 3.18 A (*d*) 2.5 A.

49. A meter has a constant of 600 revolutions per kW hr. If the meter makes 10 revolutions in 20 secs, the load is
(*a*) 0.75 kW (*b*) 1.5 kW
(*c*) 3 kW (*d*) 6 kW.

50. In a moving iron instrument 12 A current causes a deflection of the needle by 60°. For a deflection of 15° the current required will be
(*a*) 9 A (*b*) 6 A
(*c*) 4 A (*d*) 3 A.

51. Which of the following instruments can have full scale deflection of 300° ?
(*a*) Hot wire (*b*) Induction
(*c*) Moving coil (*d*) Moving iron.

52. Which of the following does not change in an ordinary transformer ?
(*a*) Voltage (*b*) Current
(*c*) Frequency (*d*) all of the above.

53. If R_s is the resistance of secondary winding of the transformer and K is the transformation ratio, the equivalent secondary resistance as referred to primary side is given by
(*a*) KR_s^2 (*b*) K^2R_s
(*c*) R_s/K (*d*) R_s/K^2.

54. At no load the current taken by a transformer

(*a*) lags behind the applied voltage by 80° (*b*) lags behind the applied voltage by 50°

(*c*) leads the applied voltage by 50° (*d*) leads the applied voltage by 80°.

55. Which part of the transformer is subjected to maximum heating

(*a*) core (*b*) winding

(*c*) oil (*d*) frame.

56. For a transformer if the full load copper losses are A and iron losses are B, then the load at which these two losses would be equal, is given by

(*a*) $FL \times \frac{A}{B}$ (*b*) $FL \times \frac{B}{A}$

(*c*) $FL \sqrt{\frac{A}{B}}$ (*c*) $FL \sqrt{\frac{B}{A}}$.

57. A transformer has at full load iron loss of 900 W and copper loss of 1600 W. At what percent of the load the transformer will have maximum efficiency ?

(*a*) 100% (*b*) 90%

(*c*) 75% (*d*) 50%.

58. A transformer has full load copper loss of 800 W and core loss of 600 W. Total loss at no load will be approximately

(*a*) 1400 W (*b*) 1100 watt

(*c*) 1000 W (*d*) 600 watts.

59. For a transformer operating at constant load current, maximum efficiency will occur at a p.f.

(*a*) zero (*b*) unity

(*c*) 0.8 lead (*d*) 0.8 lag.

60. The efficiency of a transformer at full load 0.8 p.f. lag is 95%. The efficiency at 0.8 p.f. lead is

(*a*) 99% (*b*) 95.5%

(*c*) 95% (*d*) 90%.

61. A transformer can have zero voltage regulation at p.f.

(*a*) zero (*b*) lag

(*c*) lead (*d*) unity.

62. The leakage impedances of primary and secondary windings of a transformer are $(3 + j4)\ \Omega$ and $(1 + j1.3)\ \Omega$ respectively. It can be concluded that the transformer has

(*a*) low voltage primary (*b*) low voltage secondary

(*c*) high voltage secondary (*d*) no conclusions can be drawn.

63. In a transformer the resistance between the primary and secondary must be

(*a*) zero (*b*) 1 kΩ

(*c*) 100 kΩ (*d*) Infinite.

64. In a transformer connected to constant voltage source, increase in secondary load current will ultimately

(*a*) reduce the mutual flux in the core

(*b*) increase the mutual flux in the core

(*c*) not cause any change in the magnitude of mutual flux

(*d*) decrease the mutual flux when the secondary current is below 0.8 pf lag.

65. In a step down transformer, there is change of 10 amp in the load current, this results in change of supply current of

(*a*) 10 amps (*b*) less than 10 amps

(*c*) more than 10 amps (*d*) 20 A.

66. Negative voltage regulation is indicative that

(*a*) the load is inductive only (*b*) the load is inductive or resistive

(*c*) the load is capacitive (*d*) in all cases.

67. The maximum transmission (a.c.) voltage in India is

(*a*) 132 kV (*b*) 220 kV

(*c*) 400 kV (*d*) 500 kV.

68. The value of the resistance R_L for maximum power transfer will be

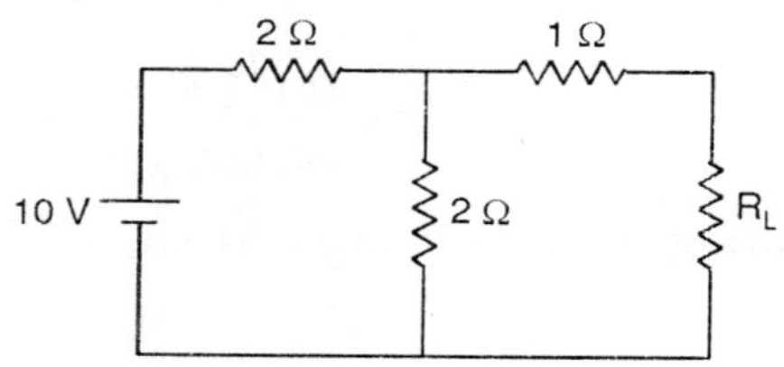

Fig. 13

(*a*) 5 Ω (*b*) 2 Ω

(*c*) 6 Ω (*d*) 3 Ω.

69. The Thevenin's equivalent resistance when the current is to be obtained through 2 Ω resistance is

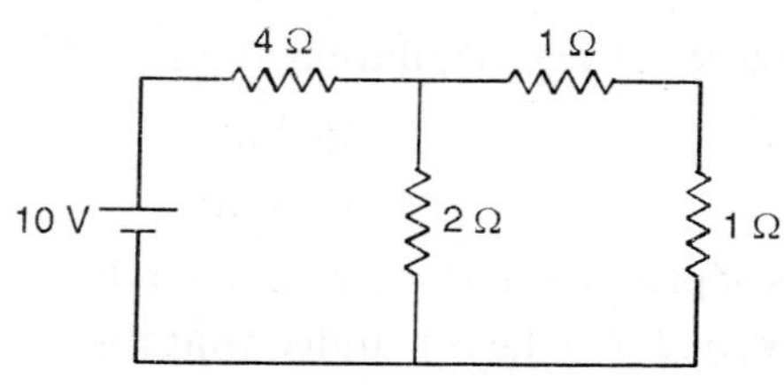

Fig. 14

(*a*) 5 Ω (*b*) 2 Ω

(*c*) 1.33 Ω (*d*) 1.5 Ω.

70. If an induction motor is started with DOL starter and the starting torque is 300 Nm, if started with star/delta starter, the starting torque is

(*a*) 300 Nm (*b*) 600 Nm
(*c*) 900 Nm (*d*) 100 Nm.

71. When oil in a transformer decomposes the major gas that it produces is
(*a*) oxygen (*b*) nitrogen
(*c*) hydrogen (*d*) none of these.

72. The torque-speed characteristic of a d.c shunt motor is
(*a*) a rectangular hyperbola (*b*) a drooping straight line
(*c*) a parabola (*d*) none of the above.

73. A d.c. series motor for large armature current, the torque is proportional to
(*a*) I_a (*b*) I_a^2
(*c*) $\sqrt{I_a}$ (*d*) $I_a^{2/3}$.

74. In case of series resonance the supply voltage depends upon the voltage across the
(*a*) inductance (*b*) capacitance
(*c*) resistance (*d*) none of these.

75. The direction of rotation of a d.c. shunt motor is reversed by
(*a*) reversing armature connection
(*b*) interchanging the armature and field connection
(*c*) adding resistance to the field circuit
(*d*) reversing supply terminals.

76. The armature of a d.c. m/c is laminated
(*a*) to reduce the hysteresis losses
(*b*) to reduce the eddy current losses
(*c*) to reduce the inductance of the armature
(*d*) to reduce the mass of the armature.

77. If the field of a d.c. shunt motor is opened
(*a*) It will continue to run at its rated speed
(*b*) the speed of the motor will become very high
(*c*) the motor will stop
(*d*) the speed of motor will decrease.

78. What will happen if the supply terminals of d.c. shunt motor are interchanged ?
(*a*) Motor will stop
(*b*) The motor will run in the same original direction
(*c*) The direction of rotation will reverse
(*d*) Motor will run at a speed lower than the normal speed in the same direction.

79. If the armature current of a d.c. series motor is increased keeping the field flux constant, then the developed torque
(*a*) increases proportionally (*b*) decreases in the inverse ratio
(*c*) remains constant (*d*) increases $\propto I^2$.

80. The core losses in a d.c. m/c occur in

(*a*) the armature only (*b*) the pole faces only

(*c*) the yoke only (*d*) both the armature and pole faces.

81. A transformer has a core loss of 64 watts and copper loss of 144 watt when it is carrying 20% overload current. The load at which this transformer will operate at the maximum efficiency is

(*a*) 80% (*b*) 66%

(*c*) 120% (*d*) 44%.

82. The short circuit test in a transformer is used to determine

(*a*) iron loss at any load (*b*) copper loss at any load

(*c*) hysteresis loss (*d*) Eddy current loss.

83. A 5-ohm load is connected to the secondary of a 1100/110 V power transformer. The effective load on 1100 V bus bar is

(*a*) 50 ohm (*b*) 500 ohm

(*c*) 0.5 ohm (*d*) 0.05 ohm.

84. A transformer has percentage resistance and reactances of 1% and 4% respectively. It's regulation at p.f. 0.8 lag and lead are respectively

(*a*) 3.2% and – 1.6% (*b*) 6% and – 4%

(*c*) 4.8% and – 3.2% (*d*) none of these.

85. A synchronous motor can run at

(*a*) a leading p.f. (*b*) unity p.f.

(*c*) lag, lead or unity p.f. (*d*) zero p.f.

86. If the field of synchronous motor is under excited, the p.f. will be

(*a*) lagging (*b*) leading

(*c*) unity (*d*) none of these is necessary.

87. When a synchronous motor is started, the field winding is initially

(*a*) short circuited (*b*) excited by a d.c. source

(*c*) open circuited (*d*) none of these.

88. For a given load the armature current of a synchronous motor is minimum when the p.f. is

(*a*) unity (*b*) slightly less than unity and lag

(*c*) slightly less than unity and lead (*d*) none of these.

89. The largest size alternator used in India is

(*a*) 100 MW (*b*) 210 MW

(*c*) 500 MW (*c*) 1000 MW.

90. The power drawn by 3-phase induction motor when first run in star connection and then in delta is

(*a*) same (*b*) $\sqrt{3}$ times that of star

(*c*) 3 times that of star (*d*) $\frac{1}{\sqrt{3}}$ times that of star.

91. A 3-phase induction motor draws a current of 20 A from mains if started by direct switching. If an auto transforms with 50% tapping is used for starting the current drawn from the mains is

(*a*) 20 A (*b*) 10 A

(*c*) 5 A (*d*) 40 A.

92. A change of 5% in supply voltage to a 3-phase IM will produce the approx change in the torque of

(*a*) 5% (*b*) 7.5%

(*c*) 10% (*d*) 15%.

93. A 50 Hz 10 pole 3-phase IM is used to drive a 60 Hz 3-phase alternator. The no. of poles of the alternator is

(*a*) 10 (*b*) 6

(*c*) 12 (*d*) 16.

94. The mechanical load in an induction motor can be represented by

(*a*) $\frac{R_{2e}}{s} - R_{2e}$ (*b*) $R_{2e} - \frac{s}{R_{2e}}$

(*c*) $\frac{1}{s(R_{2e} - 1)}$ (*d*) $R_{2e}\left(\frac{1}{s} + 1\right)$

95. In an induction motor the rotor input is 600 W and slip is 4%. The rotor copper loss is

(*a*) 700 W (*b*) 600 W

(*c*) 650 W (*d*) 24 watts.

96. In an induction motor the ratio of the rotor output to rotor input is 0.96. The percentage slip is

(*a*) 0.04% (*b*) 4%

(*c*) 6% (*d*) 12%.

97. In a 3-phase IM, the maximum torque

(*a*) is proportional to R_{2e} (*b*) is proportional to R^2_{2e}

(*c*) does not depend upon R_{2e} (*d*) none of these.

98. If a synchronous m/c is overexcited it takes lagging vars from the system when it is operated as

(*a*) Sychronous motor (*b*) Synchronous generator

(*c*) Synchronous motor as well as generator (*d*) None of the above.

99. If a synchronous motor is underexcited it takes lagging vars from the system when it is operated as

(*a*) Synchronous motor (*b*) Synchronous generator

(*c*) Synchronous motor as well as synchronous generator

(*d*) None of the above.

100. The frequency of voltage generated by an alternator having 8 poles and rotating at 250 rpm is

(*a*) 60 Hz (*b*) 50 Hz

(*c*) 25 Hz (*d*) $16\frac{2}{3}$ Hz.

Answers

1. (*a*)	**2.** (*c*)	**3.** (*d*)	**4.** (*d*)	**5.** (*c*)	**6.** (*d*)
7. (*c*)	**8.** (*d*)	**9.** (*d*)	**10.** (*a*)	**11.** (*c*)	**12.** (*c*)
13. (*b*)	**14.** (*b*)	**15.** (*d*)	**16.** (*c*)	**17.** (*c*)	**18.** (*b*)
19. (*c*)	**20.** (*c*)	**21.** (*c*)	**22.** (*c*)	**23.** (*d*)	**24.** (*c*)
25. (*c*)	**26.** (*b*)	**27.** (*d*)	**28.** (*c*)	**29.** (*b*)	**30.** (*c*)
31. (*b*)	**32.** (*a*)	**33.** (*b*)	**34.** (*a*)	**35.** (*d*)	**36.** (*d*)
37. (*c*)	**38.** (*c*)	**39.** (*b*)	**40.** (*c*)	**41.** (*c*)	**42.** (*d*)
43. (*d*)	**44.** (*d*)	**45.** (*d*)	**46.** (*d*)	**47.** (*c*)	**48.** (*a*)
49. (*c*)	**50.** (*b*)	**51.** (*b*)	**52.** (*c*)	**53.** (*d*)	**54.** (*a*)
55. (*c*)	**56.** (*d*)	**57.** (*c*)	**58.** (*d*)	**59.** (*b*)	**60.** (*c*)
61. (*c*)	**62.** (*b*)	**63.** (*d*)	**64.** (*c*)	**65.** (*b*)	**66.** (*c*)
67. (*c*)	**68.** (*b*)	**69.** (*c*)	**70.** (*d*)	**71.** (*c*)	**72.** (*b*)
73. (*a*)	**74.** (*c*)	**75.** (*a*)	**76.** (*b*)	**77.** (*b*)	**78.** (*b*)
79. (*a*)	**80.** (*d*)	**81.** (*a*)	**82.** (*b*)	**83.** (*b*)	**84.** (*a*)
85. (*c*)	**86.** (*a*)	**87.** (*a*)	**88.** (*a*)	**89.** (*c*)	**90.** (*c*)
91. (*b*)	**92.** (*c*)	**93.** (*c*)	**94.** (*a*)	**95.** (*d*)	**96.** (*b*)
97. (*a*)	**98.** (*d*)	**99.** (*c*)	**100.** (*d*).		

APPENDIX

Magnetic Circuits

MAGNETIC CIRCUITS

Oersted in 1819 discovered that a current flowing in a straight conductor deflected the needle of a compass and it shows that the paths of magnetic force are concentric circle around the conductor. Similarly if a current is passed through circular (loop) conductor the magnetic flux are found to be concentric to the circular conductor and the magnetic flux density at the centre of the circular conductor is given by (loop)

$$B = \frac{\mu I}{2r}$$

where B is the magnetic flux density in Wb/m^2 or in Tesla, I the current flowing through the circular loop of the conductor and r the radius of the circular loop, μ the permeability of the material around which the loop is wound.

$$\mu = \mu_0 = 4\pi \times 10^{-7} \text{ H/m}$$

if the coil is wound on a non-magnetic material or it is air cored and $\mu = \mu_0 \mu_r$ where μ_r is the relative permeability of the magnetic material. The magnetic flux density is related to magnetic field intensity through μ as given below.

$$B = \mu_0 \mu_r H$$

where H is the magnetic field intensity and its units are Ampere turns per meter. μ_r has no units as it is a relative permeability, μ is permeability of the material and its units are Henry/m.

As far as analysis of magnetic circuits is concerned, it is analogous to electric circuits. The analogous quantities are listed below.

Electric Circuits	*Magnetic Circuits*
e.m.f. (Volts)	mmf (amp. Turns)
Current (Ampere)	Flux (Webers)
Resistance, ohm	Reluctance, R
Conductance S	Permeability Henry/m
Electric field intensity, E V/m	Magnetic field Intensity, H AT/m
Electric flux density D coulomb/m^2	Magnetic flux density Wb/m^2

Simple magnetic circuits can be analysed based on Ohm's law in electric circuits according to Ohm's law

$$\frac{V}{I} = R$$

where V is the e.m.f., I the current and R the resistance. Picking up analogous quantities in magnetic circuits from the relations given above we have

$$\frac{mmf}{flux} = \text{Reluctance}$$

$$\frac{NI}{\phi} = \text{Reluctance}$$

where N is the number of turns in the circular coil.

Now $\phi = BA$ where A is the area of cross section of the core on which the coil is wound.

Now $$B = \mu_0 \mu_r H$$

and $H = \dfrac{NI}{l}$ where l is the length of the magnetic path.

Hence reluctance $$R = \frac{NI}{\mu_0\mu_r \dfrac{NI}{l} . A}$$

$$= \frac{l}{\mu_0\mu_r A} \text{ AT/Wb}$$

Series Magnetic Circuit

If there are one or more than one magnetic paths connected in series there could even be an air gap in the path of the same flux (same current in a series electric circuit), the net reluctance is the sum total of the individual magnetic paths.

Parallel Magnetic Circuits

Analogous to electric circuit, the magnetic circuit will be analysed using Kirchhoff's laws as shown in Fig. 1.

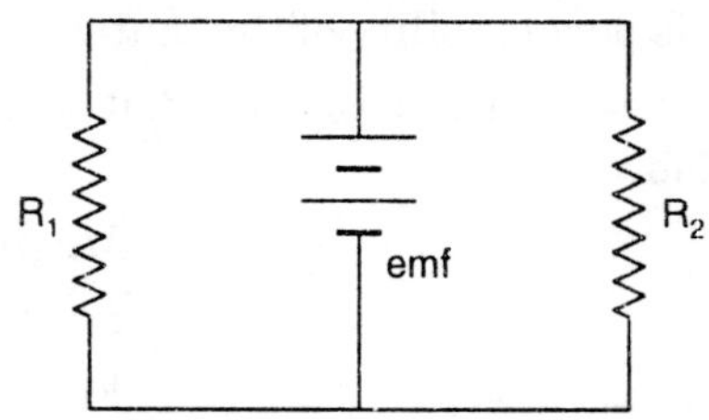

Fig. 1. Analogous parallel electric circuits.

Let us take a few examples to illustrate the application of what we have studied about the magnetic circuits.

Example 1. An iron ring of mean length 50 cms has an air gap of 1 mm and a winding of 200 turns. If the relative permeability of air is 300 when a current of 1 A flows through the coil, find the flux density.

Solution.

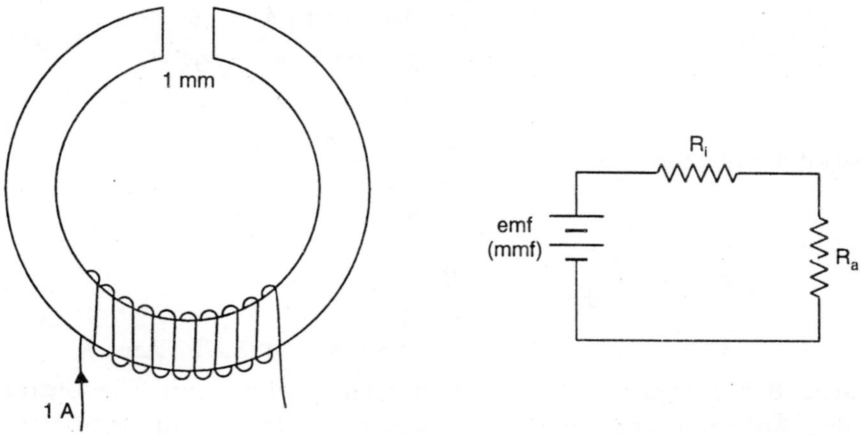

Fig. 2. (a) Magnetic Circuit (b) Electric analog.

The reluctance of magnetic path R,

$$= \frac{50 \times 10^{-2}}{4\pi \times 10^{-7} \times 300\, A}$$

$$= \frac{50 \times 1000}{12\pi\, A} = \frac{1326}{A}$$

Now reluctance of air gap R_2

$$R_2 = \frac{1 \times 10^{-3}}{4\pi \times 10^{-7} \times 1 \times A} = \frac{796}{A} \quad \text{as} \quad \mu_r = 1 \text{ for air}$$

$$\text{Total reluctance} = \frac{1326}{A} + \frac{796}{A} = \frac{2122}{A}$$

$$\phi = \frac{MMF}{\text{Reluctance}} = \frac{200 \times 1A}{2122}$$

$$= 94.2 \times 10^{-3} A$$

or $$B = \frac{\phi}{A} = 94.2 \text{ mwb/m}^2$$ **Ans.**

Example 2. A cast steel electromagnet has an air gap of length 2 mm and an iron path of length 30 cms. Find the number of ampere turns necessary to produce a flux density of 0.8 wb/m^2 in the air gap. Assume H = 700 AT/m for the flux density of 0.8 wb/m^2 for the cast steel.

Solution. Here again there are two reluctances in series, one due to cast steel (say R_1) and the other due to air gap (say R_2)

$$R_1 = \frac{30 \times 10^{-2}}{4\pi \times 10^{-7}\, \mu_r\, A}$$

Since $$B = \mu_0 \mu_r H$$

$$0.8 = 4\pi \times 10^{-7} \mu_r \times 700$$

$$\mu_r = 910$$

Hence $R_1 = \dfrac{30 \times 10^{-2}}{4\pi \times 10^{-7} \times 910\, A} = \dfrac{262.4}{A}$

and $R_2 = \dfrac{2 \times 10^{-3}}{4\pi \times 10^{-7} \times 1 \times A} = \dfrac{1592}{A}$

Hence total reluctance $= \dfrac{1592 + 262.4}{A} = \dfrac{1854.4}{A}$

Now $\phi = \dfrac{NI}{R}$

$$0.8 \times A = \frac{NI \,.\, A}{1854.4}$$

or $NI = 1854.4 \times 0.8 \simeq 1485$ AT **Ans.**

Example 3. Fig. 3 shows a three limb electromagnetic circuit. The central limbs has an air gap of 2 mm. Assume $\mu_r = 800$, determine the current in the 600 turn exciting coil to establish a flux of 100×10^{-3} mwb in the air gap.

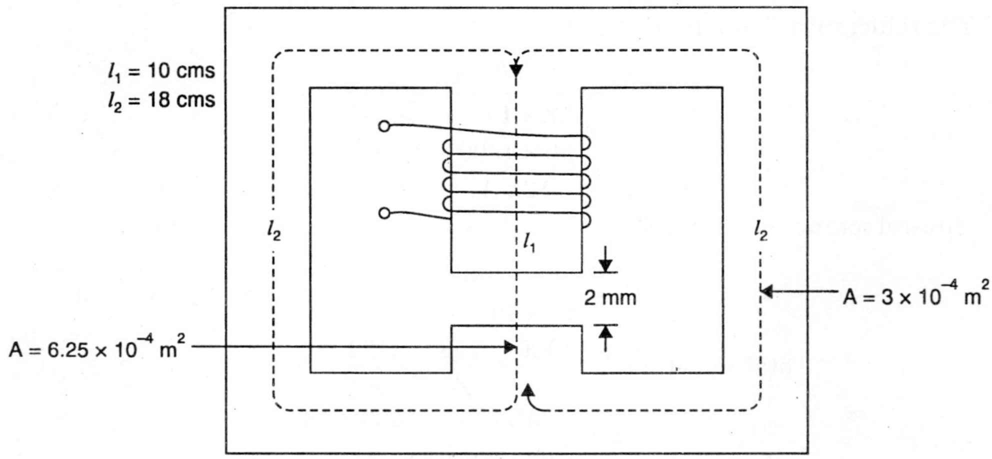

Fig. 3

Solution. Since the flux in the air gap is 100×10^{-3} mwb, the flux in the outer limbs will be 50×10^{-3} mwb. The analogous electric circuit is as shown.

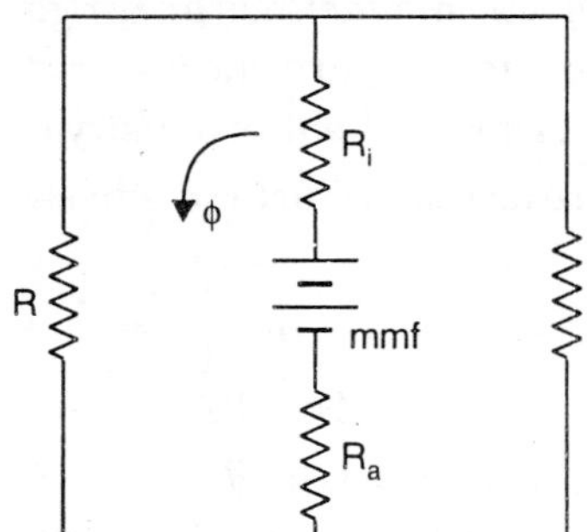

Fig. 4

$$R_a = \frac{2 \times 10^{-3}}{4\pi \times 10^{-7} \times 6.25 \times 10^{-4}}$$

$$= 25.47 \times 10^5$$

$$R_i = \frac{10 \times 10^{-2}}{4\pi \times 10^{-7} \times 800 \times 6.25 \times 10^{-4}}$$

$$= 159200 = 1.592 \times 10^5$$

Total reluctance $= 27.062 \times 10^5$

Hence $mmf = \phi R = 100 \times 10^{-6} \times 27.062 \times 10^5$

$$= 27.06 \times 10 = 270.6$$

Reluctance of one of the parallel paths

$$= \frac{18 \times 10^{-2}}{4\pi \times 10^{-7} \times 800 \times 3 \times 10^{-4}}$$

$$= \frac{18 \times 10^9}{4\pi \times 800 \times 3} = 5.97 \times 10^5$$

Hence *mmf* required for one of the parallel paths is $50 \times 10^{-6} \times 5.97 \times 10^5$

$$= 29.85$$

Hence total *mmf* $270.6 + 29.85 = 300.45\,AT$

$$\text{The current} = \frac{300.45}{600} \simeq 0.5 \text{ A} \quad \textbf{Ans.}$$

We have taken one of the parallel paths as, to find out *mmf* (e.m.f. in electrical analog) we need complete one loop.

Example 4. A 680 turn coil is wound on the central limb of the cast steel frame shown. A total flux of 1.6 mwb is required in the gap. Find the current required. Dimensions are in cms. Assume $H = 900$ A T/m for $B = 1$ Wb/m^2.

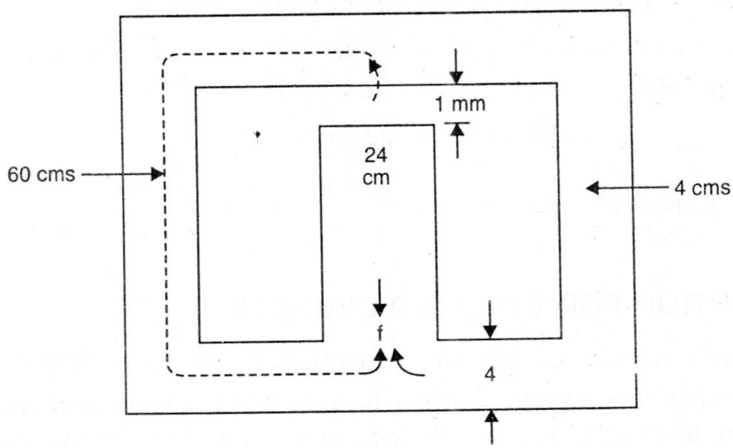

Fig. 5

Solution.

$$B = \mu_0 \mu_r H$$

$$1 = 4\pi \times 10^{-7} \mu_r \times 900$$

or $$\mu_r = \frac{10^5}{4\pi \times 9} = 885$$

Reluctance for the air gap $= \dfrac{1 \times 10^{-3}}{4\pi \times 10^{-7} \times 1 \times 16 \times 10^{-4}}$

$$= \frac{10^8}{4\pi \times 16} = 497610 \text{ AT/Wb}$$

Reluctance of the central limb $= \dfrac{24 \times 10^{-2}}{4\pi \times 10^{-7} \times 885 \times 16 \times 10^{-4}}$

$$= \frac{24 \times 10^9}{4\pi \times 885 \times 16} = 13.49 \times 10^4 \text{ AT/Wb}$$

Total reluctance of the central limb

$$= 63.25 \times 10^4 \text{ AT/Wb}$$

MMf required for the central limb

$$= 63.25 \times 10^4 \times 1.6 \times 10^{-3}$$

$$= 1012 \text{ AT}$$

Now reluctance of one of the outer limbs

$$13.49 \times 10^4 \times \frac{60}{24} = 33.72 \times 10^4$$

Hence AT required $= 33.72 \times 0.8 \times 10^{-3} \times 10^4$

$$= 269.8$$

$$\text{Total AT required} = 1012 + 269.8 = 1281.8$$

Since the two paths are identical it is necessary to consider *mmf* for only one of them. Hence total *mmf* required

$$1012 + 269.8 = 1281.8$$

$$\text{Hence current required} = \frac{1281.8}{680} = 1.9 \text{ Amp.}$$

DOT CONVENTION FOR COUPLED CIRCUITS

The significance of dot convention is explained with the help of Fig. 6.

Suppose a time varying voltage source is connected across the primary of the transformer and at any given instant the voltage source has the polarity shown and current $i\ (t)$ is in the direction shown by the arrow and is increasing with time entering the dotted terminal, this current induces a voltage in the secondary which is positive at the dotted terminal. Conversely if the current in the primary is increasing and is leaving the dotted terminal this will induce the voltage polarity in the secondary such that the undotted end will become positive and hence dotted terminals of both the windings will be negative simultaneously. The following experimental procedure can be used to establish dotted ends of the transformer windings. Mark a dot arbitrarily on one end of the primary winding and connect a d.c. source with positive terminal connected to the dotted end and the negative terminal to the undotted end of the winding through a switch. Connect a MC voltmeter across the secondary winding. The end of the secondary winding which goes positive momentarily on closing the switch on the primary side as measured by the voltmeter, is the terminal to be dotted.

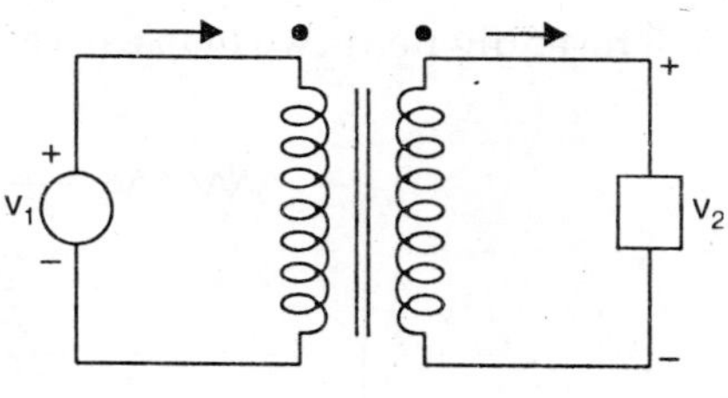

Fig. 6. A circuit for dot convention.

If there are more than two windings, similar procedure can be followed for identifying relationship between each pair of windings. For each pair of winding different forms of dots (• ❖ ■ ❁) should be used to avoid any confusion.

Suppose we are required to write loop equations for a mutually coupled circuit shown in Fig. 7.

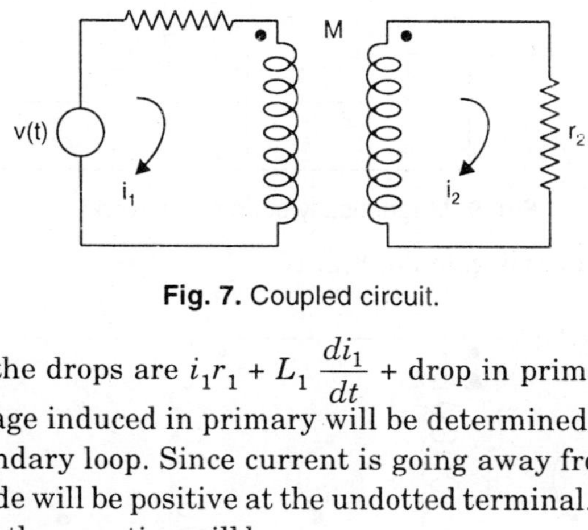

Fig. 7. Coupled circuit.

For primary loop the drops are $i_1 r_1 + L_1 \dfrac{di_1}{dt}$ + drop in primary loop due to current in secondary loop. The voltage induced in primary will be determined by the direction of i_2 with respect to its dot in secondary loop. Since current is going away from the dot, the polarity of voltage on the primary side will be positive at the undotted terminal. Therefore, for current i_1 it will be rise in voltage and the equation will be

$$i_1 r_1 + L_1\, di_1/dt - M\frac{di_2}{dt} = v(t)$$

and for the secondary loop, similarly

$$i_2 r_2 + L_2 \frac{di_2}{dt} - M\frac{di_1}{dt} = 0$$

The equivalent two loops are as shown in Fig. 8.

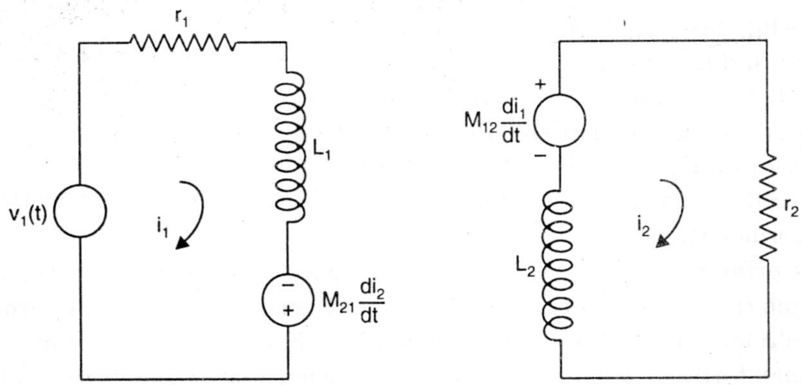

Fig. 8. Equivalent of Fig. 7.

Consider Fig. 9 which is a magnetically coupled network.

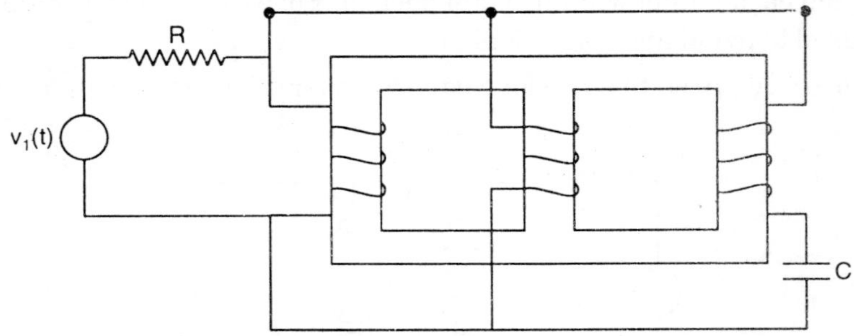

Fig. 9. Magnetically coupled network.

The equivalent with dots is given in Fig. 10.

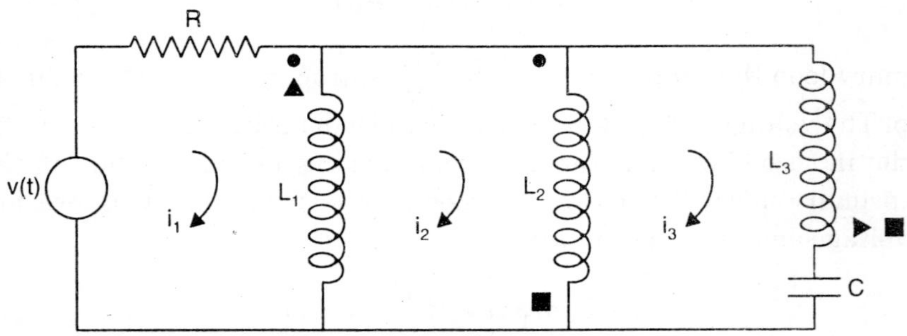

Fig. 10. Equivalent of Fig. 9.

Since all the three coils are mutually coupled, each coil will induce voltage in the other two coils, the magnitude will depend upon the current flowing through the coil and the polarity its relative direction of flow with respect to dot of the coil.

Voltage induced in coil 1

Due to coil 2, the current through coil 2 is $(i_2 - i_3)$ entering the circular dot, therefore the voltage induced in coil 1 will be with positive polarity at the dot. Due to coil 3, since the current i_3 is leaving the triangular dot, the polarity will be –ve at the corresponding dot in coil 1.

Voltage induced in coil 2

Due to coil 1, the current $(i_1 - i_2)$ enters the dot, therefore positive polarity at the dot of coil 2.

Due to coil 3, the current i_3 leaves the dot (square), the polarity +ve will be at the undotted terminal of coil 2.

Voltage induced in coil 3

Due to coil 1 the current entering the dot is $(i_1 - i_2)$ hence the positive polarity at the dotted terminal of coil 3. Due to coil 2 current $(i_2 - i_3)$ leaves the dot (square), the positive polarity will be at the undotted terminal. Hence the equivalent circuit given as follows :

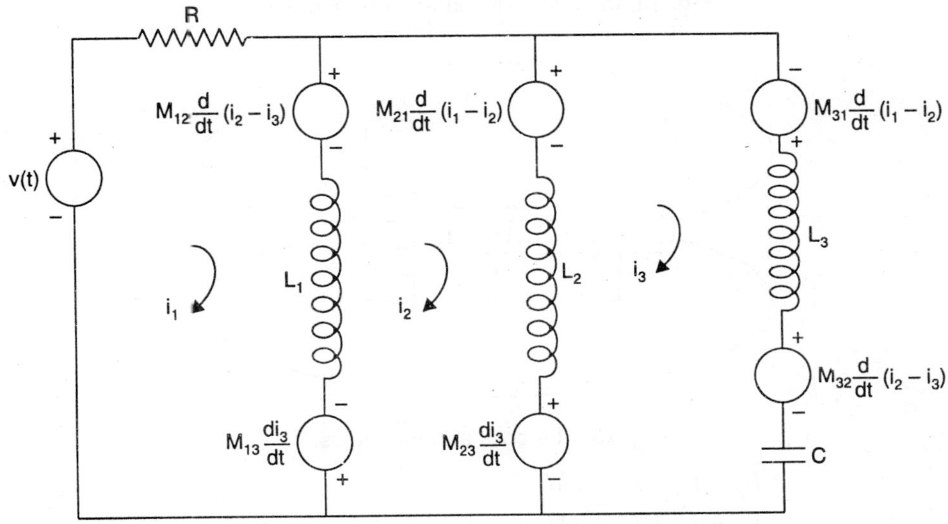

Fig. 11. Equivalent of Fig. 9. with voltage sources.

The loop equations can be written as follows :

Loop 1

$$Ri_1 + M_{12}\frac{d(i_2 - i_3)}{dt} + L_1\frac{d(i_1 - i_2)}{dt} - M_{13}\frac{di_3}{dt} = v(t)$$

Loop 2

$$M_{13}\frac{di_3}{dt} + L_1\frac{d(i_2 - i_1)}{dt} - M_{12}\frac{d(i_2 - i_3)}{dt} + M_{21}\frac{d(i_1 - i_2)}{dt} + L_2\frac{d(i_2 - i_3)}{dt} + M_{23}\frac{di_3}{dt} = 0$$

Loop 3

$$L_2\frac{d(i_3 - i_2)}{dt} - M_{23}\frac{di_3}{dt} - M_{21}\frac{d(i_1 - i_2)}{dt} - M_{31}\frac{d(i_1 - i_2)}{dt}$$

$$+ L_3\frac{di_3}{dt} + M_{32}\frac{d(i_2 - i_3)}{dt} + \frac{1}{C}\int i_3\, dt = 0$$

The above example explains clearly how voltage sources (corresponding to induced voltages) are inserted in the coils which are magnetically coupled and the dots have been assigned.

When the two coils are to be interconnected it is important to know whether the mutual inductance M is aiding or opposing. However, it is known that the effect on each of the two coils must be the same. Refer to Fig. 12(a) where the mutual inductance is aiding their self inductances whereas in (*b*) it is opposing. If L_1 and L_2 are the self inductances of the two coils and M the mutual inductance, the total inductance say if is L_A in case of Fig. 12(*a*) and L_B that in Fig. 12(*b*) then,

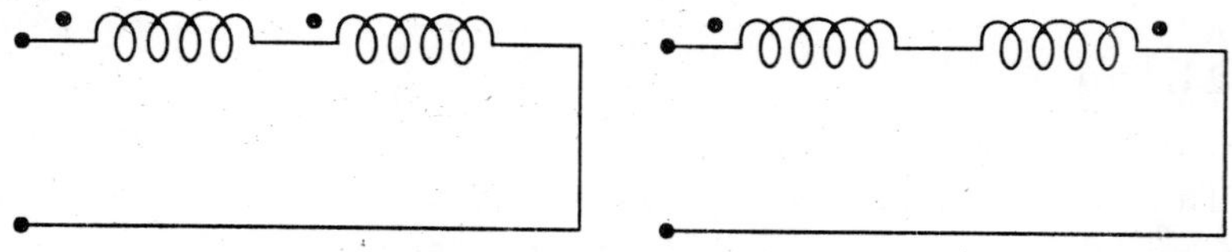

Fig. 12. (a) Two coils aid and (b) Oppose.

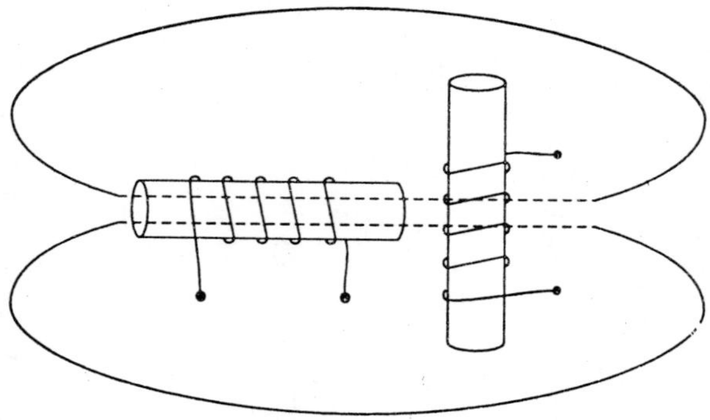

Fig. 13. The coils at right angles.

$$L_A = L_1 + L_2 + 2M$$

and
$$L_B = L_1 + L_2 - 2M$$

Therefore,
$$M = \frac{L_A - L_B}{4}$$

This method provides a very convenient way of determining the value of mutual inductance between the coils.

The co-efficient of coupling and the mutual inductance between the two coils shown in Fig. 13 are both zero. Although the flux of one coil appears to pass through the other coil, it does not actually link the turns of the other coil.

Therefore, the linkages are zero and the value of M is zero. It is only when one coil is rotated with respect to other that a linking occurs.

Example 5. Determine the inductance of the individual winding and the equivalent inductance when mutual inductance is 8H.

Since the flux through the two coils opposes each other, the mutual inductance is subtractive from each coil and hence

$$L_1 = 15 - 8 = 7\text{H}$$

and
$$L_2 = 16 - 8 = 8\text{H}$$

Therefore equivalent inductance between the terminals

$$L = \frac{L_1 L_2}{L_1 + L_2} = \frac{7 \times 8}{15} = 3.73\text{H}$$

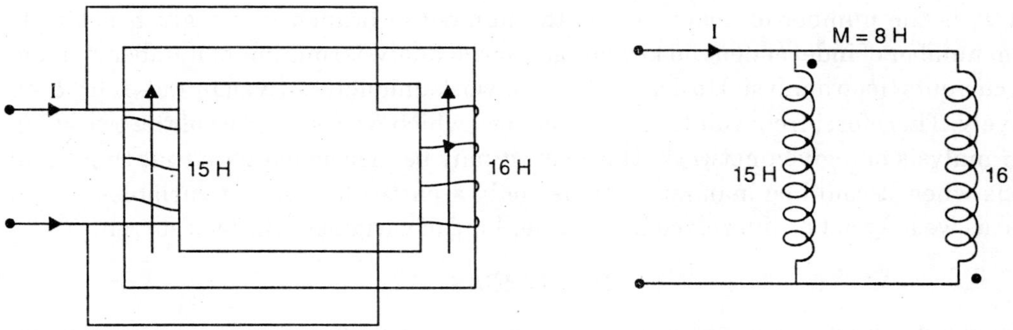

Fig. 14

Example 6. Determine the inductance between the terminals for a three coil system shown below:

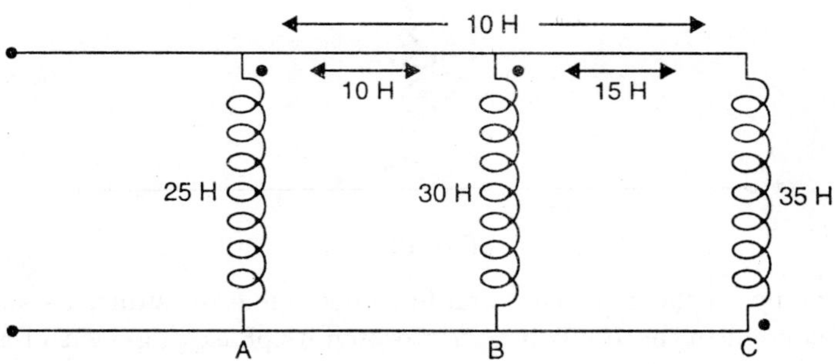

Fig. 15

From the Fig. 15

$$L_A = 25 + 10 - 10 = 25\text{H}$$

$$L_B = 30 + 10 - 15 = 25\text{H}$$

$$L_C = 35 - 15 - 10 = 10\text{H}$$

Therefore, $$\frac{1}{L_{eq}} = \frac{1}{25} + \frac{1}{25} + \frac{1}{10} = \frac{2+2+5}{50}$$

or $$L_{eq} = \frac{50}{9} = 5.55\text{H}$$

A given network will always have a unique solution in terms of its nodal voltages and loop currents. It is, therefore, required to write down a set of linearly independent simultaneous equations equal to the number of unknowns, and smaller the number of unknowns fewer the linear simultaneous equations and less time consuming will be the solution of these equations. It would be seen that sometimes it is more convenient to use nodal voltages as independent variables, sometimes loop currents as independent variables and yet sometimes a suitable combination of both from the point of view of minimum independent variables required to describe the network and hence the number of equations to seek solution of the network.

If 'n' is the number of nodes and 'b' the number of branches in a given network, the minimum number of independent nodal voltages (nodal analysis) minimum number of independent loop currents (loop analysis) to describe the network uniquely is given by $(n - 1)$ or $(b - n + 1)$ respectively. Therefore, from solution point of view, whichever is smaller of the two should be used for analysis of a given network. However, it may be mentioned here that there could be situations when it could be mandatory to use only a particular kind of variables (viz. nodal voltages) as we may not be interested in the other kind of variables (viz. loop currents).

PROBLEMS

1. Three coils are interconnected with winding sense shown by dots. The self inductance of the coils are 2 H each and the mutual inductance 1 H each. Determine L_{eq} between

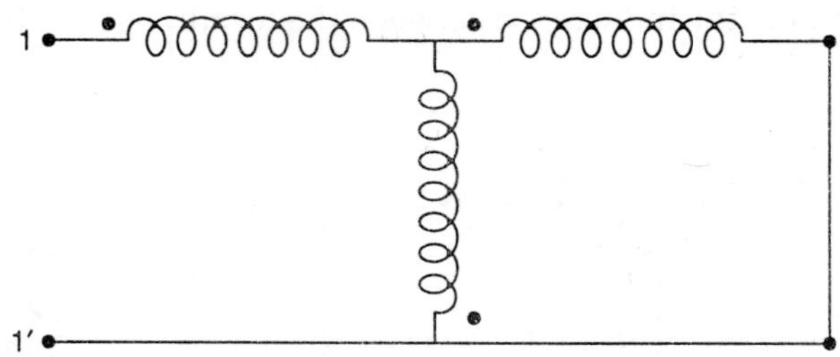

Fig. 16

2. Two coils are wound on the same core and have negligible resistance as shown in the Fig. 17. Determine the current in the 1000 ohm resistor and its phase angle with respect to the applied voltage of 100 volt at a frequency of $\frac{5000}{2\pi}$ Hz when the coils act (i) in the same sense (ii) in opposite sense.

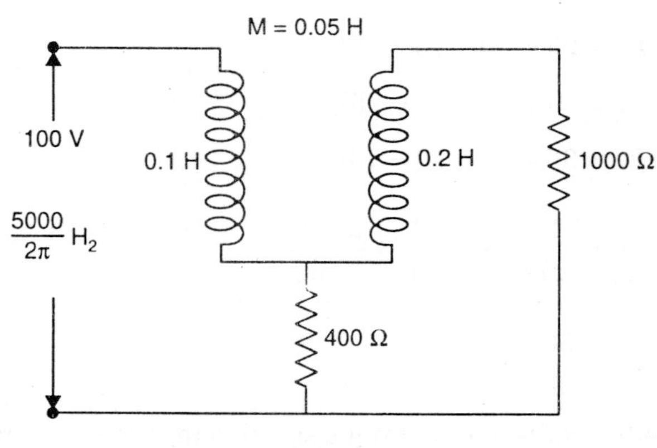

Fig. 17

3. A cast steel ring has a circular cross sections 3 cm in diameter and a mean circumference of 80 cm. The ring is uniformly wound with a coil of 600 turns. (a) Estimate the current required to produce a flux of 0.5 mWb in the ring (b) if a saw cut 2 mm wide is made in the ring, find approximately the flux produced by the current found in (a). (c) Find the current which will give the same flux as in (a). Assume $H = 650$ AT/m for $B = 0.707$ wb/m^2.

4. A ring of cast steel has an external diameter of 24 cms and a square section of 3 cm side. Inside and across the ring an ordinary steel bar, 18 × 3 × 0.4 cm is fitted with negligible gap. Calculate the number of ampere turns to be applied to one half of the ring to produce a flux density of 1 wb/m^2 in the other half.
5. An iron ring with a mean circumference of 140 cm and cross section 12 cm^2 is wound with 500 turns of wire. When the exciting current is 2 A, the flux is found to be 1.2 mwb. Determine relative permeability of iron.
6. An iron magnetic circuit has a uniform cross sectional area of 5 cm^2 and a length of 25 cm. A coil of 120 turns is wound uniformly over the magnetic circuit. When the current in the coil is 1.5 A, the total flux is 0.3 mwb and when the current is 5 A, the total flux is 0.6 mwb. For each value of current calculate H and μ_r.

REFERENCES

1. Langsdorf A.S., Theory of alternating current Machinery, McGraw Hill 1955.
2. Wadhwa C.L., Network Analysis, New Age International Publishers 2004.
3. McPherson George & Laramore Robert D., An Introduction to Electrical Machines & Transformers John Wiley & Sons 1990.
4. Fitzgerald A.E. and Kingsley C., Electric Machinery 2nd edition McGraw Hill 1961.
5. V. Del Toro, Principles of Electrical Engineering Prentice Hall.
6. Golding E.W., Electrical Measurements and Measuring Instruments SIR ISAC PITMAN & SONS 1960.
7. Hayt W.H. Jr. and Jack E. Kemmerly, Engineering Circuit Analysis McGraw Hill.

Index

H

I

K

L

M

N

O

P

Q

R

S

T

V

W